Recent Advances in Rock Mass Engineering

Recent Advances in Rock Mass Engineering

Guest Editors

Qibin Lin
Rihong Cao
Jingjing Meng

MDPI

Basel • Beijing • Wuhan • Barcelona • Belgrade • Novi Sad • Cluj • Manchester

Guest Editors
Qibin Lin
School of Resources
Environment and Safety
Engineering
University of South China
Hengyang
China

Rihong Cao
School of Resources and
Safety Engineering
Central South University
Changsha
China

Jingjing Meng
School of Infrastructure
Engineering
Nanchang University
Nanchang
China

Editorial Office
MDPI AG
Grosspeteranlage 5
4052 Basel, Switzerland

This is a reprint of the Special Issue, published open access by the journal *Applied Sciences* (ISSN 2076-3417), freely accessible at: https://www.mdpi.com/journal/applsci/special_issues/79FS3E3866.

For citation purposes, cite each article independently as indicated on the article page online and as indicated below:

Lastname, A.A.; Lastname, B.B. Article Title. *Journal Name* **Year**, *Volume Number*, Page Range.

ISBN 978-3-7258-7422-4 (Hbk)
ISBN 978-3-7258-7423-1 (PDF)
https://doi.org/10.3390/books978-3-7258-7423-1

Contents

Preface

Rock mass engineering plays a fundamental role in the safe and sustainable development of underground and surface engineering projects, particularly as engineering activities extend into deeper, more complex, and more challenging geological environments. In recent years, rapid progress in experimental techniques, numerical modeling, and theoretical analysis has significantly enhanced our understanding of rock mass behavior, while also revealing new scientific and engineering challenges.

This Reprint, Recent Advances in Rock Mass Engineering, aims to provide a focused platform for presenting recent methodological developments and engineering applications related to the mechanical behavior, failure mechanisms, and stability control of rock masses. The contributions included in this Reprint address a broad spectrum of topics, ranging from laboratory-based investigations and numerical simulations to field observations and engineering case studies. Together, they highlight the interplay between fundamental rock mechanics theory and practical engineering implementation.

This Reprint is intended for researchers, graduate students, and professional engineers working in rock mechanics, geotechnical engineering, mining engineering, and underground space development. By bringing together diverse perspectives and advanced research outcomes, this Reprint aims to promote knowledge exchange, stimulate further scientific inquiry, and support the design and construction of safer and more resilient rock engineering systems.

Qibin Lin, Rihong Cao, and Jingjing Meng
Guest Editors

applied sciences

Editorial

Recent Advances in Rock Mass Engineering

Qibin Lin [1,*], Rihong Cao [2] and Jingjing Meng [3]

1 School of Resources Environment and Safety Engineering, University of South China, Hengyang 421001, China
2 School of Resources and Safety Engineering, Central South University, Changsha 410083, China
3 School of Infrastructure Engineering, Nanchang University, Nanchang 330031, China
* Correspondence: qblin@usc.edu.cn; Tel.: +86-15273110027

Rock mass engineering serves as a critical foundation for infrastructure construction and resource development. Research and practice in this field are increasingly challenged by complex deep geological environments, heterogeneous geological structures, and heightened demands for safety and sustainability [1]. With urbanization accelerating and the extraction of energy and mineral resources reaching greater depths, rock mass engineering is being applied more extensively in tunnels, mines, underground spaces, slopes, and high dam structures. The scale, depth, and complexity of these projects continue to grow [2]. Traditional empirical design methods, which largely rely on historical engineering experience and simplified assumptions, often fall short in providing sufficient reliability and safety assurance when confronted with complex rock mass structures, multi-directional stress states, groundwater effects, and disturbances such as seismic activity or mining operations [3,4]. Modern rock mass engineering requires designs that not only ensure safety but also consider economic efficiency, construction speed, and environmental sustainability, placing greater demands on rock mechanics theory, experimental techniques, and numerical simulation.

Recent years have witnessed rapid progress in rock mechanics, particularly regarding experimental techniques, numerical simulation, constitutive modeling, and the integration of artificial intelligence (AI). These advances provide new theoretical foundations and technical pathways for predicting rock mass stability, optimizing support designs, and controlling geological hazards. For instance, developments in experimental techniques, such as triaxial compression, rock fatigue testing, dynamic loading, and micro-observation methods, enable precise observation and quantitative analysis of micro-crack evolution, damage mechanisms, and macroscopic mechanical responses in rocks [5]. In numerical simulations, approaches based on the finite and discrete element methods and multi-physics coupling models allow engineers to predict rock mass responses, analyze potential failure modes, and optimize support and construction schemes during the design phase [6]. The advancement of AI technologies offers new tools for data processing, intelligent monitoring, and prediction in rock mass engineering. For example, deep learning can analyze rock microstructures, crack distributions, and failure evolution, enabling rapid and accurate assessment of complex rock mass behaviors [7,8].

This Special Issue focuses on key cutting-edge topics in the field in recent years. It brings together multidimensional research encompassing laboratory testing, theoretical analysis, numerical simulation, intelligent algorithms, and field monitoring, aiming to comprehensively present the latest progress in rock mechanics and engineering for addressing practical challenges. This issue particularly focuses on the mechanisms of rock mass failure, the optimization of support structure design, the development of intelligent

https://doi.org/10.3390/app152312752

monitoring technologies, and the integrated applications of these aspects in tunneling, mining, slope engineering, and underground space development. It is intended to provide a high-level platform for academic and engineering exchange and collaboration, promoting high-quality development in rock mass engineering.

In geological engineering, rock masses are often composed of different rock layers. Interfaces between layers represent natural weak planes that are prone to initiate and propagate interfacial cracks, whose evolution directly affects engineering stability [9]. Research on the failure mechanisms of composite rock masses under dynamic loading not only helps reveal the internal crack propagation laws under mining-induced disturbances but also provides a key scientific basis for optimizing stope layout, designing support systems, and preventing rock bursts and dynamic failures [10]. Studies on the mechanical behavior of rocks containing pre-existing holes further reveal how hole geometry and location govern rock failure modes. Integrated research conducted using uniaxial compression tests, digital image correlation technology, and theoretical analysis shows that stress disturbances induced by holes significantly alter crack initiation locations, propagation paths, and damage evolution characteristics, providing reliable support for the design of underground structures and the evaluation of the stability of surrounding rock [11]. Regarding crack initiation and propagation mechanisms, multi-physical field monitoring combining acoustic emission and CT scanning enables 3D visualization and quantitative characterization of crack nucleation, while theoretical analyses reveal the intrinsic relationship between nucleation behavior and rocks' ultimate failure modes [12]. In studies of layered rock mass failure, fracture mechanics analyses clearly explain how increasing notch length enhances stress concentration effects, leading to reductions in shear strength and stiffness [13]. This theoretical achievement effectively addresses the difficulty of precisely defining crack propagation mechanisms in layered rock under shear loading, providing a solid theoretical basis for predicting the stability of layered surrounding rock and designing supports in tunnels, mines, and other engineering projects.

With the promotion of green mining concepts, microwave-assisted rock breaking has emerged as an efficient new technology. However, the effects of microwave heating on water-bearing rock masses and their influence on mechanical properties are not yet fully understood. Research in this area is of significant practical importance for the efficient extraction of deep hard rock [14]. The rapid development of numerical simulation technology has facilitated a shift in rock mass modeling from traditional homogeneous rock mass to more realistic gradient rock mass structures, offering new theoretical frameworks for slope stability analysis and the simulation of underground engineering responses [15]. Concurrently, studies on joint network structures indicate there is a significant nonlinear coupling effect between joint strength parameters and connectivity, and this effect plays a dominant role in the evolution of overall rock mass strength [16].

The application of AI in rock mass engineering also shows broad prospects. Researchers have developed a transferable, scalable, and interpretable intelligent fracture recognition framework, providing key technical support for digital rock physics and intelligent geological engineering [17]. Multimodal recognition methods based on deep learning and image-processing technology can automatically identify fractures, mineral distributions, and borehole structures within rock-sample CT images and achieve 3D reconstruction and modeling, offering new tools for the refined characterization of complex rock masses [18]. Furthermore, using deep networks such as ResNet50 for identifying microstructural damage in heat-treated rocks enables automatic discrimination of thermal damage and visualization of the micro-damage evolution process, providing innovative methods for geothermal development and post-fire engineering assessment [19]. Combining discrete element simulation with deep learning, systematic studies have been conducted on the evolution of granular material fabric under different loading conditions and its impact on macroscopic mechanical behavior, demonstrating good physical interpretability and engineering applicability [20].

Rock mass engineering is transitioning from an "empirical–static" paradigm to a "data–dynamic" one. Comprehensive results from experiments, numerical simulations, and field monitoring indicate that only through deep integration of rock micro-damage mechanisms, multi-source monitoring data, and intelligent prediction models can the safe, low-carbon, and efficient construction and operation of underground engineering truly be achieved. This transformation requires researchers to possess interdisciplinary knowledge spanning rock mechanics, structural mechanics, computational science, and data science as well as more flexible design concepts and real-time monitoring methods in engineering practice. We anticipate that these research outcomes will serve as a shared language for the academic and engineering communities, stimulate further interdisciplinary collaboration, and encourage continued exploration of deeper, larger, and more complex challenges in rock mass engineering, thereby revealing new directions for future development.

Author Contributions: Conceptualization, Q.L.; writing—original draft preparation, Q.L.; writing—review and editing, R.C. and J.M. All authors have read and agreed to the published version of the manuscript.

Funding: This research was funded by the National Natural Science Foundation of China (52404087), the Hunan Provincial Natural Science Foundation of China (2024JJ6383), and the Outstanding Youth Project of Hunan Provincial Education Department (23B0444).

Conflicts of Interest: The authors declare no conflicts of interest.

References

1. Xie, S.; Li, J.; Wang, S.; Li, X. An experimental investigation on the dynamic shear characteristics of wet joints. *Int. J. Rock. Mech. Min. Sci.* **2025**, *194*, 106193. [CrossRef]
2. Wang, L.; Yang, D.; Zhang, Y.; Li, W.; Kang, Z.; Zhao, Y. Research on the reaction mechanism and modification distance of oil shale during high-temperature water vapor pyrolysis. *Energy* **2022**, *261*, 125213. [CrossRef]
3. Liu, Z.; Cui, G.; Zhou, C.; Lan, C. Impact mechanism of fabric changes in different classes of redbeds under static water on their degradation of physical properties. *Rock. Mech. Bull.* **2025**, *4*, 100194. [CrossRef]
4. Fu, Q.; Yang, J.; Gao, Y.; Li, C.; Song, H.; Liu, Y.; Wu, W. Combined blasting for protection of gob-side roadway with thick and hard roof. *J. Rock. Mech. Geotech. Eng.* **2024**, *16*, 3165–3180. [CrossRef]
5. Lin, Q.; Zhang, S.; Lin, H.; Zhang, K.; Fan, W.; Huang, C.; Shao, Z.; Lan, M. Failure behavior of jointed rock masses containing a circular hole under compressive-shear load: Insights from DIC technique. *Theor. Appl. Fract. Mech.* **2025**, *139*, 105089. [CrossRef]
6. Ma, Q.; Liu, X.; Tan, Y.; Wang, R.; Xie, W.; Wang, E.; Liu, X.; Shang, J. Experimental study of loading system stiffness effects on mechanical characteristics and kinetic energy calculation of coal specimens. *Rock. Mech. Rock. Eng.* **2024**, *57*, 9941–9957. [CrossRef]
7. Jinge, Z.; Yujing, J.; Sunhao, Z.; Dongqi, S.; Zhenjiao, S.; Hongbin, C. Data-driven machine learning approaches for predicting the shear strength of rock joints. *Rock. Mech. Bull.* **2025**, *4*, 100209. [CrossRef]
8. Zhu, D.; Yu, B.; Wang, D.; Zhang, Y. Fusion of finite element and machine learning methods to predict rock shear strength parameters. *J. Geophys. Eng.* **2024**, *21*, 1183–1193. [CrossRef]
9. Alneasan, M.; Behnia, M.; Bagherpour, R. Analytical investigations of interface crack growth between two dissimilar rock layers under compression and tension. *Eng. Geol.* **2019**, *259*, 105188. [CrossRef]
10. Xue, H.; Wang, Y.; Yang, W.; Zhang, P.; Xiao, H.; Zhang, Y.; Zhang, Y. Experimental and numerical study of damage evolution and fracture characteristics of three-layer composite rocks under dynamic loading. *Appl. Sci.* **2025**, *15*, 10369. [CrossRef]
11. Zhou, Y.; Kong, D.; Zuo, Y.; Wen, Z.; Xu, M.; Chen, F.; Zhang, Y. Damage characteristics and fracture evolution laws for prefabricated hole rock specimens. *Theor. Appl. Fract. Mech.* **2025**, *136*, 104805. [CrossRef]
12. Wang, C.; Zhou, B.; Zhu, C.; Li, C.; Sun, L. Crack nucleation mechanism of rock fracture. *Eng. Fract. Mech.* **2025**, *325*, 111328. [CrossRef]
13. Mohammadi, S.D.; Mortezaei, R.; Sarfarazi, V.; Moayedi Far, A. Crack growth properties and AE phenomena in notched hard/soft layered rock under punch shear test; experimental test and numerical simulation. *Results Eng.* **2025**, *27*, 106652. [CrossRef]

14. Wang, P.; Lin, Y.; Chen, D.; Yin, T. Dynamic mode I fracture toughness and damage mechanism of dry and saturated sandstone subject to microwave radiation. *Appl. Sci.* **2025**, *15*, 9500. [CrossRef]
15. Sarfarazi, V.; Torabi-Kaveh, M.; Moayedi Far, A. Effects of weathering depth and thickness on rock failure: Experimental approach and particle flow code simulation. *J. Rock. Mech. Geotech. Eng.* **2024**, *16*, 4638–4653. [CrossRef]
16. Tang, Q.; Xie, W.; Jing, S.; Wang, X.; Su, Z. Experimental and numerical investigation on the mechanical behavior of rock-like material with complex discrete joints. *Rock. Mech. Rock. Eng.* **2024**, *57*, 4493–4511. [CrossRef]
17. Li, M.; Grasselli, G. Segmenting identified fracture families from 3D fracture networks in montney rock using a deep learning-based method. *J. Rock Mech. Geotech. Eng.* **2025**, *17*, 6120–6129. [CrossRef]
18. He, C.; Sadeghpour, H.; Shi, Y.; Mishra, B.; Roshankhah, S. Mapping distribution of fractures and minerals in rock samples using res-VGG-UNet and threshold segmentation methods. *Comput. Geotech.* **2024**, *175*, 106675. [CrossRef]
19. Gao, Y.; Yu, Z.; Yin, Q.; Sui, H.; Feng, T.; Liu, Y. Deep-learning analysis of microstructural deterioration in rocks exposed to high temperatures. *J. Rock Mech. Geotech. Eng.* **2025**, *17*, 6279–6292. [CrossRef]
20. Irani, N.; Salimi, M.; Golestaneh, P.; Tafili, M.; Wichtmann, T.; Lederer, J. Deep learning-based analysis of true triaxial DEM simulations: Role of fabric and particle aspect ratio. *Comput. Geotech.* **2024**, *173*, 106529. [CrossRef]

applied sciences

MDPI

Article

Experimental and Numerical Study of Damage Evolution and Fracture Characteristics of Three-Layer Composite Rocks Under Dynamic Loading

Huajun Xue [1,2,*], Yanbing Wang [3], Weihong Yang [1,2], Pengda Zhang [3], Hui Xiao [1,2], Yaoyao Zhang [3] and Yuanjian Zhang [1,2]

1 Inspection and Certification Co., Ltd., MCC Group, Beijing 100088, China; cumtbzhehang@163.com (W.Y.); mccxiaohui2025@163.com (H.X.); yuanjianzhang2024@163.com (Y.Z.)

2 Central Research Institute of Building and Construction Co., Ltd., MCC Group, Beijing 100088, China

3 School of Mechanics and Civil Engineering, China University of Mining and Technology-Beijing, Beijing 100083, China; zpdc18236117952@163.com (Y.W.); sqt2400604097@student.cumtb.edu.cn (P.Z.); yrssub2022@163.com (Y.Z.)

* Correspondence: xuehuajun7792@sina.com; Tel.: +86-18618325042

Abstract

In order to study the damage evolution and fracture characteristics of rock with different composite modes in three layers under dynamic loading, rock specimens with different composite modes were made by using three materials: sandstone, marble and granite. The dynamic fracture impact test was carried out by using the Hopkinson pressure bar impact loading system, the voltage signal on the Hopkinson pressure bar was calculated and processed, and the crack propagation mode of the specimen was captured by using a high-speed camera, and the stress wave characteristics, stress time–history relationship and energy change characteristics of rocks with different composite modes were studied. At the same time, combined with Distinct Lattice Spring Model numerical simulation, the fracture process of the specimen was inverted, and the changes in stress intensity factor, stress change and load–displacement change in monitoring point were analyzed to compare the dynamic fracture behavior differences between different composite rocks. The results show that the dynamic fracture process captured by the high-speed camera has a good fit with the crack propagation process simulated by numerical simulation. When marble is used as the upper material, the energy transmittance is larger, and the transmission energy ratio between sandstone and granite is basically the same due to the large difference in hardness. When the comprehensive hardness of the specimen is the same, the smaller the hardness of the material at the cracking position, the faster the cracking will be, and the smaller the hardness of the second layer of the specimen at the cracking position, the faster the cracking speed of the specimen. In terms of dynamic fracture toughness, for specimens with little difference in hardness, when the impact end material is sandstone, the dynamic fracture extreme value of the specimen is lower, and when the sandstone material is used as the impact end material, it is more likely to crack. When the first layer of material is the same, the dynamic fracture toughness of the specimen with less hardness of the second layer of material is smaller, and the easier the crack development is.

Keywords: rock mechanics; SHPB; composite rocks; dynamic fracture; damage evolution

1. Introduction

The rapid advancement of economic development has led to an expansion in the scale of geotechnical engineering works covering both surface and subsurface projects, resulting in increasingly complex and varied rock mechanics challenges [1]. These engineering issues are closely related to rock instability under impact loads and the propagation characteristics of stress waves in flawed rocks, which are pertinent to activities such as shaft excavation, mining operations, engineering blasting, dam foundation construction, and refuge excavation [2]. As a natural material susceptible to temperature fluctuations, weathering, tectonic movements, and other factors, rock formations inherently possess certain natural defects [3,4]. These defects can manifest as cracks, laminations, joints, and other irregularities, contributing to the inhomogeneity and anisotropy of the rock material [5]. Consequently, this leads to complex deformation behaviors and mechanisms of failure. Failure to accurately identify the defects in the mechanical properties of rocks may hinder efforts to prevent and manage rock-related disasters, complicating the assurance of long-term stability in underground engineering structures [6–12]. Thus, it is crucial to investigate the mechanical properties and characteristics of rocks with bedding defects. Current research works predominantly focus on the overall damage mechanisms of rock and the damage mechanisms associated with pre-existing fissures within the rock body. However, there is a notable lack of studies addressing the damage mechanisms of rock materials under various composite conditions, particularly in multi-layered contexts. Therefore, efforts to study the dynamic mechanical properties of three-layer composite rocks under dynamic loading conditions hold an immense value and relevance for both academic and engineering applications.

The research works on the static mechanical properties of composite rock materials were conducted by some scholars. Ding [13] investigated the modal transformation of stress waves at the interfaces of layered composite rocks. This process was reproduced through finite element analysis, and the progression of crack development was also examined. Regarding the sandstones and mudstones from deep dense reservoirs, layered composite rocks were created using similar material modeling tests. Uniaxial compression tests were performed, supplemented by an acoustic emission (AE) system and a digital image correlation (DIC) system, to determine the strength, elastic modulus, and other physical-mechanical parameters of the layered composite rocks. Xu Ke [14] studied the influence of fracture position on the mechanical properties of layered composite rocks under different confining pressures. A numerical model of composite rock samples with prefabricated single fractures was established using the particle flow software PFC2D, and the stress–strain, volume change, and related stress thresholds of the composite rock samples were analyzed. The study found that the increase in confining pressure changes the dominant factor of rock sample failure from fracture position to confining pressure, and the fracture in hard rock weakens the rock sample strength the least. Dong [15] carried out uniaxial compression tests on two varieties of rock-like specimens with differing layer counts. The evolution of the apparent strain field in these specimens was analyzed using digital image correlation (DIC). By applying energy dissipation theory and a damage evolution model, acoustic emission technology was employed to investigate the damage evolution process of the specimens. Park et al. [16] investigated gypsum-like composite rocks with three closed fractures via uniaxial compression, noting that wing (tensile) and secondary (shear) cracks initiate from fracture tips, with their initiation stress increasing with fracture inclination, spacing, and overlap rate, and coalescence modes linked to geometric parameters. Sagong et al. [17] tested rock-like composites with 3 and 16 fractures, finding crack types consistent with fewer-fracture specimens; multi-fracture specimens showed lower crack initiation and coalescence stress, with coalescence mostly in a "columnar" mode dependent on geometric parameters.

The research works on the dynamic mechanical properties of composite rock materials were conducted by some scholars. Li Jin [18] transformed raw materials from deep rock formations into various types of layered composite rock specimens in consideration of the acidic environment and the effects of different angles at the interface layers, leading to experimental investigations into the mechanical properties and damage mechanisms of these layered composite rocks. Zhao Guoyan [19] employed a split Hopkinson pressure bar (SHPB) to perform impact tests on composite sandstone featuring non-penetrating fractures of varying counts and depths. This research established correlations between dynamic peak stress, dynamic modulus of elasticity, damage modes, fractal dimensions of the damaged specimens, energy dissipated per unit volume, and the energy absorption rate in relation to the fissure characteristics. Additionally, Li Dejian [20] employed the finite element method-discrete element method (FDEM) for simulating Brazilian disc splitting experiments on composite coal rock specimens with beddings. The study identified four distinct splitting damage modes that emerged with increasing angles of the laminations and analyzed the tensile strength and energy dissipation values. Guéry et al. [21] established a micromechanical model of three-phase composite geomaterials composed of clay matrix, calcite and quartz, considering the non-associated dilatant elastoplasticity of clay matrix and elastic damage of calcite (including microcrack closure effect). Through nonlinear homogenization, they revealed the regulatory mechanism of mechanical properties of each component on the dynamic response of composite materials.Bikong et al. [22] proposed a micro-macro model of clay rock, pointing out that subcritical propagation of anisotropic microcracks in clay matrix is the key mechanism for its time-dependent deformation. Using two-step homogenization and considering microcrack interaction and mineral inclusions, they found that mineral composition and microcrack distribution significantly affect the dynamic creep characteristics of composite rocks.

At present, most existing research works focus on the overall rock failure mechanism and the failure mechanism of pre-fissured rock masses in defective rocks; however, there are relatively few studies on the failure mechanism of rock materials under different multi-layer composite modes. Therefore, the basic and applied research on the dynamic mechanical properties of three-layer composite rocks under dynamic loading holds significant research value and engineering significance. In this work, a Hopkinson pressure bar impact loading system was operated to conduct dynamic fracture impact tests on three-layer composite rock specimens. A high-speed camera was used to capture the crack propagation patterns of the specimens. The research aims to investigate the characteristics of stress waves, the relationship between stress and time, and the energy variations in the rock under various composite modes. Through the numerical simulation using DLSM, efforts were made to invert the fracture process of the specimens and analyze the variations in the stress intensity factor, the stress changes at monitoring points, and the load–displacement changes. This further facilitated a comparison of the difference in dynamic fracture behavior among different composite rocks, and a study of the damage evolution and fracture characteristics of three layers of rocks with different composite modes under dynamic loading.

2. Test Program and Fabrication of Specimens

2.1. Test Programand Apparatus

A split Hopkinson pressure bar (SHPB) system (Figure 1) was adopted, which has widely recognized reliability in dynamic loading tests. The system consists of four parts: a loading drive system, a pressure bar system, a data acquisition system, and an image acquisition system. The functions and parameters of each part are as follows:

(1) Loading drive system: including launch chamber, high-pressure gas chamber, pressure control valve, high-pressure nitrogen cylinder, etc.
(2) Pressure bar system: composed of impact bar, incident bar, transmission bar, absorption device, etc.
(3) Data acquisition system: composed of strain gauges, junction box, ultra-dynamic strain gauge, oscilloscope, computer, etc.
(4) Image acquisition system: high-speed camera, supplementary lighting device, etc.

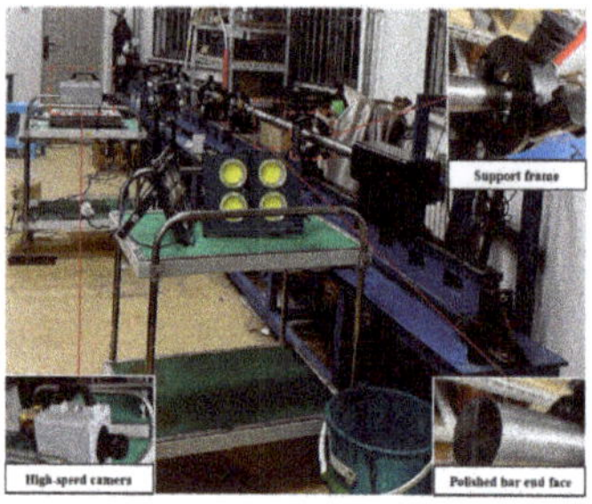

Figure 1. SHPB test system.

The elastic bars in the pressure bar system are made of high-strength alloy steel. Taking the bar made of 40rC alloy steel as an example, its elastic limit can reach 800 MPa, and under an impact pressure of 800 MPa, an impact velocity of 40 m/s can be achieved. To ensure the smooth transmission of stress pulses at the contact surfaces, the two end faces of the elastic bars were polished to make the contact surfaces flat, smooth, and well-contacted. The bar ends are shown in Figure 1. To meet the requirement of fixing the pressure bar and replacing elastic bars of different diameters and materials at any time, a support frame as shown in Figure 1 is mostly used. Figure 1 shows the high-speed camera for recording the test process, with a maximum frame rate of 100,000 frames per second. Higher resolution requires higher lighting conditions. Due to the limited power of the supplementary lighting device, in this study, when using a professional high-speed camera, the image resolution was set to VGA (896 $\times$ 848) and the shooting speed was 100,000 frames per second.

To explore the dynamic fracture characteristics of the three-layer composite specimen, six unique experimental programs were prepared. Figure 2 presents a physical depiction of the three-layer composite rock specimen, whereas Figure 3 gives a schematic representation of the same. For example, the specimen illustrated in Figure 2c is referred to as the sandstone-marble-granite (SDH) which consists of an upper layer of sandstone, a middle layer of marble, and a lower layer of granite. This configuration incorporates various rock materials in each layer, with the two layers adhered together using epoxy resin. The beddings are aligned at an angle of 0°, with the lower beddings spaced 6 mm from the tip of cut seam, and the space between the two layers also measuring 6 mm. The corresponding schematic for this physical specimen is depicted in Figure 3c, where the diagonal area represents sandstone, the square area denotes marble, and the grid area denotes marble, while the white area denotes granite.

Each specimen was subjected to loading in the direction of the incident bar impacting the specimen while in a clamped state, with the contact point located at the top of the half disk. Given that the transmissive bar end of the SHPB system utilized in the experiment is a conventional plane, it would not be feasible to directly apply loading via three-point bending. Therefore, a specially designed three-point bending support was affixed to the end of the transmissive bar to accommodate the loading requirements of the Notched Semi-Circle Bending (NSCB) specimen.

(**a**) Marble-Sandstone-Granite (DSH) (**b**) Marble-Granite-Sandstone (DHS) (**c**) Sandstone-Marble-Granite (SDH)

(**d**) Sandstone-Granite-Marble (SHD) (**e**) Granite-Marble-Sandstone (HDS) (**f**) Granite-Sandstone-Marble (HSD)

Figure 2. Physical drawing of three-layer composite rock specimen.

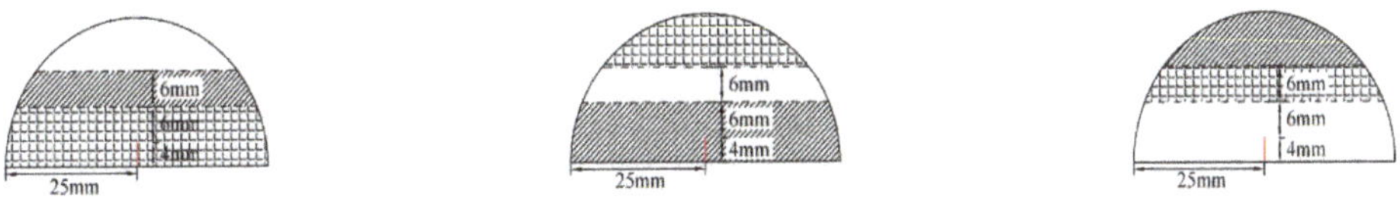

(**a**) Marble-Sandstone-Granite (DSH) (**b**) Marble-Granite-Sandstone (DHS) (**c**) Sandstone-Marble-Granite (SDH)

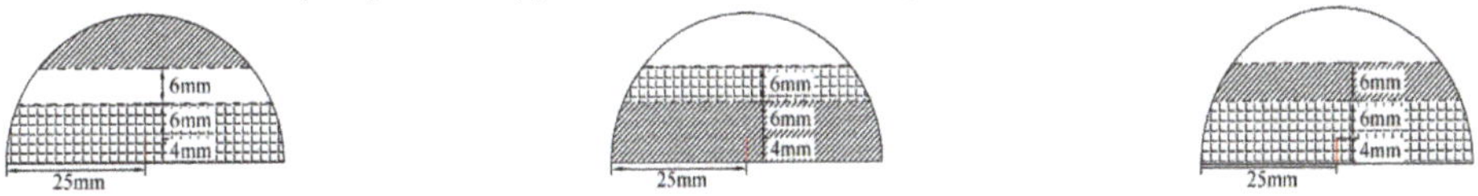

(**d**) Sandstone-Granite-Marble (SHD) (**e**) Granite-Marble-Sandstone (HDS) (**f**) Granite-Sandstone-Marble (HSD)

Figure 3. Schematic diagram of three-layer composite specimen.

2.2. Fabrication of Specimens

Following the extraction of the cylindrical core from the rock, the core was shaped into a standard disc specimen in accordance with the dimensions recommended by the International Society for Rock Mechanics for disc experiments, specifically maintaining a diameter-to-thickness ratio of 2:1 [23]. The dimensions of the disc specimen measured 50 mm in diameter and 25 mm in thickness. Subsequently, the half-disc specimen was prepared with pre-existing cracks and transformed into the NSCB specimen, as illustrated in Figure 1. Each half-disc specimen was then cut at the bottom center. The bottom center of each half-disc was processed through cutting, with a prefabricated crack of 4 mm in depth located along the centerline of the half-disc, forming a reasonable ratio with the specimen's diameter and thickness. This configuration not only ensures that the crack tip can effectively initiate the propagation of the main crack but also avoids excessive reduction in the overall strength of the specimen due to an overly deep cut. The crack had a width of 0.3 mm, and its tip was sharpened using a diamond wire saw, ensuring that the width of the crack tip was precisely controlled to 0.1 mm. The fundamental physico-mechanical parameters of the three types of rock materials are set out in Table 1.

Table 1. Physical and mechanical parameters of test materials.

Rock Material	Density (kg/m^3)	Poisson's Ratio	Elastic Modulus (GPa)	Protodyakonov's Hardness	UCS (MPa)
Marble	2500	0.3	3.0	8	90.5
Sandstone	2600	0.24	4.6	6	62.3
Granite	3000	0.2	18.4	12	150.1

3. Test Results and Analysis

3.1. Analysis of Specimen Damage Process

Figure 4a–f illustrate the dynamic fracture expansion graphs of six groups of three-layer composite specimens captured by a high-speed camera. The fracture processes of these groups exhibit remarkable similarity; specifically, cracks initiate at the tip of cut seam and expand vertically upward. The cracks display only minor lateral displacements, influenced by variations in hardness across the laminar surfaces, which cause changes in the direction of crack propagation before reverting to the original direction for continued expansion. The three-layer composite specimens exhibit two modes of damage: crack propagation and localized extrusion. Additionally, both primary and secondary cracks emerge at the pre-existing cracks and at the interfaces of the interlayer structure. Figure 4a depicts the marble-sandstone-granite composite specimen (DSH), while Figure 4f presents the granite-sandstone-marble composite specimen (HSD). A notable similarity between these specimens is the presence of sandstone as the intermediate layer. However, the materials at the incident and exit ends differ, with one specimen featuring marble and the other granite. A comparative analysis of the damage forms reveals variations from the bottom to the top of the specimens. Within the three layers, the width of the cracks varies from the initiation point to the crack tip, with both sandstone specimens being susceptible to slight deviations in crack direction at the material interface. This phenomenon is attributed to the differing hardness of the two rock types, resulting in a minor degree of stress concentration at the bedding surface.

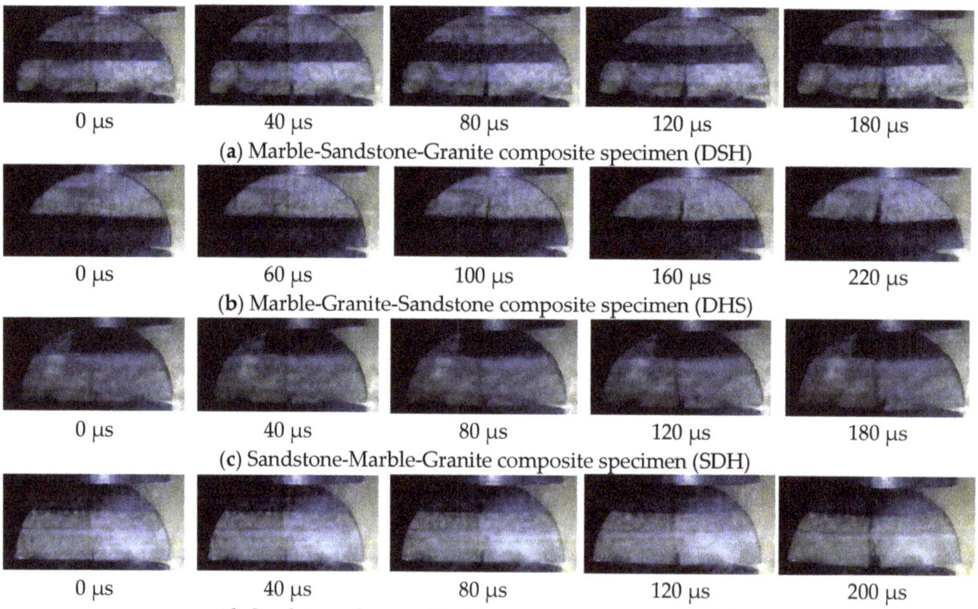

(**a**) Marble-Sandstone-Granite composite specimen (DSH)
(**b**) Marble-Granite-Sandstone composite specimen (DHS)
(**c**) Sandstone-Marble-Granite composite specimen (SDH)
(**d**) Sandstone-Granite-Marble composite specimen (SHD)

Figure 4. *Cont.*

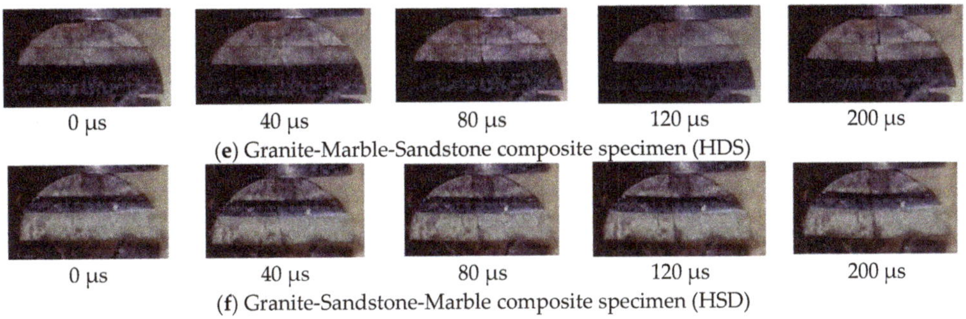

Figure 4. Schematic diagram of three-layer composite specimen (captured by high-speed camera).

A comparison of the damage duration of the specimens reveals that the total duration for crack propagation is greater in HSD than in DSH. Furthermore, the initiation of cracks occurs earlier in HSD compared to DSH. This can be attributed to the presence of the same weak surface of bedding in both specimens. At this stage, the propagation of cracks is influenced by the material properties beyond the weak surface of bedding. The overall extent of material damage can be assessed by examining the destruction of the upper and lower layers, which also experiences an additional stress concentration effect due to material hardness, ranked in a descending order: Granite > Marble > Sandstone. Consequently, the greater hardness of the HSD bedding surface results in a more pronounced offset damage effect. This principle is similarly applicable to specimens with identical interlayer materials, such as HDS and SDH, as well as SHD and DHS. When the interlayer material remains constant, a higher hardness at the incident end correlates with an increased duration of crack extension and an earlier onset of crack initiation.

3.2. Stress Wave Characterization

Figures 5 and 6 illustrate the stress waveform curves of the three-layer composite specimens. The incident waveforms for the six composite specimens exhibit a high degree of similarity. Notably, the DSH and DHS specimens demonstrate a slower response during the initial phase of the descending section, while the remaining specimens show comparable behavior in this segment. The ascending sections of all six specimens are also largely consistent. By observing the transmitted waveforms, the SDH, DSH, and DHS specimens show resembling waveforms, characterized by smooth sine curves that align closely, with their peaks nearly equating. In contrast, the transmitted wave forms of the other three specimens exhibit greater irregularity, featuring more fluctuations and higher peaks compared to the first three specimens. Regarding the transmitted wave forms, the peak value of the SDH specimen is lower than that of the DSH specimen. When marble is utilized as the impacted end material for the composite specimen, it results in improved wave impedance matching, generating a greater number of transmitted waves [24]. Similarly, the peak value of the HSD specimen is less than that of the SHD specimen, indicating that the wave impedance effect of sandstone surpasses that of granite. Consequently, the wave impedance matching of the composite specimen with marble is superior to that of the specimen using granite as the impacted end material. This leads to the conclusion that among the three materials, marble exhibits the highest wave impedance effect, while granite has the lowest, implying that the DHS specimen has a larger peak value than the HDS specimen. From the perspective of inherent material properties, differences in elastic modulus also contribute to the aforementioned variations in stress wave characteristics. As shown in Table 1, the elastic moduli of the three rock types follow the order: granite > sandstone > marble. The elastic modulus directly affects the propagation velocity of stress waves in materials, while wave impedance, a key indicator of stress wave transmission efficiency, has a matching degree that depends on wave velocity differences between materials. The ascending segments of the incident wave curves for the six specimen groups are nearly overlapping, indicating consistent initial transmission laws of stress waves at the interface between the pressure bar and the specimen during the initial impact stage. However, the early descending segments of DSH and DHS are gentler, reflecting that the lower elastic modulus of marble results in a more gradual attenuation rate of stress waves.

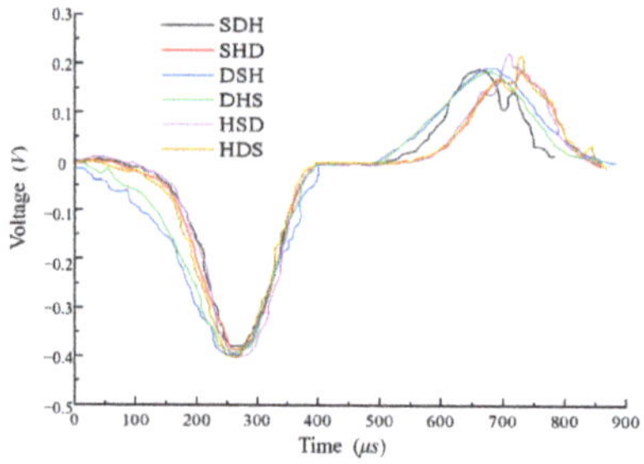

Figure 5. Incident wave stress waveform of three-layer composite specimen.

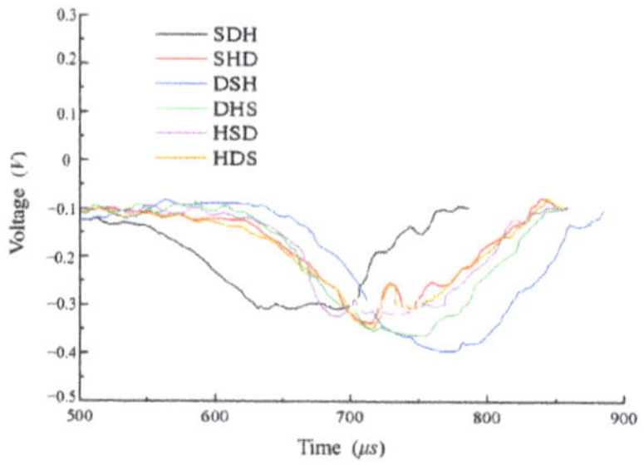

Figure 6. The stress waveforms of the three-layer composite specimen.

3.3. Analysis of Stress Characteristic Curve and Peak Intensity Change

Figure 7 illustrates the time–history curve of the stress intensity factor for the three-layer composite specimen. The overall trend of the stress intensity factor is consistent across the six specimens, beginning at a value of zero. Initially, the curve fluctuates around this zero value before gradually increasing over time, reaching a peak after approx. 100 μs of fluctuation, followed by a decline back to zero after a certain period of time. Among the various specimen combinations, the SDH specimen demonstrates the earliest rise and peak, indicating that it achieves fracture toughness first. This phenomenon can be attributed to the order of material impact on the specimen: sandstone, marble, and granite, arranged from lowest to highest hardness (hardness: sandstone < marble < granite). The SHD specimen follows closely behind the SDH specimen in the ascending phase, with the curves of SDH and SHD positioned to the left of the curves of the other specimens, suggesting that, when the overall hardness of the specimens is equivalent, a lower hardness at the crack location results in faster cracking. Furthermore, the SDH specimen is positioned to the left of the SHD curve, indicating that when both the overall hardness and crack location are the same, a lower hardness in the second layer of the specimen leads to quicker cracking initiation. This trend is also observed in the other groups of specimens.

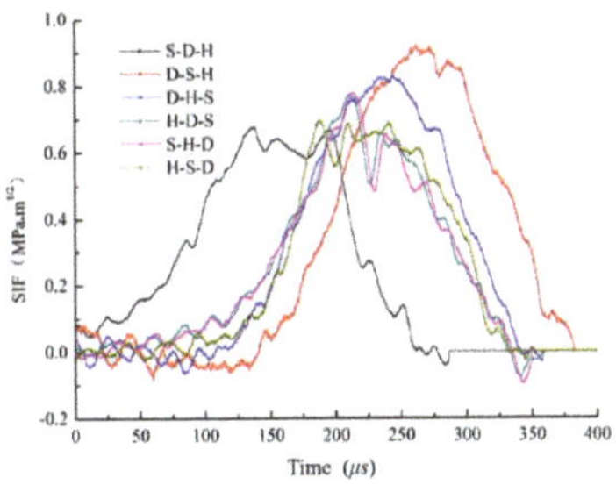

Figure 7. The Stress intensity factor time–history curve of the three-layer composite specimen.

3.4. Analysis of Dynamic Fracture Toughness

Table 2 presents the dynamic fracture toughness values of the specimens, which are derived from the curve illustrated in Figure 5 and plotted as the dynamic fracture toughness curve for the three-layer composite specimens in Figure 8. A comparison of the data reveals that the dynamic fracture toughness of the SDH and SHD specimens is lower. This suggestis that when the impact section material is sandstone, the dynamic fracture extremum of the specimen decreases. This indicates that sandstone materials are more prone to fracturing when utilized as impact section materials, particularly in respect of hardness. When comparing the SDH and SHD specimens, both of which feature sandstone as the impact end material, it is noted that the dynamic impact toughness of the SDH specimen is lower than that of the SHD specimen. The hardness of the two-layer materials follows the order: SDH < SHD. When the first layer of material remains constant, a decrease in the hardness of the second layer correlates with reduced dynamic fracture toughness, facilitating crack propagation. This trend is also observed in the remaining groups of combined specimens. However, the data for the HSD specimen deviates from this pattern. According to the aforementioned pattern, the dynamic fracture toughness of the HSD specimen should rank second. However, the significant hardness disparity between granite and sandstone leads to increased susceptibility to damage at the interface of these materials. Consequently, the specimens with substantial hardness differences are not covered by the aforementioned pattern.

Table 2. Dynamic fracture toughness of three-layer composite laminates.

Test Specimen	Dynamic Fracture Toughness/MPa·m$^{1/2}$
SDH	0.67605
DSH	0.92144
DHS	0.83249
HDS	0.77472
SHD	0.78014
HSD	0.69563

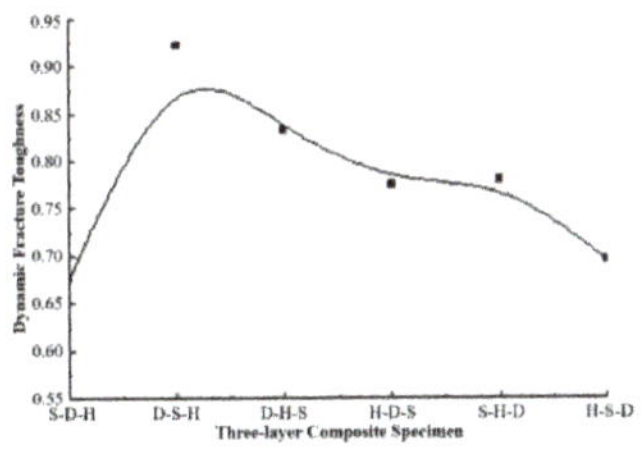

Figure 8. Comparison of dynamic fracture toughness of three-layer composite laminates.

3.5. Characterization of Energy Evolution in the Impact Process

The propagation and energy dissipation characteristics of stress waves in rocks under dynamic impact loading are crucial for revealing the dynamic response mechanisms of composite rocks. According to the energy transmission law when stress waves pass through rock joints, the energy during impact can be divided into incident energy E_I, reflected energy E_R, transmitted energy E_T, and dissipated energy E_D, which satisfy the energy conservation relationship:

$$E_I = E_R + E_T + E_D \tag{1}$$

For three-layer composite rocks, different material combinations and bedding interface properties significantly affect the efficiency of energy transmission and dissipation. Table 3 presents the energy parameters of six three-layer composite specimens (SDH, DSH, SHD, HSD, DHS, HDS) during impact, including specific values of incident energy, reflected energy, transmitted energy, and dissipated energy, as well as their proportions relative to the incident energy.

Table 3. Energy-to-energy ratio of composite specimen.

Specimen Type		Incident Energy E_I/J	Reflected Energy E_R/J	Transmitted Energy E_T/J	Dissipated Energy E_D/J	E_R/E_I	E_T/E_I	E_D/E_I
Three-layer composite	SDH	30.79	6.75	8.99	15.04	21.93%	29.19%	48.86%
	DSH	40.08	9.90	16.13	14.04	24.71%	40.23%	35.04%
	SHD	31.03	7.23	8.58	15.22	23.30%	27.64%	49.04%
	HSD	33.04	7.73	9.36	15.94	23.40%	28.35%	48.24%
	DHS	36.57	9.06	12.65	14.86	24.77%	34.58%	40.63%
	HDS	31.12	6.97	8.86	15.29	22.39%	28.47%	49.13%

Figure 9 illustrates the energy ratio curves for the three-layer composite specimens. Firstly, regarding the reflected energy ratio, all specimens maintain a relatively consistent level, due to the fact that the six specimens are constructed from the same three materials, with variations only in the arrangement of the upper, middle, and lower layers. As a result, there are only minor differences in the reflected energy ratio. In terms of the transmittance energy ratio, the specimen composed of marble-sandstone-granite (DSH) exhibits a higher value compared to the sandstone-marble-granite (SDH), sandstone-granite-marble (SHD), and granite-sandstone-marble (HSD) specimens. Notably, there is no significant difference between the sandstone-granite-marble (SHD) and granite-sandstone-marble (HSD) specimens. Furthermore, the marble-granite-sandstone (DHS) specimen demonstrates a greater transmittance than the granite-marble-sandstone (HDS) specimen. This indicates that when marble is used as the impacting material, the transmittance energy is enhanced. The transmittance energy ratios for sandstone and granite are largely similar due to their comparable hardness. Regarding the dissipated energy ratio, the sandstone-marble-granite (SDH) specimen surpasses the marble-sandstone-granite (DSH) specimen, while no significant difference is observed between the sandstone-granite-marble (SHD) and granite-sandstone-marble (HSD) specimens. Additionally, the granite-marble-sandstone (HDS) specimen exhibits a higher dissipated energy ratio than the marble-granite-sandstone (DHS) specimen.

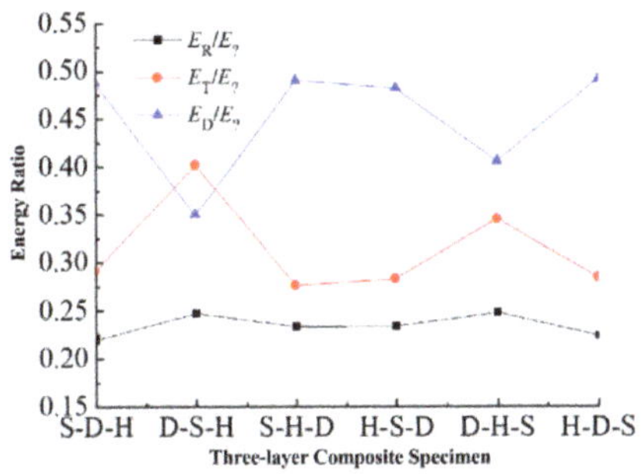

Figure 9. The energy ratio curve of three-layer composite specimen.

4. DLSM Numerical Simulation Analysis of Stress Wave Transmission and Stress Damage in Three-Layer Composite Rock Materials

4.1. Modeling

DLSM numerical simulation reproduces the discontinuous failure characteristics of rock materials through the particle discrete element method, while incorporating the finite element concept to handle stress wave propagation problems. The model of the three-layer composite specimen is illustrated in Figure 10. Initially, a straight-cut groove half-disc model identical to the overall dimensions of the test specimen was constructed [25]. The model measures a diameter of 50 mm and a thickness of 25 mm, with the diameter of the spherical particles set at 25 mm. The prefabricated cut seam measures 4 mm in length and 0.5 mm in width. To simulate the beddings present in the test specimen, a layer of spherical particles with varying parameters of the rock matrix was set, with a thickness of 0.5 mm. The parameters of the rocks were selected from the data measured in the actual experiment, as shown in Table 1. Parameters such as particle stiffness and bond strength refer to the conventional values used in discrete element simulations of rock-like materials.

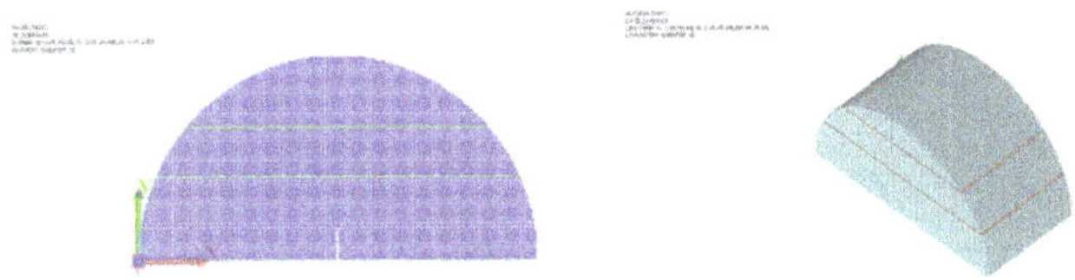

(**a**) 2D Hoek-Brown model (**b**) 3D block model

Figure 10. Rock test composite model.

The boundary conditions of the model are depicted in Figure 11, illustrating the clamped and strained conditions of the specimen on the Hopkinson bar during testing. Since the transmissive bar was not secured when the incident bar struck the specimen, the boundary conditions for the left and right supports were established as velocity loads with a value of 0. The boundary conditions at the loading point correspond to the dynamic load experienced by the specimen due to the impact of the incident bar, which was measured using the velocity acquisition instrument of the Hopkinson bar during the test. Based on the impact velocity recorded by the Hopkinson rod velocity acquisition instrument, the boundary condition at the loading point is also defined as a velocity load. To standardize the simulated dynamic loading conditions, all models with varying bedding spacing are set to 6000 mm/s, with reference to the average impact velocity observed in the test.

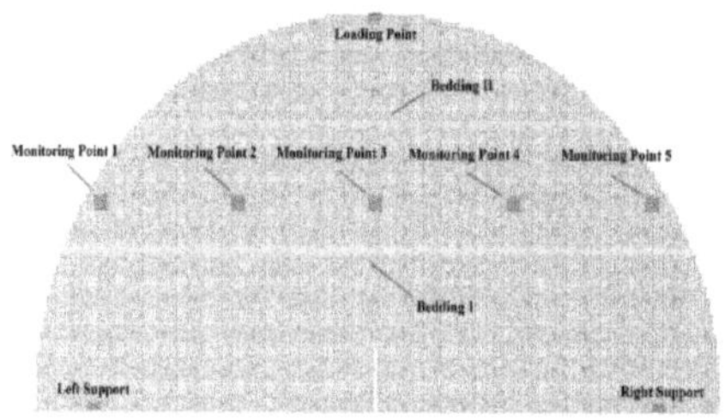

Figure 11. Setting of model monitoring points.

To examine the impact of varying layer spacing on the stress and energy evolution characteristics of the rock matrix situated between two beddings, five monitoring points were established within this interval. The relevant stress and strain parameters for these monitoring points were derived from numerical simulations [26]. It is important to note that the positions of the monitoring points remain consistent across different layer spacing models, with the coordinates for each monitoring point given in Table 4.

Table 4. Coordinates of monitoring points.

Coordinate	X	Y	Z
Monitoring Point 1	5	13	12.5
Monitoring Point 2	15	13	12.5
Monitoring Point 3	25	13	12.5
Monitoring Point 4	35	13	12.5
Monitoring Point 5	45	13	12.5

4.2. *Dynamic Fracture Process Analysis*

Figure 12 illustrates the fracture simulation process of the three-layer composite specimen. Initially, a comparative analysis of the test and simulation of the fracture process reveals that the sandstone-marble-granite (SDH) specimen exhibits a relatively similar crack propagation pattern in both scenarios. The crack initiates at the tip of cut seam and extends vertically upward. Upon reaching the bedding, the crack continues to propagate along the bedding in an upward direction. After a specific distance, the crack expands once more in the direction vertical to the bedding until it reaches the top of the specimen. A notable distinction is that the crack in the simulated specimen appears smoother, with fewer "jagged teeth" along its edge. In contrast, the edge of the crack in the test specimen is rougher and features more counts of jagged teeth. This discrepancy arises from the heterogeneous nature of the actual test specimen, whereas the simulated specimen is composed of homogeneous material, which does not produce jagged cracks during the tearing process of spherical particles. In comparison to the simulated results, the test and simulation results for the marble-sandstone-granite (DSH) specimen also demonstrate a high degree of similarity. In this case, the cracks do not initiate from the tip of cut seam, but rather begin at the lower end of the bedding and propagate upward along the bedding, ultimately penetrating perpendicularly near the upper end of the bedding and expanding to the top of the specimen. The difference between the two lies in the later stages of crack development. The test results indicate that the cracks initially expand in a straight line before abruptly bending upward, while the simulated results show a bending upward of propagation. Significant discrepancies were found when comparing the test and simulation results of sandstone-granite-marble (SHD) specimens. Firstly, in the experimental results, the cracks did not propagate from the tip of cut seam; instead, they initiated damage directly from the lower end of the bedding. In contrast, the simulation results indicated that cracking initiated at the tip of cut seam. Secondly, there is a notable difference in the location of the cracks that penetrate the upper end of the bedding. The experimental results showed penetration occurring from the centerline of the specimen, whereas the simulation results indicated penetration from a position closer to the upper end of the bedding.

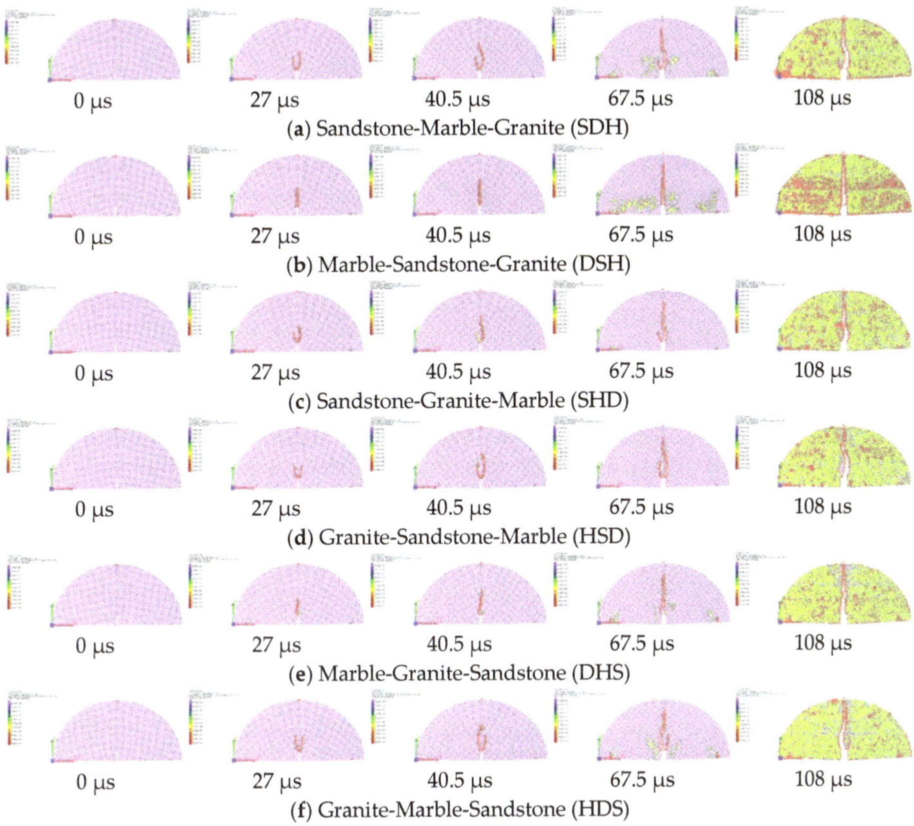

Figure 12. Dynamic fracture simulation of three-layer composite specimen.

4.3. Analysis of Crack Propagation Displacement Fitting Curve

Figure 13 illustrates the displacement load curve resulting from the simulation of the three-layer composite specimen. It is evident that the slope of the ascending section of the curve remains consistent when the upper material is identical. Notably, the steepest slope is observed when the upper material is granite, while the shallowest slope is observed with sandstone as the upper material. Regarding the peak load, there is no significant distinction between the SDH and SHD configuration. Specifically, when the upper material is sandstone, the slope of the ascending section remains unchanged. Although the peak load does not exhibit considerable variation with changes in the middle- and lower-layer materials, the specimen with the highest peak load is identified as HSD, followed by HDS (which is slightly lower), and DSH which is slightly higher than DHS. This outcome contradicts the earlier assumptions regarding the behavior of the previous two specimen groups, where it was anticipated that the peak load would be greater with sandstone as the middle layer compared to marble or granite.

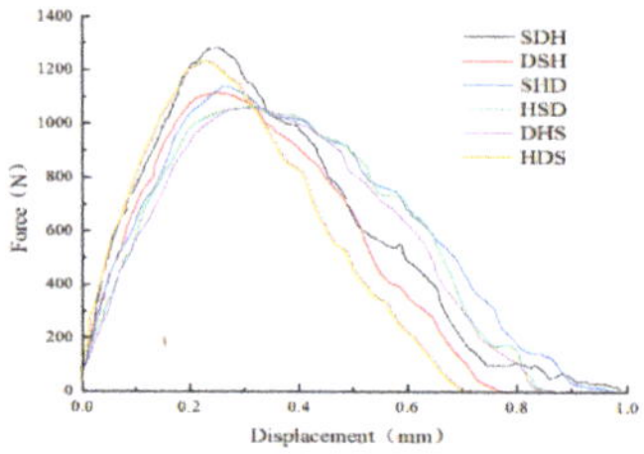

Figure 13. The displacement–load curve obtained from the simulation process.

This deviation is mainly related to the synergistic effect of the mechanical properties of the upper layer material and the hardness difference between layers. On one hand, since the elastic modulus and hardness of granite are significantly higher than those of sandstone and marble, when granite is used as the upper layer material, its high strength can significantly enhance the overall stiffness of the specimen, resulting in generally higher load peaks for HSD and HDS. This "bearing-dominant effect" of the upper layer material outweighs the influence of the middle layer material properties on the load peak. On the other hand, the middle layer of HSD is sandstone, with granite and marble as the upper and lower layers, respectively, and the interlayer hardness difference is relatively moderate, leading to weak stress concentration effect and difficulty in early failure of the bedding interface, thus supporting a higher load peak. In contrast, the upper layers of DSH and DHS are marble, whose hardness is lower than that of granite, resulting in lower overall stiffness. Even if the middle layer is sandstone, it is difficult to offset the impact of insufficient strength of the upper layer material, leading to their load peaks being lower than that of HSD. In addition, although there is a hardness difference between granite and sandstone in HSD, the high strength of the upper granite can inhibit early interface instability, further enhancing its bearing capacity, which ultimately leads to the deviation from the expected law.

4.4. Analysis of Stress–Strain Characteristics at Monitoring Points

Figure 14 presents the stress and strain curves obtained from numerical simulations at monitoring point 3. Monitoring points 1 and 5 are positioned symmetrically relative to the midline of cut seam, resulting in nearly equal forces on both sides. Consequently, the stress and strain curves for these points are largely similar. The average stress and strain values for points 1 and 5 are calculated to derive their respective curves, and a similar approach is applied to obtain the stress and strain curves for points 2 and 4. The stress and strain time–history curves for points 2 and 4 are derived using the same method [27].

Figure 14a illustrates the stress and strain curves recorded at point 3, where the initial layer consists of sandstone in the simulation. The stress-time curve begins to increase gradually around 12 μs, indicating that the stress wave has penetrated through layer II and is accumulating energy within the rock matrix situated between the two layers. At approx. 25 μs, the stress curve experiences a rapid ascent, with the combined curves of the two layers exhibiting varying rates of ascending. Notably, the curve corresponding to a 6-mm spacing between the layers shows the least steep rise, suggesting that the pore space between the layers is being densified at a faster rate. Once the layers reach their densification limit, the underlying rock matrix begins to experience stress, resulting in an increase in pressure at the monitoring point. As the bedding attains the compaction threshold, the stress on the rock matrix below escalates swiftly. The curves for the specimens with two different material combinations reached their respective stress peaks at 43 μs and 45 μs, measuring 18.20 MPa and 19.80 MPa, respectively, with a change ratio of 8.79% as observed in relation to the order of the marble-granite combination.

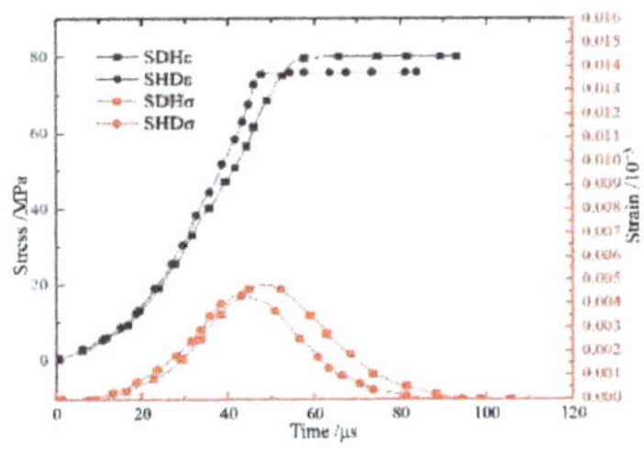

(a) First-layer sandstone

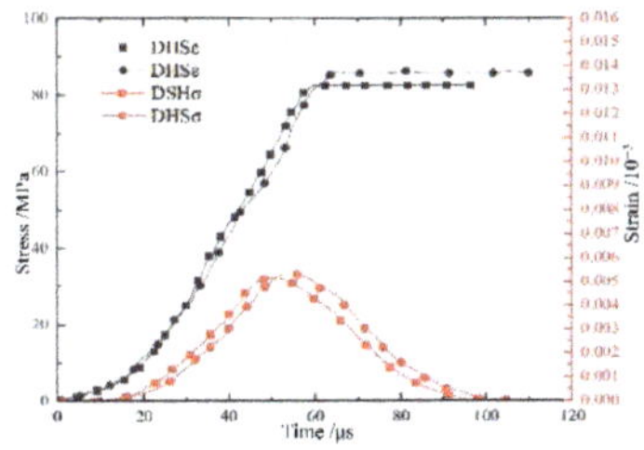

(b) First-layer marble

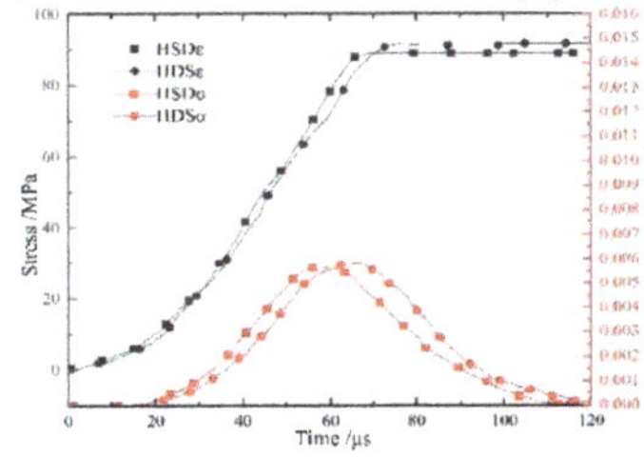

(c) First-layer granite

Figure 14. Stress–strain time–history curve at monitoring point 3.

The strain-time curve depicted in the figure is categorized into three stages. In the initial stage, spanning from 0 to 25 μs, the impact load begins to exert force on the tip of the specimen, primarily affecting the densification of bedding II. During this period, the stress conduction is relatively poor, resulting in a gradual increase in the strain curve. As the bedding is compacted to the strain limit, the curve transitions into the second phase. In the course of simulation, once each section of the specimen attains the compaction limit, stress conduction becomes more uniform, leading to a rapid increase in strain at monitoring point 3. As the spacing between the beddings widens, bedding II approaches the end of the impact load, allowing for a more rapid compaction due to the dynamic load. This process drives elastic strain into the rock matrix situated between the two beddings, causing the upward slope of the curve in phase II to rise in correlation with the increased spacing of the bedding. The rate of increase in the curve gradually diminishes in the later stages, then transitions into phase III, where cracks begin to propagate near the monitoring point. This results in the rapid destabilization of the rock matrix between the beddings, the elastic deformation capacity approximates the limit. The two sets of curves reach their respective strain peaks at 48 μs and 50 μs, respectively, measuring 0.013×10^{-3} and 0.014×10^{-3}. The observations indicate that both the peak stress and peak strain at monitoring point 3 exhibit similar patterns of change, with their peak values and the time points of these peaks showing consistent trends over time. The peak stress point is identified at the moment when crack propagation reaches monitoring point 3. At this point, which is situated near the rock matrix, significant compression pushes the material to its elastic deformation limit. This results in the formation of numerous fine cracks and damage, leading to a maximum bearing capacity, after which the stress begins to decline rapidly. The greater spacing of the bedding causes bedding II to endure earlier compression due to the impact load, so the rock matrix between the beddings becomes crushed. Consequently, the overall strength of this section is more susceptible to premature failure. The strain peak value observed at monitoring point 3 exhibits a clear relationship with the bedding material. If the first layer is composed of sandstone, a minimal difference in hardness between the first and second layers results in an earlier appearance of the stress–strain peak value, particularly when the hardness of the second or third layers is lower. This phenomenon occurs because the initial impact load is more effectively absorbed by the first layer, leading to damage in the bedding matrix. The lower the hardness of the material, the less energy is required for its destruction. Additionally, when the bedding surfaces are in close proximity, the impact load can drive the two laminar rock substrates to fail prematurely, thereby diminishing the overall strength of the section. The time required for cracks to penetrate bedding I is reduced, allowing cracks to reach the monitoring point more swiftly [28], which destabilizes the rock matrix in that area and results in reduced strain under identical stress conditions.

The stress–strain characteristics of the three-layer composite bedding specimen are influenced by the overall hardness of the materials utilized. This specimen comprises three materials, and it is observed that a lower hardness in the material located near the impact end results in a reduced overall hardness of the specimen. Consequently, this leads to a diminished peak stress–strain value, with the time taken to achieve peak stress and strain occurring earlier. Additionally, the time to reach peak strain and peak stress is also influenced by the differences in material hardness. In the case where the first layer is granite, followed by marble, the aforementioned trend is followed. However, when the first layer consists of sandstone, the significant disparity in hardness between sandstone and granite, along with the brittleness of granite, results in the peak SHD occurring earlier than the SDH. This can be attributed to the contrasting hardness and the inherent brittleness of granite.

Figure 15a–c illustrate the time–history curves of stress and strain for sandstone monitoring points 2 and 4 in the course of simulation. The stress curves initially exhibit a gradual loading phase, followed by a rapid increase after 25 μs, ultimately reaching a peak stress. This peak coincides with the moment when the crack propagates to bedding II, resulting in destructive separation between bedding II and the rock matrix situated between the two beddings. Consequently, the internal bonding force diminishes, leading to a reduction in stress transmission from the loading point to the section, and a rapid decline in stress commences. In comparison to Figure 15d–f, under identical material conditions, the stress and strain peaks at monitoring points 2 and 4 are elevated by 13.13% and 14.77%, respectively, when compared to monitoring point 1. Additionally, the stress peaks at monitoring points 1 and 5 are increased by 22.82% and 23.63% relative to those at monitoring points 2 and 4. This indicates that the farther a monitoring point is from the centerline, the higher the peak stress that can be attained at that location. The ratio of peak stress increase relative to the integrated hardness of the specimen is most evident at monitoring point 3, while the correlation between peak stress and integrated hardness diminishes as the distance from the centerline of the cut seam increases. The maximum stress observed at the monitoring point is associated with the residual strength at that location. A lower integrated hardness of the specimen facilitates the susceptibility of bedding II to the compressive effects of the impact load, which can lead to the failure of the rock matrix situated between the beddings. Additionally, as the distance from the centerline increases, the overall strength of the rock matrix near the monitoring point is less influenced by the damage caused by extrusion, allowing for a higher peak stress to be attained. The curve exhibits a gradual increase during stage I. In comparison to the curve at monitoring point 1, the duration of stage II for each curve is extended, indicating that the peak strain point occurs later. When the crack propagates to monitoring point 1, monitoring points 2 and 4 remain intact, and their residual strength enables the strain to continue increasing. As the crack reaches bedding II, the rock matrix between the two beddings progressively loses its bond with bedding II, which coincides with further degradation of the specimen. The rock matrix near monitoring points 2 and 4 ultimately loses its load-bearing capacity and attains the peak strain point. A closer examination of Figure 15a–c indicates that the peak strain at monitoring points 1 and 5 occurs later than at the first two points, suggesting that the farther one is from the centerline, the slower the destruction rate of the adjacent rock matrix and the greater the residual strength.

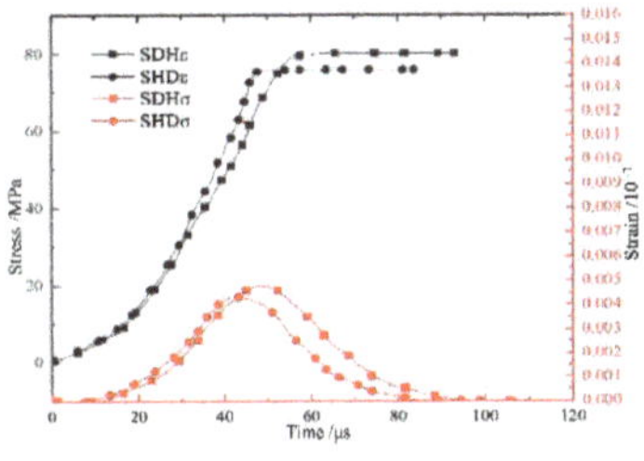

(**a**) First-layer sandstone (Monitoring points 2 and 4)

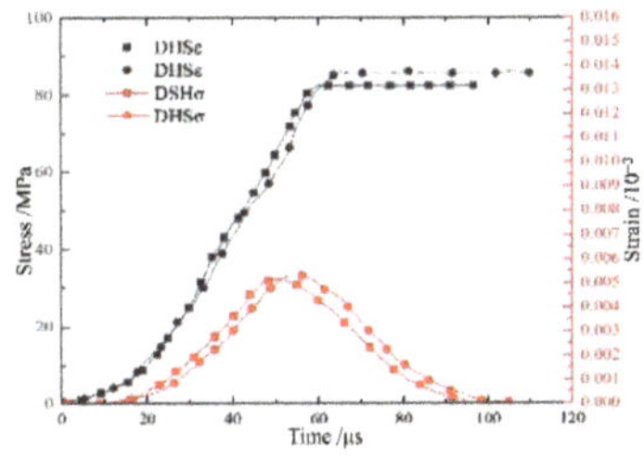

(**b**) First-layer marble (Monitoring points 2 and 4)

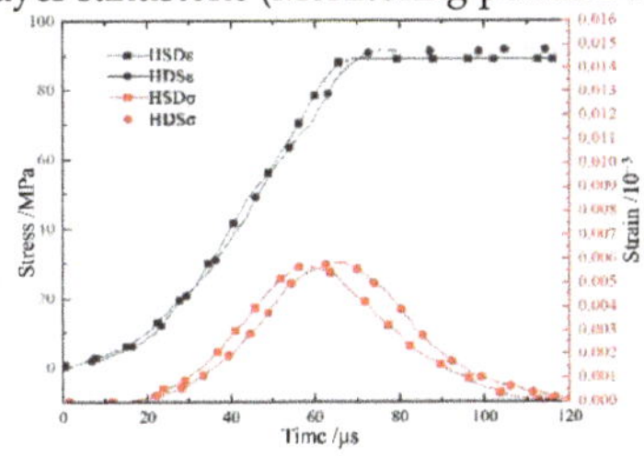

(**c**) First-layer granite (Monitoring points 2 and 4)

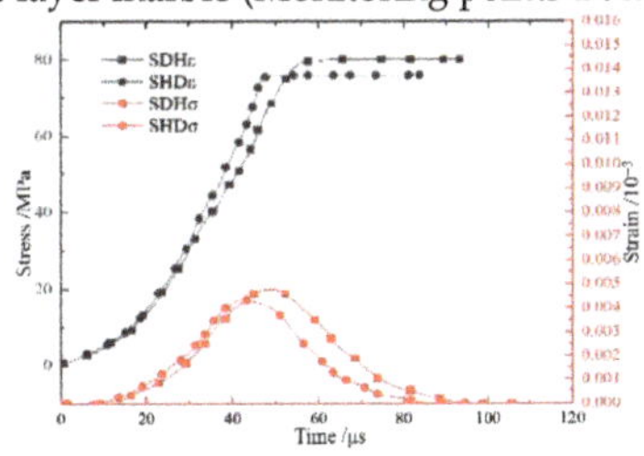

(**d**) First-layer sandstone (Monitoring points 1 and 5)

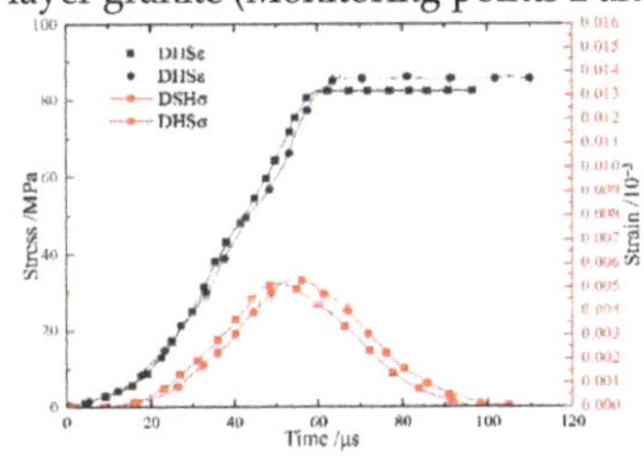

(**e**) First-layer marble (Monitoring points 1 and 5)

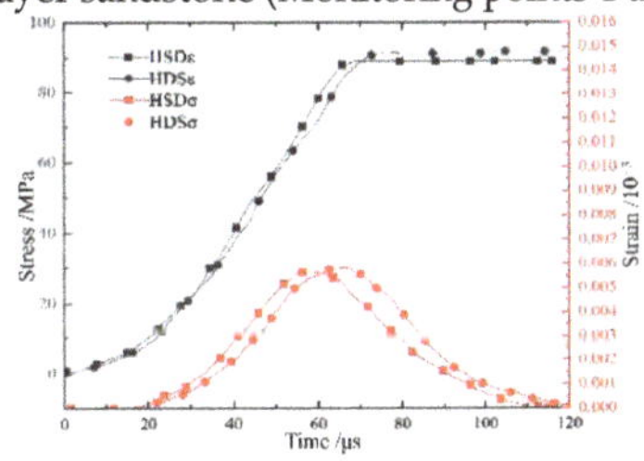

(**f**) First-layer granite (Monitoring points 1 and 5)

Figure 15. Stress–strain time–history curves at monitoring points 2, 4 and 1, 5.

5. Conclusions

(1) The fracture mechanism of the composite specimen primarily initiates at the tip of cut seam, subsequently propagating along the upper surface and penetrating the bedding structure. When marble is utilized as the impact end material, it tends to generate secondary cracks due to stress concentration rather than originating from the tip of cut seam. Furthermore, the fracture morphology observed in the three-layer composite specimens exhibits a notable similarity, without significant distinctions.

(2) When subjected to identical impact conditions, the incident wave forms across various composite specimens are nearly identical. However, the peak value of the transmitted wave varies with the position switch of each layer. It has been observed that marble demonstrates superior wave impedance matching when utilized as the material subjected to the impact. The ratio of reflected energy in the three-layer composite specimens remains relatively constant, while the transmitted energy ratio indicates that when marble serves as the upper material, the energy transmission ratio is greater. In contrast, the transmitted energy ratios for sandstone and granite are largely comparable due to their substantial hardness differences.

(3) When the integrated hardness of a specimen remains constant, a lower hardness at the crack initiation site leads to a more rapid onset of crack initiation. If the crack initiation site is identical, a reduced hardness in the second layer of the specimen results in quicker crack propagation. Regarding dynamic fracture toughness, specimens exhibiting minimal hardness variation demonstrate that when the impact material is sandstone, the dynamic fracture peak value is diminished, causing these specimens more susceptible to cracking. For specimens where the first layer material is uniform, those with a softer second layer material exhibit lower dynamic fracture toughness, facilitating crack expansion. Conversely, specimens with a significant hardness disparity experience faster damage, which deviates from this observed trend.

(4) The dynamic fracture behavior simulated by DLSM aligns closely with experimental findings. The load–displacement curve indicates that the peak load for specimens with granite as the upper layer is significantly greater than that of other specimens. Among specimens sharing the same upper material, variations in peak load are evident. Specifically, when the lower material is marble or granite, the peak load surpasses that of specimens with sandstone as the lower layer. However, when the upper material is either marble or granite, no substantial difference is observed between these specimens and those with sandstone as the lower layer. Notably, the three-layer composite specimen exhibiting the highest peak load is HSD, followed closely by HDS, while DSH exhibits a slight increase over DHS.

(5) The stress–strain characteristics of the three-layer composite laminated specimen are influenced by the overall hardness of the material. Specifically, a lower hardness in the material composition located near the impact end results in a reduced overall hardness of the specimen, which in turn leads to a diminished peak stress–strain value and a delayed time to achieve peak stress and strain. The time to reach peak strain and peak stress is also contingent upon the differences in material hardness. Where the first layer is composed of granite, followed by marble, the aforementioned trend is followed. However, when the first layer consists of sandstone, the significant hardness disparity between sandstone and granite, along with the brittleness of granite, alters this dynamic. Consequently, the time to reach the peak for the sandstone layer occurs prior to that of the granite layer.

(6) The findings of this study hold significant application prospects for practical engineering. In geotechnical engineering involving dynamic loading such as tunnel excavation, mine blasting, and hydraulic engineering, for composite rock masses composed of multiple layers of different rocks, their damage evolution and fracture characteristics under impact loading can be predicted based on the properties (e.g., hardness, elastic modulus) of each layer of rock. This provides a theoretical basis for optimizing engineering design and formulating reasonable construction schemes.

Author Contributions: Conceptualization, H.X. (Huajun Xue); methodology, Y.W.; software, H.X. (Hui Xiao), Y.Z. (Yuanjian Zhang); formal analysis, W.Y.; data curation, P.Z.; writing—review and editing, Y.Z. (Yaoyao Zhang), P.Z. All authors have read and agreed to the published version of the manuscript.

Funding: This work was financially supported by the National Key Research and Development Program of China (2023YFC3805900); the Key Research and Development Project of Metallurgical Corporation of China Ltd. (YCC20 Kt01); the Special Project of Inspection and Certification Co., Ltd., MCC (2024ZYZX04).

Informed Consent Statement: Not applicable.

Data Availability Statement: All data that support the findings of this study are included in this manuscript.

Acknowledgments: We would like to express our gratitude to Inspection and Certification Co., Ltd., MCC, Central Research Institute of Building and Construction Co., Ltd., MCC Group, and the School of Mechanics and Civil Engineering for providing the experimental platforms and equipment that supported the smooth implementation of our tests and numerical simulations.

Conflicts of Interest: Authors Huajun Xue, Weihong Yang, Hui Xiao and Yuanjian Zhang are employed by the company Inspection and Certification and Central Research Institute of Building and Construction Co., Ltd., MCC Group. The remaining authors declare that the research was conducted in the absence of any commercial or financial relationships that could be construed as a potential conflict of interest.

References

1. Hopkinson, J. On the rupture of iron wire by a blow. *Proc. Manch. Lit. Philos. Soc.* **1872**, *11*, 40–45.
2. Hopkinson, J. Further experiments on the rupture of iron wire. *Proc. Manch. Lit. Philos. Soc.* **1872**, *1*, 119–121.
3. Al-aasm, I.S.; Mrad, C.; Packard, J. Fluid compartmentalization of Devonian and Mississippian dolostones, western Canada sedimentary basin: Petrologic and geochemical evidence from fracture mineralization. *Can. J. Earth Sci.* **2019**, *56*, 265–305. [CrossRef]
4. Lenton, A.; Mcinnes, K.L.; O'grady, J.G. Marine projections of warming and ocean acidification in the australasian region. *Aust. Meteorol. Oceanogr. J.* **2015**, *65*, S1–S28. [CrossRef]
5. Lei, Q.H.; Gao, K. A numerical study of stress variability in heterogeneous fractured rocks. *Int. J. Rock Mech. Min. Sci.* **2019**, *113*, 121–133. [CrossRef]
6. Odling, N.E. Scaling and connectivity of joint systems in sandstones from western Norway. *J. Struct. Geol.* **1997**, *19*, 1257–1271. [CrossRef]
7. Shi, H.; Song, L.; Wang, D. Strength model for damaged sandstone based on power-law distribution of crack length. *J. China Univ. Min. Technol.* **2022**, *51*, 263–272.
8. Shi, H.; Zhang, H.; Song, L.; Wu, Y. Variation of strata pressure and axial bolt load at a coal mine face under the effect of a fault. *Arch. Min. Sci.* **2019**, *64*, 351–374.
9. Li, D.; Wan, Q.; Zhu, Q.; Hu, C. Experimental study on mechanical properties and failure behaviour of fractured rocks under different loading methods. *J. Min. Saf. Eng.* **2021**, *38*, 1025–1035.
10. Zhang, M.; Wang, W.; Wang, Q.; Zhang, S. Dynamic failure process and strain-damage evolution law of sandstone based on SHPB experiments. *Explos. Shock Waves* **2021**, *41*, 37–50.
11. Guo, D.M.; Yan, P.Y.; Zhang, Y.S.; Zhang, W.; Ding, Y.Y.; Zhao, L.M. Experimental research on the sprayed concrete-surrounding rock combined body subjected to cyclic impact loadings. *J. Vib. Shock* **2019**, *38*, 105–111.
12. Xie, B.J.; Wang, X.Y.; Lv, P.Y. Dynamic properties of bedding coal and rock and the SHPB testing for its impact damage. *J. Vib. Shock* **2017**, *36*, 117–124.
13. Ding, X.; Yang, Y.; Zhou, W.; An, W.; Li, J.; Ebelia, M. The law of blast stress wave propagation and fracture development in soft and hard composite rock. *Sci. Rep.* **2022**, *12*, 17120. [CrossRef] [PubMed]
14. Xu, K.; Xiao, T.L.; Zhao, Y.F.; Shen, H.C. Effect of fracture location on mechanical properties of composite rock samples under different surrounding pressure. *Min. R D* **2023**, *43*, 67–75.
15. Dong, T.; Cao, P.; Lin, Q.; Liu, Z.; Xiao, F.; Zhang, Z. Fracture evolution of artificial composite rocks containing interface flaws under uniaxial compression. *Theor. Appl. Fract. Mech.* **2022**, *120*, 103401. [CrossRef]
16. Park, C.H.; Bobet, A. Crack initiation, propagation and coalescence from frictional flaws in uniaxial compression. *Eng. Fract. Mech.* **2010**, *77*, 2727–2748. [CrossRef]
17. Sagong, M.; Bobet, A. Coalescence of multiple flaws in a rock-model material in uniaxial compression. *Int. J. Rock Mech. Min. Sci.* **2002**, *39*, 229–241. [CrossRef]
18. Li, J. Study on Mechanical Properties and Failure Characteristics of Layered Composite Rock Under Acid Environment. Ph.D. Thesis, East China Jiaotong University, Nanchang, China, 2018.
19. Zhao, G.Y.; Li, Z.Y.; Wu, H.; Wang, E.J.; Liu, L.L. Dynamic failure characteristics of sandstone with non-penetrating cracks. *Rock Soil Mech.* **2019**, *40*, 73–81.
20. Li, D.J.; Qi, H.; Li, C.X.; Feng, J.L. Brazilian disc splitting tests and numerical simulations on coal samples containing bedding planes. *J. Min. Sci. Technol.* **2020**, *5*, 150–159.
21. Guéry, A.A.C.; Cormery, F.; Shao, J.F.; Kondo, D. A micromechanical model of elastoplastic and damage behavior of a cohesive geomaterial. *Int. J. Solids Struct.* **2008**, *45*, 1406–1429. [CrossRef]
22. Bikong, C.; Hoxha, D.; Shao, J.F. A micro-macro model for time-dependent behavior of clayey rocks due to anisotropic propagation of microcracks. *Int. J. Plast.* **2015**, *69*, 73–88. [CrossRef]

23. Zhou, Z.L.; Li, D.Y.; Ma, G.W.; Li, J.C. Failure of rock under dynamic compressive loading. *J. Cent. South Univ.* **2008**, *15*, 339–343. [CrossRef]
24. Zhang, M.T.; Wang, W.; Zhang, S.Y. Numerical Simulation of Failure Process and Damage of Gray Sandstone under Impact Loading. *Blasting* **2020**, *37*, 46–54.
25. Yang, S.Q.; Li, Y.; Huang, Y.H.; Tang, W.L. Particle flow analysis of macroscopic and microscopic mechanical properties of Brazilian disc containing a hole under splitting test. *J. China Univ. Min. Technol.* **2019**, *48*, 984–992.
26. Xie, L.; Zhu, W.; Wang, S.H.; Niu, L.L. Three-Dimensional parallel computing on failure process of rock specimen with a pre-existing circular opening. *Chin. J. Geotech. Eng.* **2011**, *33*, 1447–1455.
27. Yang, S.Q.; Lu, C.H.; Qu, T. Investigations of Crack Expansion in Marble Having a Single Pre-existing Hole: Experiment and Simulations. *J. China Univ. Min. Technol.* **2009**, *38*, 774–781.
28. Sha, R.D.; Liang, Z.Z.; Qian, X.Q. Numerical Simulation of Rock Specimens Failure Mode with a Single Hole Under Dynamic Loading. *J. Water Resour. Archit. Eng.* **2019**, *17*, 65–70+92.

applied sciences

MDPI

Article

Dynamic Mode I Fracture Toughness and Damage Mechanism of Dry and Saturated Sandstone Subject to Microwave Radiation

Pin Wang [1,2,*], Yinqi Lin [1], Duo Chen [1] and Tubing Yin [2]

1 School of Resources, Environment and Materials, Guangxi University, Nanning 530004, China; 2339010144@st.gxu.edu.cn (Y.L.); 18346926860@163.com (D.C.)
2 School of Resources and Safety Engineering, Central South University, Changsha 410083, China; tubing_yin@mail.csu.edu.cn
* Correspondence: pinwang8023@gmail.com

Abstract

Microwave-assisted rock fragmentation has been considered as one of the most promising technologies in rock excavation, but due to the fact that excavation is usually carried out in water-rich environments, understanding the dynamic fracture properties of rocks with different water contents after microwave irradiation is thus desirable. This study employed an enhanced split Hopkinson pressure bar (SHPB) system to perform dynamic fracture tests on pre-cracked semi-circular bending (SCB) specimens. It systematically explores the changes in the mechanical properties of sandstone under both dry and saturated conditions after exposure to 700 W of microwave radiation for 10 min. Infrared thermal imaging was utilized to capture the temperature distribution across the specimens, while digital image correlation (DIC) and high-speed photography were used to simultaneously record the crack propagation process. Based on the principle of energy conservation, the analysis of energy dissipation during fracture was performed, and the micro-damage evolution mechanism of the material was revealed through scanning electron microscopy (SEM). The results demonstrated that saturated sandstone exhibited a more rapid heating response and significantly lower dynamic fracture toughness and fracture energy compared to dry samples after microwave irradiation. These findings indicate that water saturation amplifies the weakening effect induced by microwaves, making the rock more susceptible to low-stress fractures. The underlying damage mechanisms of microwave radiation on water-bearing sandstone were interpreted with the theory of pore water pressure and structural thermal stresses.

Keywords: microwave-assisted rock breaking; fracture process zone; energy dissipation; digital image correlation

1. Introduction

With the rapid growth of the world population and the increasing demand for mineral resources from industrialization, mining companies have been forced to improve production efficiency to meet the growing demand for mineral resources. Given the continuous depletion of high-grade mineral deposits on or near the surface, the high demand for mineral resources is forcing advances in mining and exploration technologies to achieve more sustainable, automated, continuous, and deep mining rock breaking techniques [1,2]. Continuous mining technology has been identified as a key factor in improving mining production efficiency [3–5].

The efficient fragmentation of hard rock and maintaining the long-term stability of the surrounding rock are significant challenges encountered during underground mineral resource exploitation. The traditional drilling and blasting method is unable to meet the requirements of green continuous mining in deep hard rock mines due to its intermittency and risk during mining operations, low energy utilization rate, and the potential for rock damage and water inrush disasters [6–8]. Relevant scholars have proposed rock breaking methods such as mechanical cutting, high-pressure water jets, microwave or laser beams, and high-temperature plasma jets [9–14] and conducted extensive theoretical research and laboratory experiments. Among them, mechanical cutting has the advantages of continuous operation, high construction quality, minimal excavation-induced damage, high safety, and economical operation. It underwent significant development in the latter half of the last century. Notably, the successful development and application of full-section tunnel boring machines (TBMs) along with coal mining machinery have significantly advanced comprehensive mechanized coal resource extraction, greatly enhancing both operational efficiency and safety. However, in metal mines, the extreme wear and tear on cutterheads is a significant issue due to the high strength and abrasiveness of the ore, rendering the mechanical crushing of rocks an exceptionally challenging task [15,16]. To solve this challenge, researchers have proposed a variety of assisted rock breaking methods. Microwave-assisted rock breaking has a good application prospect [17,18] and can effectively improve mechanical wear. Microwave action on hard rock will produce thermal effects, resulting in heating damage to the rock [19–21].

Fracture toughness is a key parameter in the fracture mechanics of rocks, and the current research on rock fracture toughness under microwave radiation mainly focuses on static conditions [22–25]. Zhou et al. [26] used crack extension strain gauges to measure the dynamic fracture crack extension rate of rocks and found that the crack extension rate initially increased due to the fracture process zone (FPZ) and then gradually leveled off. Lu et al. [27] and Dai et al. [28,29] observed that the uniaxial tensile and compressive strengths and point load strength of rocks decreased significantly with increasing microwave power and radiation time. Hartlieb et al. [30], through a combination of indoor experiments and numerical simulations, conclude that the microwave heating effect depends on the depth of microwave penetration and rock size, in addition to the non-homogeneity of the rock material. Qin et al. [31] studied the effects of different microwave radiation durations at a power density of 8×10^6 W/m^3 on granite samples. Xia et al. [32] conducted three-point bending impact loading tests on Laurentian granite NSCB samples using the SHPB system to determine their mode I dynamic fracture parameters, including dynamic initiation toughness, fracture energy, dynamic extension toughness, and crack extension rate. The fracture toughness of interfaces between different rocks or minerals is also crucial for assessing rock mass stability in mining operations [33].

The above studies have focused on the effect of microwaves on the fracture toughness of dry rock [34], but in practice, rock blasting is often carried out in water-rich environments. The presence of water significantly affects the physical and mechanical properties of rock [35–37]. During microwave heating, the presence of pore water enhances the microwave heating capacity of sandstone, exposing the rock to high-temperature thermal shock. Additionally, the volume expansion of water can significantly damage the rock. Dai et al. [38] found that water is a key factor in the deterioration of the rock. Microwave radiation after cooling in water can produce greater damage to the rock. Peinsitt et al. [39] irradiated sandstone, granite, and basalt under both dry and water-saturated conditions in a 2.45 GHz, 3 kW multimode cavity microwave oven and measured the uniaxial compressive strength and ultrasonic velocity of the rocks. Their results indicated that water saturation enhanced the weakening effect of microwaves on rocks, particularly on sandstone, while having minimal impact on granite and a negligible effect on basalt. Deng et al. [40] investigated the effect of water saturation on the kinetic behavior of sandstones under different porosity conditions and its mechanism and found that the porosity of sandstones increased and the kinetic strength and total energy dissipation decreased with the increase in the number of cycles. Li et al. [41] concluded from experiments that as the water content decreases, the long-term strength of the rock sample increases, and the damage becomes more severe. The presence of pore water intensifies the damage caused by microwave radiation, laying a solid foundation for the broader application of microwave-assisted rock breaking technology.

This paper aims to investigate the dynamic fracture mechanical properties of dry and saturated rocks under the action of microwave radiation. The mode I fracture toughness of NSCB samples was measured in dry and water-saturated states under the modified SHPB test system, respectively. The thermal imager was employed to analyze the changes in the heating rate and temperature distribution uniformity of the sample surface. The fracture failure process of the NSCB sample was recorded by a high-speed camera, the deformation field of the sample was analyzed and calculated by digital image correlation (DIC) technology, and the variation of the crack propagation rate and rotational angular velocity of sandstone with the loading rate was obtained. In addition, the study explores the internal mechanisms and differences in fracture toughness between dry and saturated sandstones under microwave radiation by examining the evolution of the microstructure around the tip of the prefabricated crack and changes in the size of the FPZ. This research provides insights that are more aligned with the complex geological environments and stress conditions encountered during microwave-assisted rock breaking, particularly under dynamic disturbance loads.

2. Materials and Methods

2.1. Sample Preparation

The sandstone samples used in this study were sourced from an underground metal mine in Changsha City, Hunan Province, China. These samples are characterized by high porosity and water absorption. The intact sandstone fragments exhibit a brownish-red color and a relatively uniform texture, with no visible surface cracks. X-ray diffraction analysis was used to obtain specific mass fractions of the mineral components, the main components being quartz (41.23%), feldspar (28.41%), calcite (13.04%), and muscovite (9.27%), as well as minor amounts of hematite (5.63%) and chlorite (2.42%). The grain size of the minerals ranges from 0.03 to 0.5 mm, with some larger particles reaching up to 0.6 mm, classifying the sandstone as a fine- to medium-grained sedimentary rock. The basic physical and mechanical properties of the sandstone are presented in Table 1.

Table 1. Essential physical and mechanical property parameters of sandstone (mean values and standard deviation).

	ρ (kg·m^{-3})	P (%)	C_P (m·s^{-1})	C_S (m·s^{-1})	K_{IC} (MPa·m$^{1/2}$)	E (GPa)	v
Mean values	2336.2	13.3	2521.3	1362.9	0.71	6.47	0.28
Standard deviation	±2.17	±0.25	±37.5	±13.36	±0.06	±0.19	±0.01

Note: ρ, density; P, porosity; C_P, longitudinal wave velocity; C_S, transverse wave velocity; K_{IC}, fracture toughness; E, elastic modulus; v, Poisson ratio.

To investigate the dynamic fracture behavior of the selected sandstone, cores were extracted from the same sandstone rock mass using a diamond drill bit with an internal diameter of 50 mm. During the coring process, the direction of the drill bit was maintained consistently to avoid the influence of rock material anisotropy on the test results. Cylindrical cores were cut into 20 mm thick discs using a rock cutter, and then each disc was sawn in half. Because of the saw tooth thickness, only one standard half-disc sample was obtained for one disc sample. An edge notch of 5 mm in length and 1 mm in width was machined on the half-disc sample. In order to make the notch crack tip sharp enough, a diamond circular saw was used to further sharpen and form the crack tip. The schematic diagrams of the NSCB sample and the machined sample are shown in Figure 1.

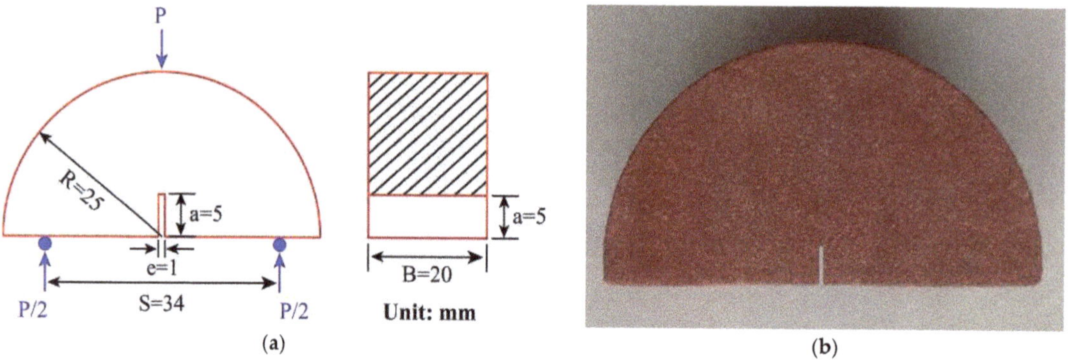

Figure 1. (**a**) Schematic diagram of the NSCB sample; (**b**) processed red sandstone NSCB sample.

A total of 36 NSCB samples were prepared and dried in a vacuum oven at 105 °C for 48 h. The dried samples were then divided into two groups: dry and water-saturated. The water saturation process followed the free absorption method, where the samples were fully submerged in pure water with a pH of 7 and a temperature of 25 °C for 48 h to achieve a saturated state, as shown in Figure 2a. During the soaking process, the sample was taken out every 1 h interval and then dried the surface of the sandstone. The water on the surface of the sandstone was weighed until the weight of the sample remained constant. By changing the weight of sandstone, the moisture content of the sample under different soaking time conditions can be calculated.

$$w = \frac{m_w - m_d}{m_d} \times 100\% \tag{1}$$

where w represents the moisture content of the sample, while m_d and m_w denote the dry weight and wet weight of the sample after soaking for a certain period, respectively. Figure 2b illustrates the variation curve of the water content of sandstone samples with immersion time. During the initial stage of immersion, the water content of the sandstone increases rapidly. However, as immersion time progresses, the growth rate of the water content gradually slows. After approximately 28 h of immersion, the weight of the sandstone no longer increases, indicating that the rock has reached a fully water-saturated state. Through regression analysis, the water content of sandstone and soaking time conform to the following functional relationship:

$$w = w_s\left(1 - e^{-at}\right) = 5.43\left(1 - e^{-0.19t}\right); \quad R^2 = 0.990 \tag{2}$$

where w_s is the saturated water content of the sample, t is the sample soaking time, and a is the fitting parameter.

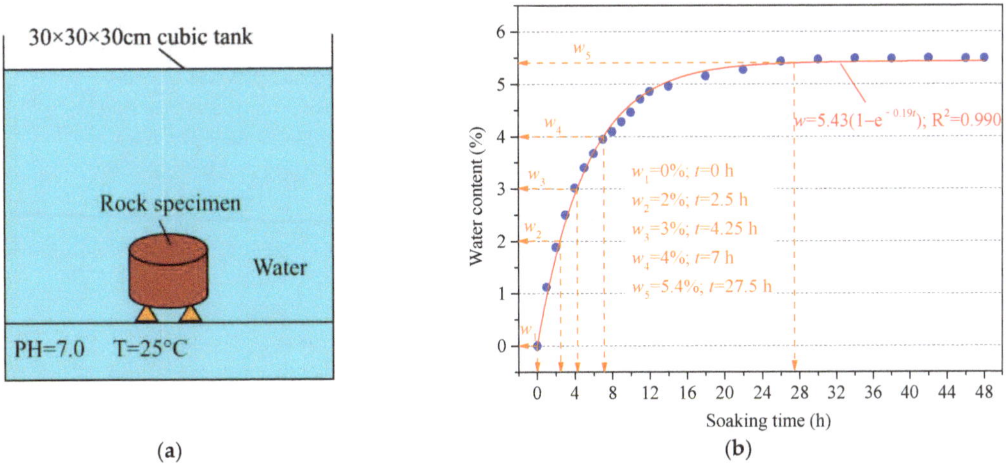

Figure 2. (**a**) Schematic diagram of the water immersion treatment method; (**b**) variation of water content vs. soaking time.

For each group of samples, three samples were used for static fracture strength tests, and fifteen samples were used for dynamic fracture tests with different loading rates.

2.2. Experimental Devices and Techniques

2.2.1. Microwave Heating System

The microwave heating equipment is a multimode cavity industrial microwave oven produced by Hunan Huaye Microwave Technology Co., Ltd., Changsha, China. Figure 3 shows its internal structure and temperature measurement method. The system mainly includes four parts: microwave heating components, control panel, temperature measurement equipment, and cooling system. The microwave generator is composed of four magnetrons with a maximum output power of 1.5 kW. The working voltage of the equipment is 380 V, the microwave frequency is 2.45 GHz, and the maximum output power is 6 kW. The microwave oven has built-in thermocouples and infrared temperature measurement equipment to monitor the surface temperature of the sample in real time.

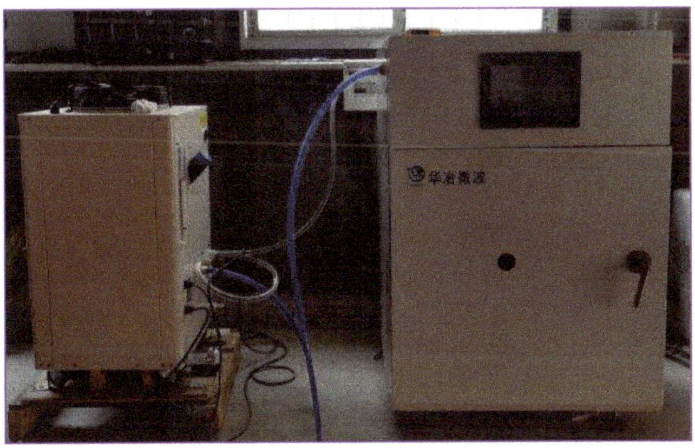

Figure 3. Multimode industrial microwave oven equipment.

In order to study the damage caused by radiation time on dry and saturated sandstone. The prepared dry and saturated sandstone samples were heated in a 700 W multimode cavity microwave oven for 10 min, and the surface temperature of the samples was measured by a full-color infrared camera with 320 × 240 pixels at intervals of 30 s. At the same time, the weight change in the samples was recorded, and the whole measurement process was completed within seconds after the end of the microwave radiation and the opening of the furnace door, so the surface temperature changes of the samples could be ignored.

2.2.2. Split Hopkinson Pressure Bar System

The SHPB test system is used to test the dynamic mechanical properties of rock. It is mainly composed of a stress wave loading device, a signal acquisition system, and a high-speed camera. The loading device drives the spindle-shaped bullet to hit the incident rod through the compression cylinder, generating a slowly rising half-sinusoidal incident stress wave, eliminating the effect of P-C oscillations, and ensuring the stress balance at both ends of the sample. The sample is clamped between the incident rod and the transmission rod and kept in a straight line with the elastic rod. After the bullet hits the incident rod, the incident wave propagates to the sample, the reflected wave and transmitted wave are generated due to the difference in wave impedance, and the residual energy is absorbed by the absorber rod. The bullet, incident rod, and transmission rod are made of high-strength 40 Cr alloy steel, with a density of 7810 kg/m^3, a Poisson's ratio of 0.28, and a longitudinal wave velocity of 5400 M/s. The signal was collected by pasting strain gauges on the incident rod and transmission rod 750 mm away from the sample. The strain signals are displayed as voltage signals on an oscilloscope, after the Wheatstone bridge and super dynamic strain gauge are amplified and filtered. The high-speed camera is triggered by the incident wave signal and synchronized with the oscilloscope to record the dynamic rock fragmentation process completely. The schematic diagram of the installation and loading method of the NSCB sample during the dynamic fracture process is shown in Figure 4. Finally, the voltage signal recorded by the strain gauge is converted into the stress signal, and the strength, strain, and strain rate of rock dynamic failure can be calculated by the following formula [42]:

$$\sigma(t) = \frac{A_e E_e}{2A_s}[\varepsilon_I(t) + \varepsilon_R(t) + \varepsilon_T(t)] \tag{3}$$

$$\varepsilon(t) = \frac{C_e}{L_s}\int_0^t [\varepsilon_I(t) - \varepsilon_R(t) - \varepsilon_T(t)]dt \tag{4}$$

$$\dot{\varepsilon}(t) = \frac{C_e}{L_s}[\varepsilon_I(t) - \varepsilon_R(t) - \varepsilon_T(t)] \tag{5}$$

where E_e, C_e, and A_e are the modulus of elasticity of the elastic bar, the longitudinal wave speed, and the cross-sectional area, L_s and A_s are the sample length and the cross-sectional area, and $\varepsilon_I(t)$, $\varepsilon_R(t)$, and $\varepsilon_T(t)$ are the incident stress, the reflected stress, and the transmitted stress, respectively.

2.2.3. Verify the Stress Balance

The stress balance between the two ends of the sample is the basic premise of the validity of the SHPB test. According to the one-dimensional stress wave propagation theory of the SHPB test, the load at both ends of the sample is as follows:

$$F_1(t) = E_e A_e[\varepsilon_I(t) + \varepsilon_R(t)] \tag{6}$$

$$F_2(t) = E_e A_e \varepsilon_T(t) \tag{7}$$

where E_e and A_e are the modulus of elasticity and cross-sectional area of the elastic bar and ε_I, ε_R, and ε_T, are the incident strain, reflected strain, and transmitted strain, respectively.

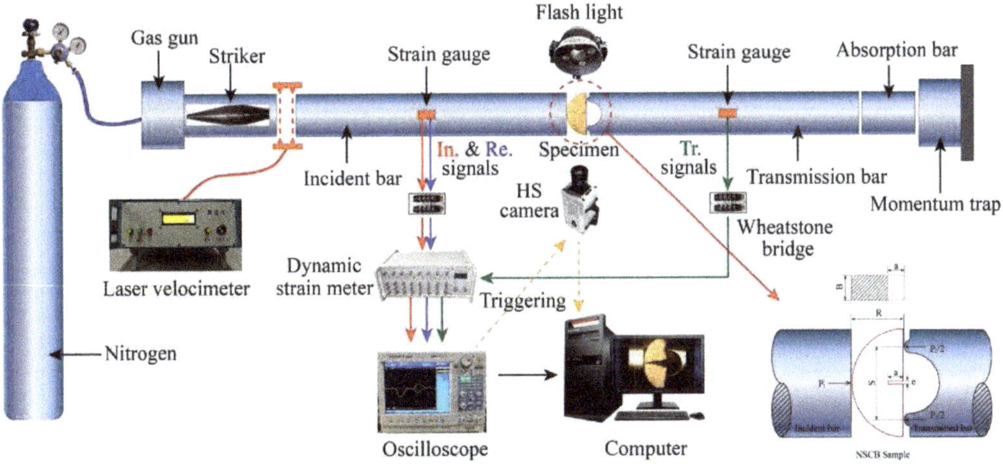

Figure 4. Schematic diagram of the NSCB sample loading method in the dynamic fracture test.

Figure 5a shows the force balance curve of a typical NSCB sandstone sample during dynamic testing. It can be observed that during the application of impact load, the superimposed stress curve and the transmitted stress curve basically coincide, that is, $F_1 = F_2$, indicating that the loading force at both ends of the sample is almost the same. Therefore, the inertia effect during dynamic loading can be ignored, and the quasi-static theory can be used to analyze the stress on the sample. The formula for calculating the mode I dynamic stress intensity factor of the NSCB sample is as follows [43,44]:

$$K_I(t) = Y' \frac{F(t)\sqrt{\pi l_n}}{2RB} \quad (8)$$

where $K_I(t)$ is the mode I dynamic fracture stress intensity factor (SIF); l_n is the length of the precast crack; R and B are the radius and thickness of the sample, respectively; and Y' is the dimensionless parameter related to the sample geometry as follows [45,46]:

$$Y' = -1.297 + 9.516\alpha - (0.47 + 16.457\alpha)\beta + (1.071 + 34.401\alpha)\beta^2 \quad (9)$$

where S is the distance between the two loading points of the transmission; $S = 34$ mm, $\alpha = \frac{S}{2R}$, and $\beta = \frac{l_n}{R}$.

During the test, the dynamic stress equilibrium was verified for each sample. Two test data sets from the dry group and one from the water-saturated group did not reach dynamic equilibrium, and thus, these three data sets were excluded from the analysis. On the other hand, Figure 5b shows a typical relationship curve of the stress strength factor with loading time, defining the slope of the approximate linear segment during the rise of the curve as the loading rate. In Figure 5b, the loading rate and fracture toughness of the sample are $\dot{K}_I = 42.94$ GPa·m$^{1/2}$/s and $K_I = 2.22$ MPa·m$^{1/2}$, respectively.

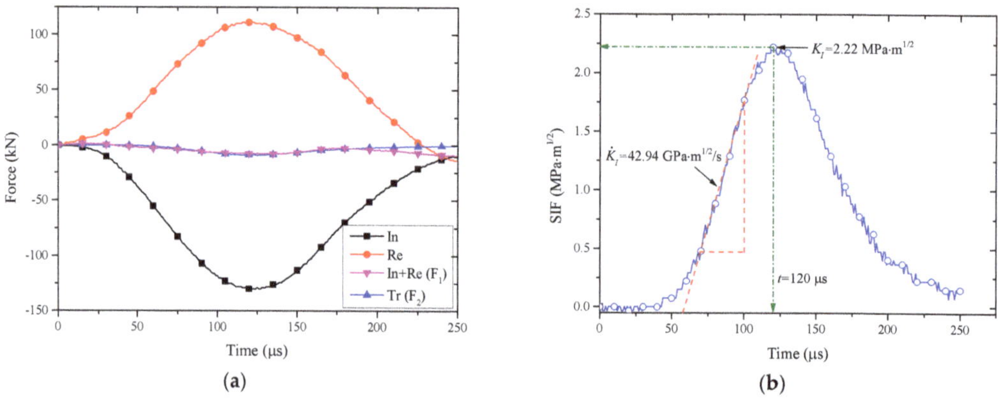

Figure 5. (**a**) Verification of dynamic stress balance at both ends of the NSCB sample; (**b**) determination method of test loading rate.

2.2.4. Digital Image Correlation Technology

This test uses a Hikvision, Hangzhou, China FASTCAM AS1.1 high-speed camera to capture the dynamic fracture process of a sandstone sample and a DIC technique to analyze and calculate the deformation field of the sample. In order to achieve synchronous recording between the high-speed camera and stress waves, the camera is connected to a super dynamic strain gauge. Once the voltage signal recorded in the incident rod rises to −34 mV, the high-speed camera starts shooting. The camera resolution was set to 256 × 244 pixels, and the capture rate was 80,000 frames per second, taking a picture every 12.5 microseconds.

The basic principle of the DIC technique relies on the assumption that the grayscale value of the image remains constant during the loading process. Therefore, strain and displacement fields are obtained using image correlation by tracking the motion of pixels in the reference and deformed images [47]. For a detailed explanation of the DIC technique, refer to the introduction by Sutton et al. [48]. To accurately capture pixels, it is necessary to prepare a speckle field with high resolution on the surface of sandstone samples. Firstly, clean the sandstone sample surface with a degreasing agent and then spray the sample surface with white matt paint to form a thin and uniform white background. After the white paint dried, select a speckle-making plate with suitable density bumps and press it on the surface of the white paint to produce randomly distributed black spots. Taking the photos taken by a high-speed camera before the deformation of the sample as the reference, and a series of deformation images during the loading process as the target, the deformation and failure process of the sample during the whole loading process is analyzed.

VIC-2D 6.0 software is used to analyze and calculate the deformation field on the specimen surface, which mainly includes the following steps: (1) import a series of speckle photos into the VIC-2D data processing software and then select the region of interest (ROI) on the surface of the sample; (2) divide the ROI into several subsets of a specified size and step size and calibrate the images based on the actual size of the sample; (3) calculate the initial displacement of these subsets and search for corresponding subsets in the target image based on the correlation algorithm; (4) calculate displacement and engineering strain based on the different positions of the subset in the reference image and the target image; and (5) obtain the strain field through differential treatment of the displacement. The working principle and process of calculating the sample deformation field using the DIC technique are illustrated in Figure 6.

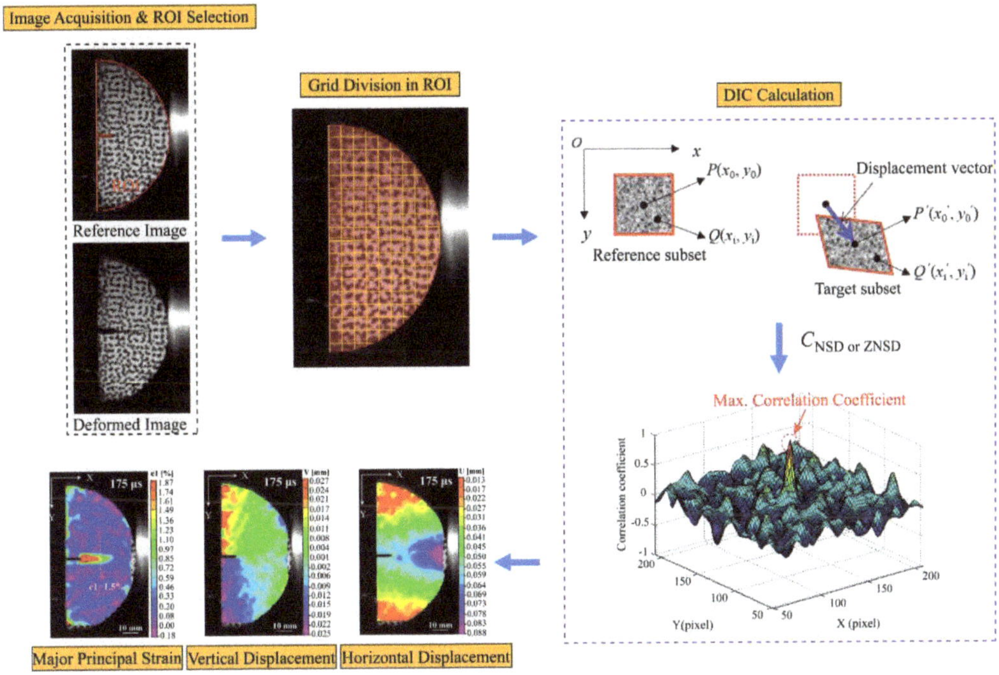

Figure 6. Schematic diagram of the working principle of the sample deformation field calculated by DIC technology.

3. Results

3.1. Temperature Distribution Characteristics

The surface temperature distribution characteristics of the water-saturated sandstone samples recorded by the thermal imager during microwave radiation are given in Figure 7. Due to the small size of the half-disc samples, the thermographic differences between dry and water-saturated sandstone samples are not particularly pronounced. However, as the duration of microwave radiation increases, the samples exhibit a pattern of higher temperatures at the center and lower temperatures at the edges. The center of the sample appears bright white, with the heat extending outward. Eventually, the highest temperature stabilizes at the center and remains unchanged. This phenomenon occurs primarily due to the differing heat dissipation conditions from the interior to the exterior of the rock, resulting in a higher temperature at the center of the rock, which gradually decreases in a circular pattern toward the edge of the sample.

As the mode I fracture of NSCB samples propagates along the prefabricated cracks, it can be observed from Figure 7 that when the microwave heating time reaches 120 s, the highest temperature begins to concentrate around the prefabricated cracks. This indicates that microwave radiation has a significant impact on the FPZ of the sandstone.

Due to the uneven temperature distribution on the surface of the sample, thermal imaging analysis software IVMS-4800 2.1.0.3 was used to quantitatively analyze the surface temperature of the sandstone sample. Figure 8 shows the maximum temperature, average temperature, and the ratio of maximum temperature to average temperature on the surface of dry and saturated sandstone samples during microwave radiation, respectively, to quantitatively analyze the effect of pore water on the microwave heating effect of sandstone semi-circular samples.

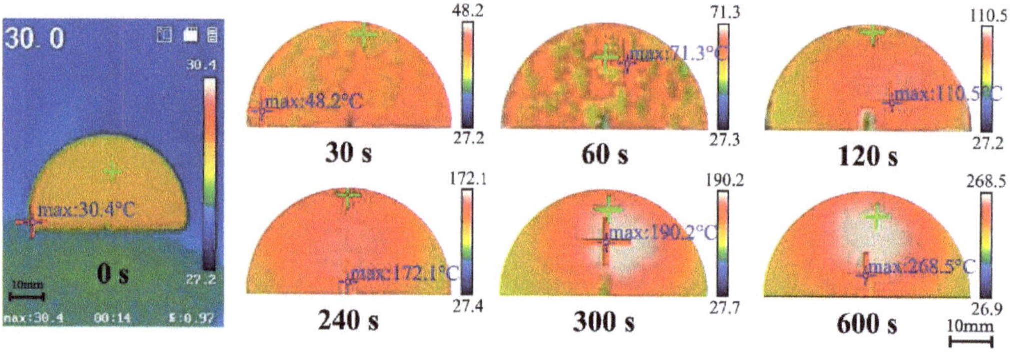

Figure 7. Thermal imaging of the temperature distribution of water-saturated sandstone samples during microwave heating.

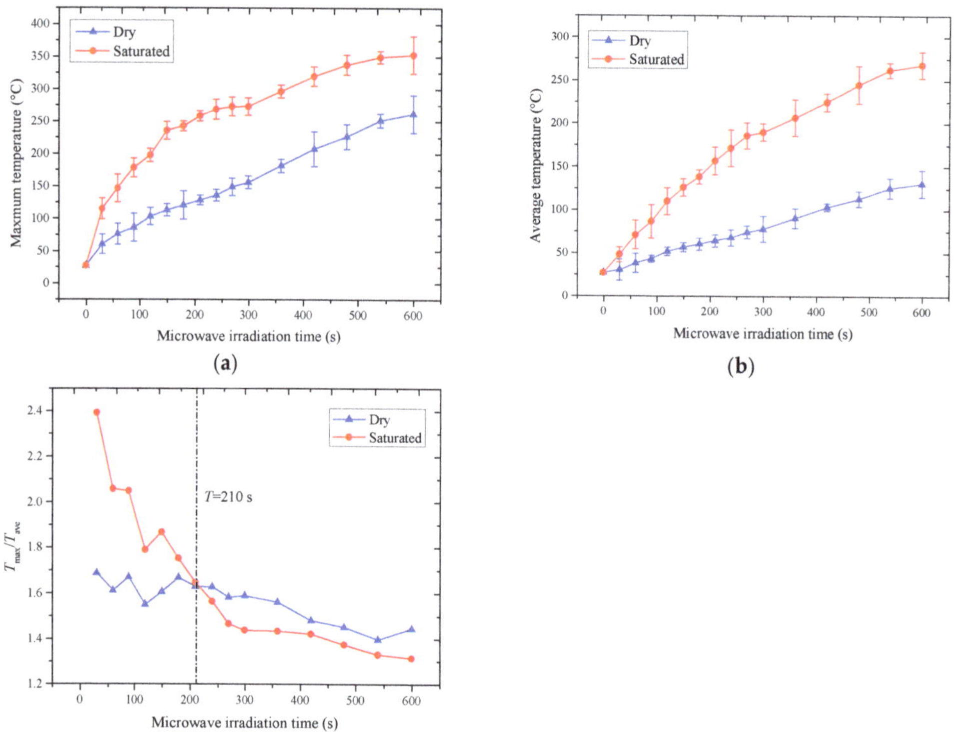

Figure 8. Variation of the maximum surface temperature: (**a**) average temperature, (**b**) variation with microwave radiation time, (**c**) maximum surface temperature of the sample and the ratio of the average temperature (T_{max}/T_{ave}) variation with microwave radiation time.

Firstly, for dry sandstone samples, both the maximum and average temperatures show an approximately linear increase in temperature with the increase in microwave radiation time. This also indicates that the dielectric heat generation capacity of dry sandstone samples does not undergo significant changes during microwave heating. In contrast, the temperature rise in the water-saturated sandstone samples can be divided into two distinct stages; the turning point of microwave radiation time appears at 300 s. In the first stage,

0–300 s, the presence of pore water leads to an increase in the dielectric properties of the sandstone, in which the water molecules are the first to absorb microwave energy and convert it into heat energy for transfer to the rock matrix, resulting in a higher heating rate of the water-saturated sandstone. In the second stage, when the microwave radiation time exceeds 300 s, the average surface temperature of the sample has approached 200 °C. At this time, the water inside the rock has been completely dried, and the main microwave absorption medium at this stage is the rock matrix, rather than the water molecules inside the sample. Since the dielectric loss factor of sandstone minerals is relatively low, the efficiency of microwave energy absorption decreases, leading to a slower rate of surface temperature rise beyond 300 s.

In addition, the ratio of the highest temperature to the average temperature on the surface of the sample (T_{max}/T_{ave}) can, to some extent, reflect the uniformity of temperature distribution. The larger the ratio between the two, the more uneven the temperature distribution on the rock surface. Figure 8c shows the variation of this ratio for both dry and water-saturated sandstone with microwave radiation time. It can be observed that both curves exhibit a rapid decline followed by stabilization as the microwave radiation time increases, which also indicates that with the extension of microwave radiation time. The heat convection and conduction within the sample in high high-temperature environment will lead to a more uniform rock temperature distribution. Another important phenomenon is that the temperature non-uniformity of saturated sandstone is higher than that of dry sandstone before 210 s, which may be due to the dielectric heat generated by the pore water inside the rock, causing the heating rate of the sample to be greater than its own thermal conduction or diffusion rate, resulting in a significant "hot spot effect" on its surface. After 210 s, the unevenness of temperature distribution on the surface of saturated sandstone tends to stabilize and is lower than that of dry sandstone samples. This may be because saturated sandstone has reached a higher temperature state at this time, and its internal heat conduction and convection are intense. On the other hand, the loss of moisture also leads to a slower heating rate of the rock, during which the heat conduction rate is greater than the microwave heating rate, resulting in a more uniform temperature distribution on the surface of the sample.

3.2. Static Physico-Mechanical Property Changes

The changes in static physical and mechanical parameters of dry and water-saturated sandstone after exposure to microwave radiation are shown in Figure 9. (1) The density of the sandstone showed little to no change after microwave radiation, whether in the dry or water-saturated state. This is likely because the temperatures during the experiment were not high enough to cause mineral decomposition. (2) Under the effect of microwave radiation, the presence of pore water has a significant pore expansion and fracturing effect on the rock, as shown by the increase in porosity of water-saturated sandstones by about 35% after experiencing microwave radiation. (3) Due to the low sensitivity of dry sandstone to microwave radiation, the damage to these samples was minimal, as evidenced by a decrease in longitudinal wave velocity and static fracture toughness of less than 5%. (4) The presence of pore water significantly aggravated the damage caused by microwave radiation. Compared to dry samples, the longitudinal wave velocity and fracture toughness of water-saturated sandstone decreased by 26% and 45%, respectively, after microwave exposure.

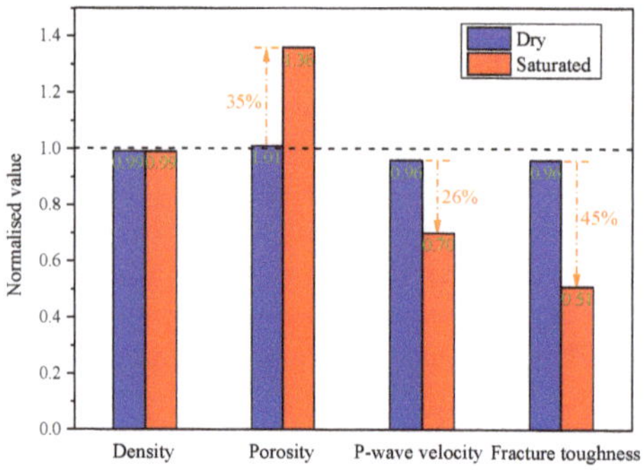

Figure 9. Normalized values of static physical and mechanical parameters of dry and water-saturated sandstone after microwave radiation action.

3.3. Correlation of Dynamic Cracking Toughness Rate

The initiation toughness measured by the NSCB method is actually the peak value of the stress intensity factor historical curve at the beginning of crack propagation, which can reflect the rock's ability to resist initiation. Therefore, fracture toughness is an important parameter for evaluating the weakening effect of microwave radiation on fractured rock masses. The dynamic fracture initiation toughness of both dry and water-saturated sandstones after microwave radiation is presented in Figure 10. It can be found that the fracture toughness of sandstone in both dry and water-saturated states shows a significant and similar loading rate dependence, and the fracture toughness of water-saturated sandstone is significantly lower than that of dry samples after microwave radiation. At a loading rate of 200 GPa·m$^{1/2}$/s, the fracture toughness of the water-saturated samples decreased by approximately 33.4% compared to the dry samples, indicating that pore water plays a positive role in microwave weakening of rock fracture toughness. In Figure 10, lowercase subscripts d and s are used to distinguish between the fracture initiation toughness of dry sandstone and water-saturated sandstone, represented as K_{Id} and K_{Is}, respectively. The fitted relationship between fracture toughness and loading rate of sandstone can be expressed as follows:

$$K_{Id} = 12.35 - 11.44 * 0.995^{\dot{K}_I} \qquad \left(R^2 = 0.988\right) \tag{10}$$

$$K_{Is} = 9.55 - 9.18 * 0.996^{\dot{K}_I} \qquad \left(R^2 = 0.989\right) \tag{11}$$

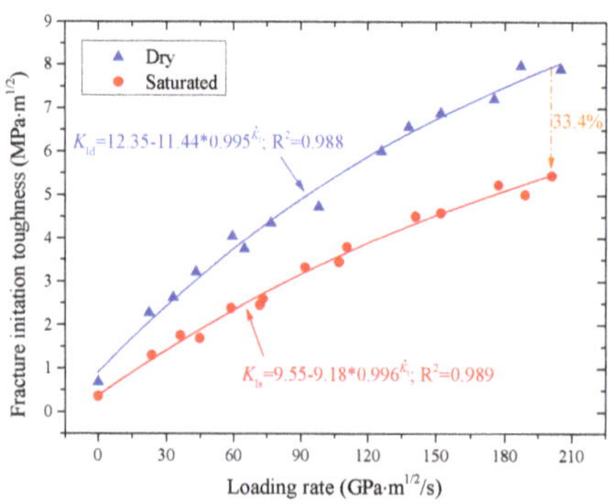

Figure 10. Relationship between sandstone cracking toughness and loading rate.

The fracture initiation, propagation, and debris movement trajectories of the NSCB samples, as captured by the high-speed camera, are shown in Figure 11. The photos taken by the high-speed camera indicate that the rock started to crack at a loading time of 162.5 μs, and the crack propagation began at the tip of the pre-existing crack and propagated along the potential failure path. However, the initial fracture shape was so small that it could only be detected through repeated comparison of the images before and after fracture initiation. The fracture propagation path of the NSCB samples became visible to the naked eye at 175 μs. As the loading time progressed, the fracture continued to propagate along the potential failure path toward the incident bar, gradually splitting the sample into two intact fragments, each resembling a fan (one-quarter of a circle). These two fragments rotated around the incident bar's loading point in opposite directions—upward and downward—using the fracture line as the central axis. The rotation angles between the two fragments were measured at different time intervals: 8.5° at 287.5 μs, 11° at 325 μs, and 17.5° at 400 μs. This indicates that the angular velocity of rotation of the fragment increased gradually during this period, reaching 581.7 rad/s and 756.3 rad/s, respectively. It should be noted that the calculation method of the angular velocity of rotation is based on the artificial measurement of the photographs taken by the high-speed camera, and there is inevitably some error in the calculation process.

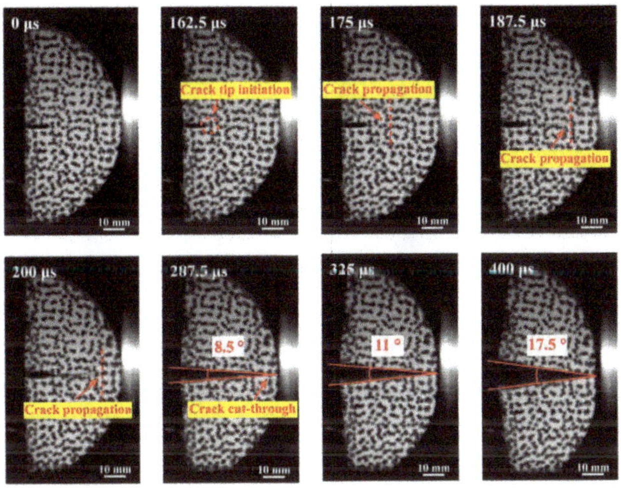

Figure 11. Photograph of the dynamic fracture process of dry sandstone taken by a high-speed camera.

3.4. Quantification of Crack Growth Rate Based on DIC

The DIC technique enables the calculation of the maximum principal strain on the surface of the NSCB sample throughout the entire loading process, providing a visualization of the evolution of horizontal and vertical displacements. As shown in Figure 12, at 162.5 μs, microcracks start to appear from the pre-existing crack tip. Horizontal strain ε_{xx} is observed at the loading end of the incident rod after the stress wave reaches the sample, and vertical strain ε_{yy} begins to form in the area around the notch tip. This is consistent with the crack initiation time determined by our high-speed camera.

Microcracks develop rapidly over time and expand along the potential fracture path toward the incident bar, eventually splitting the sample into two equal fragments. As shown in Figure 12c, all values of the vertical strain at the crack tip are positive, which also indicates that the fracture is caused by tensile stress. In addition, the crack visible to the naked eye as observed by the high-speed camera was compared with the crack extension process as indicated by the maximum principal strain obtained from the DIC processing, and the results showed that the maximum principal strain corresponding to the crack tip extension that could be identified by the high-speed camera was $e1 = 1.5\%$.

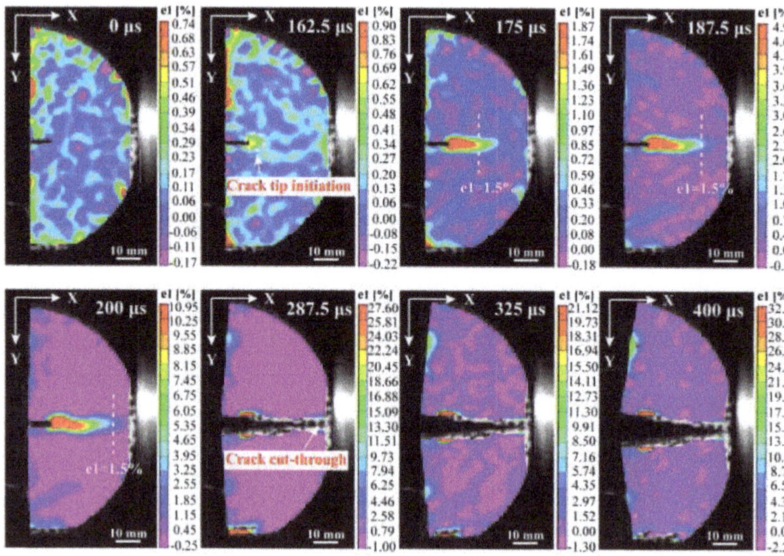

(a)

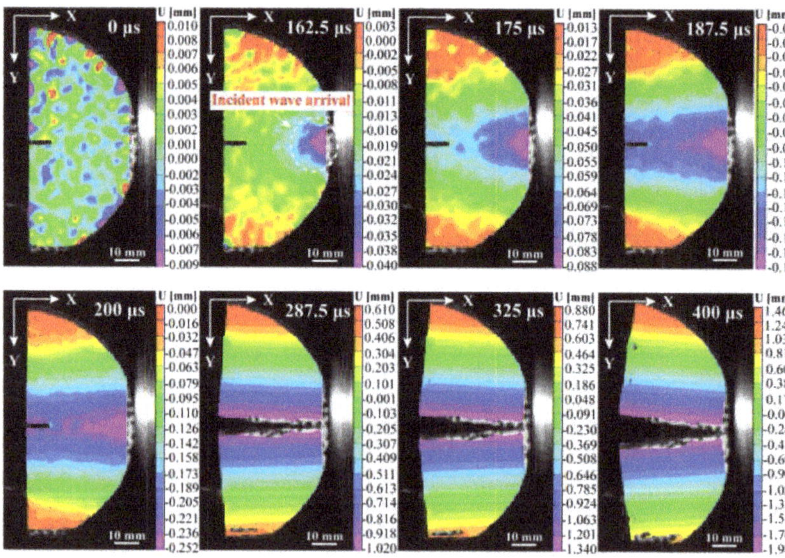

(b)

Figure 12. *Cont.*

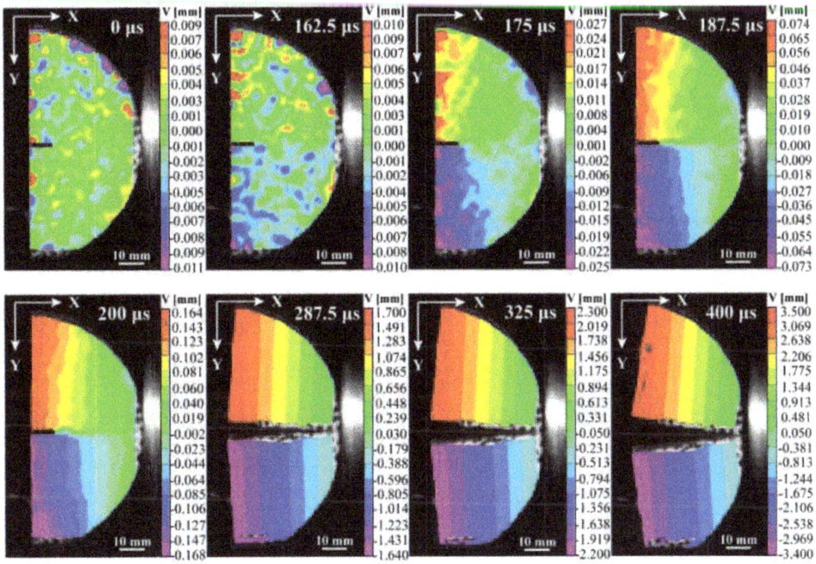

(**c**)

Figure 12. Under DIC technology, the maximum principal strain (**a**), horizontal displacement strain (**b**), and vertical displacement strain (**c**) on the surface of NSCB are shown.

The crack extension velocity is an essential parameter for characterizing the dynamic fracture process of rocks. Based on the DIC technique, the crack extension velocity is determined from the variation of the vertical displacement field of the sample during the dynamic loading process. The DIC post-processing software VIC-2D 6.0 includes a virtual displacement extensometer measurement tool. As shown in Figure 13a, two displacement pins, E0 and E1, are placed perpendicular to the crack extension trajectory, with a distance of 15 mm between them. Additionally, Zhou et al. [26], while measuring the crack extension velocity in dynamic fracture of rocks using crack extension strain gauges, found that the velocity initially increases due to the influence of the FPZ before gradually leveling off. To minimize the impact of the FPZ on measurement results, E0 was placed 1 mm from the prefabricated crack tip.

As shown in Figure 13b, the displacement extensometer values remain at 0 at the beginning of loading. As the loading time progressed, the crack passes through the displacement meter elongation meters E0 and E1 at 150 μs and 187.5 μs, respectively, and the two elongation meter values continue to increase. For a given loading time, the values of E0 and E1 are the crack opening displacement (COD) at the given position of the sample. The moment of a sudden increase in E0 and E1 data can be regarded as the fracture time at the corresponding position, and the distance between the two extensometers is 15 mm. From this, it can be calculated that the average speed of crack propagation in the sample within this propagation distance range is 400 m/s.

The variation in crack propagation speed of microwave-radiated sandstone for the loading rate was obtained, as shown in Figure 14. It can be observed that the crack propagation rate of sandstone increases with the loading rate after microwave radiation. In the loading rate range of 22.4~205 GPa·m$^{1/2}$/s, the crack expansion rates of dry and water-saturated sandstone samples after microwave radiation were 259.5~610.9 m/s and 362.6~760.4 m/s, respectively. Under the same loading rate conditions, the crack propagation speed of the saturated sample is significantly higher than that of the dry sample, and the sensitivity of the loading rate to the crack propagation speed of the saturated sample is also higher. This is because the steam water pressure generated by the saturated sample inside the rock under microwave radiation will reduce the bonding strength between particles, ultimately leading to a significant decrease in its ability to resist load and crack propagation.

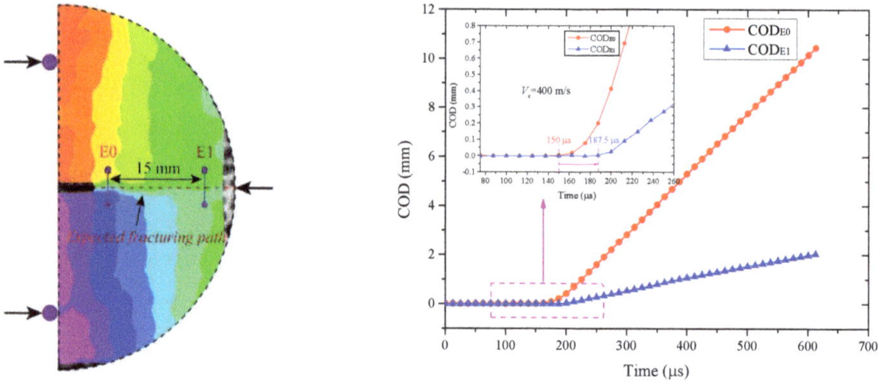

(a) (b)

Figure 13. (**a**) Schematic diagram of the displacement extensometer arrangement; (**b**) COD history of the two extensometers measured by DIC.

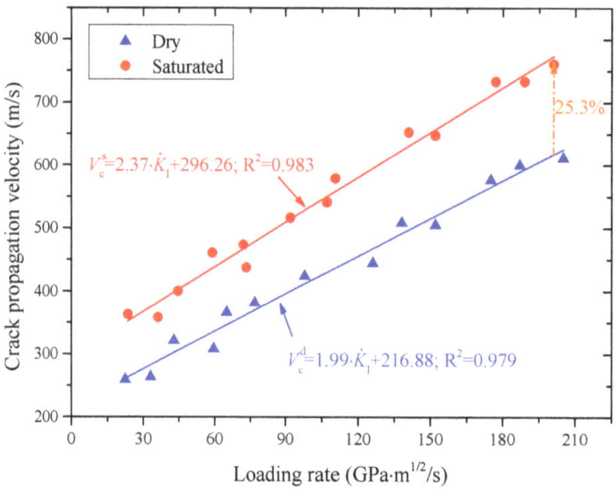

Figure 14. Crack expansion rate versus loading rate.

3.5. Energy Dissipation Characteristics of Dynamic Fracture

Under impact loading, energy is transferred in the form of stress waves. Based on elastic wave theory, the energy carried by each waveform in the SHPB system can be calculated according to the following formula [49–53]:

$$W_I = E_e A_e C_e \int_0^{\tau} \varepsilon_I(t)^2 dt \tag{12}$$

$$W_R = E_e A_e C_e \int_0^{\tau} \varepsilon_R(t)^2 dt \tag{13}$$

$$W_T = E_e A_e C_e \int_0^{\tau} \varepsilon_T(t)^2 dt \tag{14}$$

where W_I, W_R, and W_T represent the energy carried by the incident, reflected, and transmitted waves, respectively; C_e is the P-wave velocity of the elastic bar, and τ is the moment when the sample is completely fractured. According to the Law of Conservation of Energy, if the energy lost due to friction between the elastic bar and the sample is negligible, the energy absorbed by the sample can be expressed as the following equation:

$$W_S = W_I - W_R - W_T \tag{15}$$

Generally, in the NSCB dynamic fracture test, the energy absorbed by the rock sample consists of three main components [54–56]: (1) the total energy dissipated by fracture (Ω), which is used to generate new fracture surfaces and microcracks; (2) the residual kinetic energy (T) trapped in the fragments, manifested as two fragments rotating around the loading orbit during mode I fracture; and (3) the energy dissipated in other forms, including heat, electromagnetic energy, and acoustic energy. Due to the small proportion of other energies in conventional SHPB testing, they can be ignored. Thus, the total fracture dissipation energy can be approximated as follows [43]:

$$\Omega = W_S - T \tag{16}$$

The residual kinetic energy of the fragment is used primarily for the translational and rotational motion of the two fractured fragments, as calculated by the following formula [43]:

$$T = 2\left(\frac{1}{2}mv_t^2 + \frac{1}{2}I\omega^2\right) = mv_t^2 + I\omega^2 \tag{17}$$

$$I = \frac{R^2}{36\pi^2}\left(9\pi^2 + 18\pi - 128\right)m \tag{18}$$

where m is the mass of each flight fragment; I represents the inertia of rotation around the axis; ω is the rotational angular velocity; and v_t denotes the displacement caused by the barycenter of the fragment, and it can be obtained from the product of the distance between the center of mass and the loading point and the angular velocity of rotation.

In this study, the DIC method was employed to efficiently and accurately measure the angular velocity of rotation of fractured fragments ω. The fracture fragment rotation process during the crack failure is schematically illustrated in Figure 15a. Two symmetric reference points, P_1 and P_2, located on either side of the prefabricated fracture, were selected from the horizontal and vertical displacement fields using VIC-2D software. The distances L from P_1 and P_2 to the mounting point are equivalent, and the angular velocity of fragment rotation can be calculated using the formula $\omega = (\Delta d/L)/t$, where Δd is the displacement of the selected reference point in this study and $t = 12.5$ μs. Based on the displacement trajectories of P_1 and P_2, the variations of the angular velocity of rotation can be deduced from the DIC analysis, as shown in Figure 15b.

It can be observed that after the initiation of the rock crack, the angular velocity of rotation increases rapidly with loading time and then stabilizes. Moreover, the angular velocities of the two fragments are nearly equal, indicating that the rotation of the two fragments is almost symmetrical at the onset of crack initiation. As loading time continues, the rotational angular velocity experiences a slight decrease after 400 μs, eventually reaching and maintaining a relatively stable value. This minor decrease is likely due to the sliding of the sample between the elastic bars. Therefore, the average value of the first stable phase of angular velocity is selected as the average rotational angular velocity $\overline{\omega}$ of the fragments, which is obtained by averaging the angular velocity values in the dashed rectangle in Figure 15b. In Figure 15b, the average of the rotation angle with sandstone fragments is 685.6 rad/s.

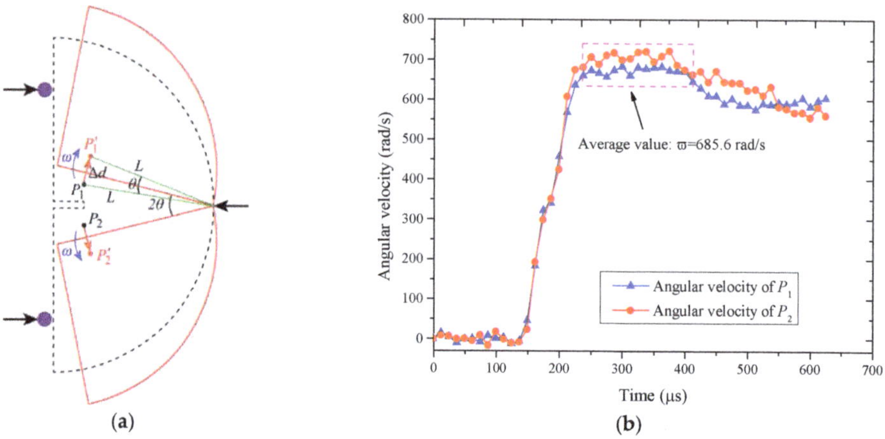

Figure 15. (**a**) Schematic diagram of the rotation of sample NSCB; (**b**) variation of rotational angular velocity with loading time.

Furthermore, the average dynamic fracture energy of the NSCB sample G_{dC}, which is the energy consumed per unit fracture area (dA), can be expressed as follows [43]:

$$G_{dc} = \frac{d\Omega}{dA} \approx \frac{W_S - T}{A_l} \tag{19}$$

In calculating the average dynamic fracture energy G_{dC}, the newly created fracture surface is calculated by $A_l = (R - a)t$. The roughness of the fracture surface and the area of the newly formed microcracks are not evaluated. Therefore, the G_{dC} obtained in this study represents the upper limit of the average fracture energy.

The correlation between energy distribution and loading rate of sandstone samples in dynamic NSCB tests is shown in Figure 16a. At the same loading rate, the residual kinetic energy of the water-saturated samples is greater than that of the dry samples. This indicates that the ability of water-saturated sandstone to resist loading and crack propagation is reduced after microwave radiation, manifested by a higher rotational angular velocity of fragments, resulting in a higher residual kinetic energy of the calculated sample.

The energy absorbed by the saturated sample W_S is less than that of the dry sample. At a loading rate of 200 GPa·m$^{1/2}$/s, the disparity between them is 20.1%. From the energy absorption values alone, it can be inferred that microwave radiation has a more pronounced damaging and fracturing effect on water-saturated sandstone, resulting in fracture damage at lower energy absorption levels.

The residual kinetic energy of the water-saturated sandstone samples is greater than that of the dry samples. This results in a significantly larger difference in the dynamic fracture energy consumed to generate new fracture surfaces and microcracks between the two states. It also indicates that the presence of pore water has a significant effect on the magnitude of the residual kinetic energy during the calculation of the dynamic fracture energy of microwave-radiated sandstones.

Based on the principle of energy conservation, the dynamic fracture energy consumed per unit fracture area is shown in Figure 16b. It can be observed that the dynamic fracture energy increases with the rise in loading rate, and the dynamic fracture energy of the water-saturated sandstone sample is significantly lower than that of the dry sandstone sample after microwave radiation. At a loading rate of 200 GPa·m$^{1/2}$/s, the difference in fracture energy G_{dC} between the dry and water-saturated sandstone samples reaches 61.04%. This also indicates that when sandstone is in a fully saturated state, microwave radiation can have a significant weakening effect on sandstone, leading to a significant decrease in the fracture energy of the rock. In the engineering application of rock excavation, the reduction in rock fracture energy indicates that only a small amount of mechanical energy is required to break the rock during mechanical excavation so as to obtain more efficient rock breaking.

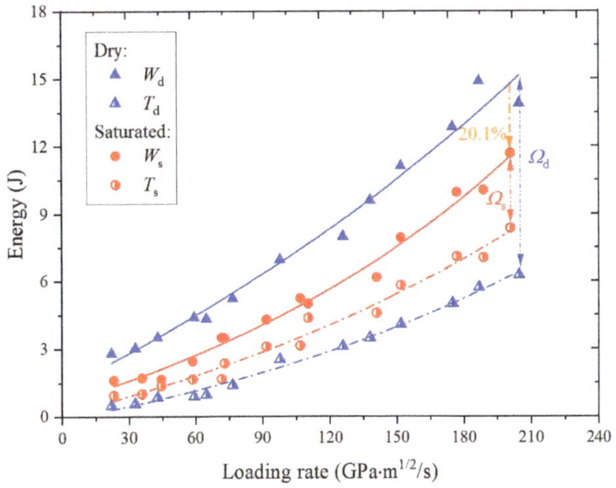

(**a**)

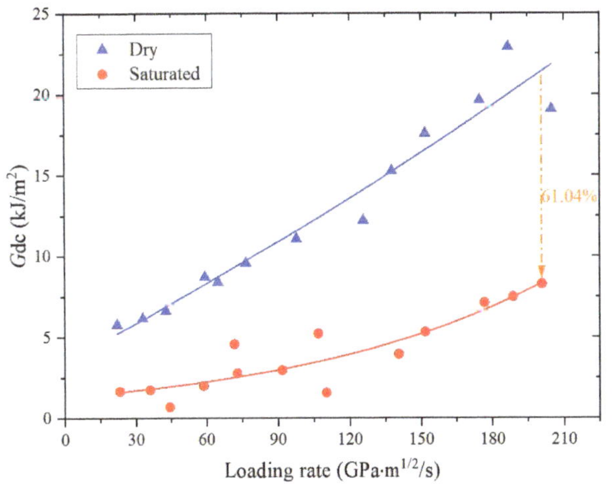

(**b**)

Figure 16. Energy distribution in relation to loading rate: (**a**) absorbed energy and residual kinetic energy of the sample; (**b**) dynamic rupture energy.

3.6. Dynamic Fracture Propagation Toughness

Dynamic fracture propagation toughness is a parameter that reflects the resistance of materials to crack propagation. When cracking occurs in an engineering rock, it is crucial to implement measures to prevent further crack expansion, ensuring the safety of the rock structure and avoiding further damage. Dynamic fracture propagation toughness K_I^P represents the critical dynamic stress intensity at a given crack propagation speed. Under high-strain-rate loading conditions, the effect of the crack expansion rate on the dynamic stress intensity factor cannot be ignored. Ravi-Chandar [57] has given a formula to calculate a particular crack expansion rate accounting for K_I^P as follows:

$$K_I^P = \sqrt{k(V_c)\frac{E \cdot G_{dC}}{1-\mu^2}} \tag{20}$$

$$k(V_c) = \sqrt{1 - \mu\frac{C_s^2 \cdot R(V_c)}{\alpha_p V_c^2}} \tag{21}$$

$$R(V_c) = 4\alpha_p\alpha_s - \left(1+\alpha_s^2\right)^2 \tag{22}$$

$$\alpha_p = \sqrt{1 - \frac{V_c^2}{C_p^2}} \tag{23}$$

$$\alpha_s = \sqrt{1 - \frac{V_c^2}{C_s^2}} \tag{24}$$

The relevant physical and mechanical parameters of the sandstone involved above are summarized in Table 1.

Figure 17 illustrates the relationship between dynamic fracture propagation toughness and the crack propagation speed of sandstone, showing a positive correlation between the two. Moreover, under the same crack propagation speed conditions, the expansion toughness of dry sandstone samples is higher. In other words, the dry sandstone must overcome a higher fracture extensional toughness if the rock is to produce the same damage rate. This observation also indicates that water-saturated sandstone sustains more significant internal damage after microwave radiation, making it more prone to low-stress fractures or rapid failure when subjected to external loads. For assisted rock breaking, a lower fracture propagation toughness means that the breaking tool can more easily penetrate the rock, leading to the formation of longer fractures. Additionally, a rock with lower propagation toughness will exhibit a faster crack expansion rate under the same cutting force, significantly improving the efficiency of rock breaking operations.

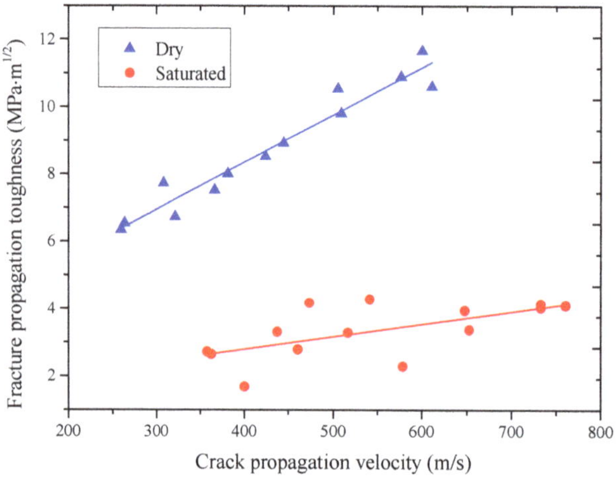

Figure 17. Relationship between dynamic fracture propagation toughness and crack extension rate.

4. Discussion

4.1. Water-Weakening Mechanisms in Microwave-Assisted Rock Breaking

Microwave heating is the process of converting microwave energy into thermal energy by utilizing the dielectric properties of materials. Based on the varying sensitivity of different materials to microwaves, they can be categorized into three distinct types: microwave transparent, microwave insulator, and microwave absorber. The sandstone material selected for this study exhibits relatively low dielectric properties typical of rock-forming minerals, classifying it as a low microwave absorber. During the drying process of sandstone via microwave radiation, most of the microwave energy passes through the sandstone specimen; only a small fraction is absorbed and converted into thermal energy within the material. Consequently, under conditions of microwave radiation, the heating rate of dry sandstone specimens remains relatively low.

Numerous studies have demonstrated that water, being a strong polar molecule, serves as an effective microwave-absorbing material. The presence of water alters the dielectric properties of rock minerals significantly. In a microwaved environment, water molecules undergo rotational orientation due to electromagnetic fields; this interaction leads to friction and collisions among neighboring molecules, resulting in pronounced heating phenomena. The heat absorbed by water subsequently propagates to surrounding rock matrices through conduction and convection processes, thereby increasing the heating rate in water-bearing sandstones.

However, as temperature rises during this process, moisture undergoes phase change, generating water vapor, which exerts pressure on inner pore walls—this contributes to crack formation and development within the rock structure, as evidenced by micromorphological characteristics observed in sandstone samples. Therefore, during microwave treatment of sandstone, pore water functions dually as both a microwave absorber and fracturing fluid; its fundamental principles and processes are illustrated in Figure 18.

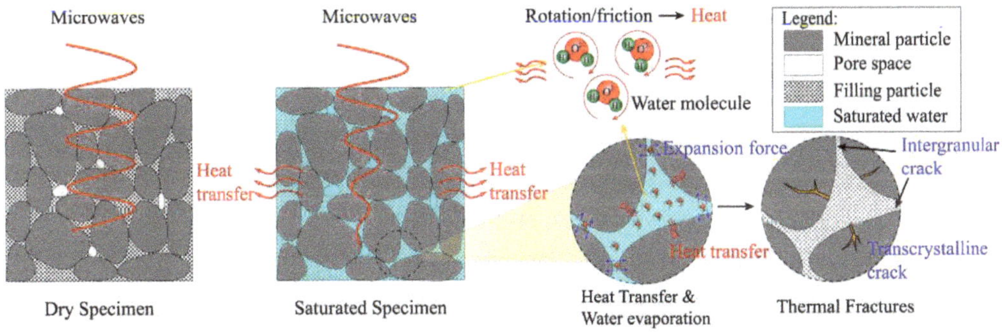

Figure 18. Schematic diagram of the microwave heating process and the weakening mechanism of water-bearing sandstone.

4.2. Microcrack Extension in Sandstone Microwave Damage

Analysis of the test results reveals that microwave radiation caused a significant decrease in the dynamic fracture toughness of sandstone. The dynamic fracture initiation and propagation toughness of water-saturated samples were notably lower than those of dry samples. The dynamic fracture energy and extensional toughness of the water-saturated sandstone samples were also lower under the same loading rate or crack extension rate. In addition, the crack propagation speed of the saturated sandstone samples also increased significantly after the microwave radiation. These experimental findings demonstrate that the presence of pore water within the rock exacerbates the damage caused by microwave radiation, making the rock more susceptible to low-stress fractures or rapid failure when subjected to external loads. This series of phenomena can be attributed to the evolution of the microstructure around the prefabricated crack tip following microwave radiation.

A rectangular area measuring 15 mm in length and 10 mm in width, situated above the precast crack tip, was selected for Scanning Electron Microscope (SEM) analysis. The microstructural changes in both dry and saturated sandstone samples after microwave radiation were observed, and the results are shown in Figure 19. The SEM images reveal that, prior to microwave treatment, the sandstone samples were internally compact, with only a few initial pores visible and no significant microcrack development, as shown in Figure 19a.

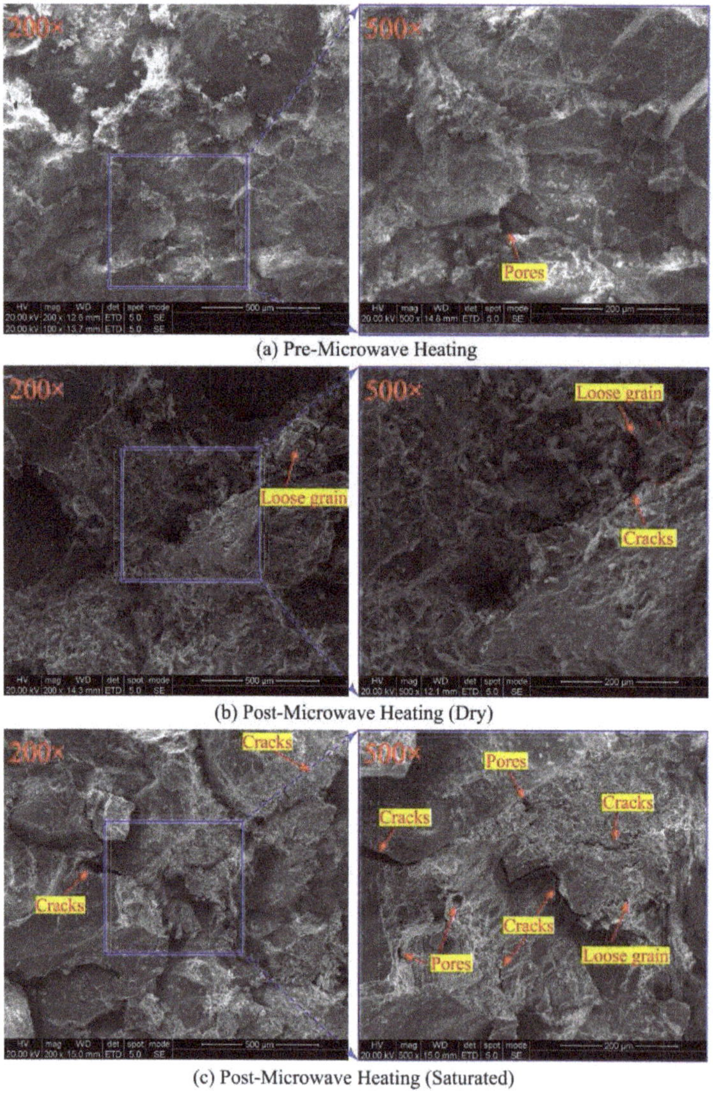

Figure 19. SEM images of the NSCB sample around the crack tip before and after microwave irradiation.

The internal microstructure of both dry and water-saturated sandstone samples changed to some extent after exposure to the same microwave power and irradiation time. For the dry sandstone sample, a small number of cracks developed along the crystal boundaries inside the rock following microwave radiation, but these cracks were relatively small, and the overall structure remained fairly dense, as shown in Figure 19b. This is because microwave-sensitive minerals within the sandstone are the first to heat up under microwave radiation, leading to significant volume expansion and a non-uniform temperature distribution inside the rock. The localized inhomogeneous deformation generated within the rock and the temperature gradients lead to its structural thermal stresses at the microscopic level. With structural thermal stresses, along-crystal cracks are generated between sandstone mineral grains. The generation of cracks will lead to an increase in the FPZ of the rock and the reduction in its ability to resist external loading and crack expansion, which is reflected in the decrease in its dynamic fracture toughness. Figure 20 illustrates the increase in sandstone crack density and the enlargement of the FPZ due to microwave radiation.

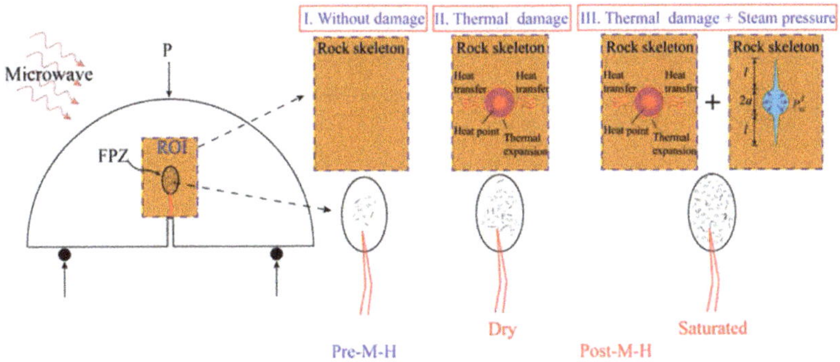

Figure 20. Schematic diagram of the increase in the FPZ size of sandstone due to increased density of cracks by microwave radiation.

In the SEM images of the water-saturated sandstone samples after microwave radiation, numerous cracks and pores can be observed along the crystals. Additionally, there are many loose particles within the rock, resulting in a rougher microstructural surface, as illustrated in Figure 19c. Compared to the dry sandstone samples with the action of microwave radiation, the size and number of microcracks inside the water-saturated sandstone are significantly increased, resulting in this difference for two main reasons as follows:

(1) The presence of pore water within sandstone enhances the rock's dielectric properties, resulting in a higher heating rate under microwave radiation. Moreover, under the same microwave radiation conditions, sandstone can reach higher temperatures. The combination of high temperature and thermal shock leads to significant thermal damage within the sandstone.

(2) Under microwave rapid heating conditions, pore water undergoes a phase transition and quickly transforms into steam, forming a binary gas mixture consisting of water vapor and air within the pores. Due to the constraint imposed by the pore size on the volumetric expansion of water molecules, expansion pressure is generated and exerted on the inner surfaces of the pores. This pressure leads to the development of additional pores within the rock structure and causes existing microcracks to expand and coalesce. Since the size of the FPZ in rocks is proportional to the density of microcracks, microwave irradiation results in a significant increase in FPZ size in saturated sandstone. This increase is more pronounced compared to that observed in dry sand-

stone samples. Consequently, the dynamic fracture toughness of saturated sandstone after microwave treatment is considerably lower than that of its dry counterpart.

In summary, during microwave heating of sandstone samples, the presence of pore water not only enhances the microwave absorption capacity of the sandstone, subjecting the rock to high temperatures and thermal shock, but also generates vapor pressure that promotes the expansion of internal pores and the propagation of microcracks. This creates a dual weakening effect, contributing to the reduction in fracture toughness of sandstone under microwave radiation. The resulting formation and agglomeration of microcracks near the crack tip will lead to an increase in the FPZ of the rock, and the anti-load capacity of the pre-cracking will be lowered. Thus, after microwave radiation, the dynamic crack initiation toughness, extensional toughness, and dynamic fracture energy of water-saturated sandstone samples decrease, while the crack propagation rate increases.

4.3. Limitations and Future Research Perspectives

Considerable efforts have been devoted to investigating the alterations in the physical and mechanical properties of rocks subjected to microwave irradiation. Nevertheless, the factors influencing the extent of microwave-induced weakening in rocks remain insufficiently explored, and relatively few studies have integrated microwave heating technology into engineering applications. Based on existing research, the following recommendations for future investigations are proposed:

(1) The NSCB semi-circular sandstone specimens used in the experiment were small in size, which facilitated laboratory control and observation. However, their thermal conductivity and crack propagation behavior differed in scale from those of large rock masses in the field. Therefore, caution should be exercised when extrapolating the conclusions to the engineering scale.

(2) Most of the studies on the response of rocks to microwave radiation have been carried out using multimode cavity microwave ovens. Compared with single-mode-cavity microwave ovens, multimode-cavity microwave ovens have the advantages of uniform heating, high energy utilization, and no risk of microwave leakage, but they have relatively low power densities and are only suitable for laboratory studies. Therefore, after fully understanding the destructive behavior of microwave irradiation on small-sized rock samples, it is necessary to use high-power single-mode-cavity microwave equipment to conduct surface radiation tests on large-sized rock bodies, thus shifting the laboratory research to field tests.

(3) In engineering practice, it is sometimes necessary to simultaneously employ mechanical cutting and microwave irradiation to achieve continuous excavation in underground engineering and tunneling operations. Consequently, further research on the dynamic mechanical properties of rocks under real-time microwave heating conditions remains essential. Additionally, addressing the issue of preventing microwave leakage during shock loading is a critical problem that must be resolved first.

(4) Microwave radiation is a multi-field coupled heating process influenced by electromagnetic, temperature, stress, and mass transfer fields. It is necessary to analyze the mechanical behavior of rocks under microwave radiation by the finite element or discrete element method to determine the relationship between the microwave-induced temperature distribution, thermal stress, and fracture density.

5. Conclusions

This study systematically elucidates the dominant role of water in the dynamic fracture behavior of sandstones under microwave radiation, where the presence of pore water leads to a dual weakening effect of high temperature and thermal shock on sandstones. For such microwave-insensitive sandstones tested in this paper, the initial water content is the decisive factor controlling the microwave fracturing effect. As long as the initial water content of the sample exceeds a specific value, microwave radiation can cause significant damage to the rock. When excavating microwave-insensitive rocks, water can be injected into the rock to increase the dielectric properties, providing an effective solution for the application of microwave fracturing technology in microwave-insensitive rocks. Therefore, we further investigated the coupled effects of water content and loading rate on the mechanical behavior of rocks under microwave radiation. Based on the SHPB test system, we investigated the dynamic mechanical behavior of water-bearing sandstone after microwave radiation, which provides key theoretical support for the application of microwave-assisted rock breaking technology in water-rich formations. The main conclusions are as follows:

(1) The heating rate of saturated sandstone samples under microwave radiation is significantly higher than that of dry sandstone. And, with the passage of microwave radiation time, the heating process of dry sandstone samples increases approximately linearly, while the heating rate of saturated sandstone samples slows down after 300 s of irradiation time.
(2) The initiation toughness of sandstone exhibits significant loading rate dependence. The fracture toughness of saturated sandstone after microwave radiation is significantly lower than that of dry samples. This indicates that pore water plays a positive role in microwave weakening of rock fracture toughness.
(3) The crack propagation speed of the saturated samples was significantly higher than that of the dry samples under the same loading rate condition, and the loading rate sensitivity of the crack propagation speed of the saturated samples was also higher. When the loading rate was 200 $GPa \cdot m^{1/2}/s$, the crack extension rate of microwave-irradiated saturated sandstone samples increased by 25.3% relative to that of dry samples.
(4) The dynamic fracture extension toughness and fracture energy of the saturated sandstone specimens were smaller than those of the dry specimens after microwave radiation, which indicated that the saturated sandstone's ability to resist crack extension was reduced after microwave radiation, and the rock was more susceptible to low stress fracture and rapid failure.

Author Contributions: P.W.: methodology, data curation, validation, resources, and writing—review editing. Y.L. and D.C.: data curation, investigation, and writing—original draft. T.Y.: supervision. All authors have read and agreed to the published version of the manuscript.

Funding: This research work was funded by the National Natural Science Foundation of China (No. 52504079 and No. 41972283); the Guangxi Natural Science Foundation of China (No. 2024GXNSFBA010427 and No. 2025GXNSFAA069611); and Guangxi Emergency Management Joint Innovation Science and Technology Project (No. 2025GXYJ009 and No. 2025GXYJ024).

Institutional Review Board Statement: Not applicable.

Informed Consent Statement: Not applicable.

Data Availability Statement: The original contributions presented in this study are included in the article. Further inquiries can be directed to the corresponding author.

Acknowledgments: We are grateful for the support provided by these funds, which laid the foundation for our experiments. We also express our gratitude to the Multi-disciplinary Integrated Innovation Experimental Teaching Center for Resource, Environment and Materials of Guangxi University for providing us with such a research environment.

Conflicts of Interest: The authors declare that they have no known competing financial interests or personal relationships that could have appeared to influence the work reported in this paper.

Abbreviations

The following abbreviations are used in this manuscript:

NSCB	notched semi-circular bend
SHPB	split Hopkinson pressure bar
TBM	tunnel boring machine
FPZ	fracture process zone
DIC	digital image correlation
ROI	region of interest
COD	crack opening displacement
SEM	scanning electron microscope

References

1. Gu, D.S.; Li, X.B. *Modern Science and Technology for Metal Deposit Mining*; Metallurgical Industry Press: Beijing, China, 2006.
2. Wang, Y.M. Opportunities and challenges to metal mine mining industry and the technical countermeasures. *Mod. Min.* **2011**, *27*, 1–14. (In Chinese) [CrossRef]
3. Li, X.B.; Cao, Z.W.; Zhou, J.; Huang, L.Q.; Wang, S.F.; Yao, J.R.; He, Z.G.; Ma, C.D.; Dong, L.J.; Zhao, G.Y. Innovation of hard rock mining methods and construction of intelligent green mines: A case study of kai yang phosphate mine. *Chin. J. Nonferrous Met.* **2019**, *29*, 2364–2380. (In Chinese) [CrossRef]
4. Li, X.B.; Huang, L.Q.; Zhou, J.; Wang, S.F.; Ma, C.D.; Chen, J.Z.; Liu, Z.X.; Li, Q.Y.; Zhao, G.Y. Review and prospect of mining technology in hard rock mines. *Chin. J. Nonferrous Met.* **2019**, *29*, 1828–1847. (In Chinese) [CrossRef]
5. Li, X.B.; Yao, J.R.; Du, K. Innovation of mining models and construction of intelligent green mine in hard rock mine: In Kai yang Phosphate Mine as an example. *Chin. J. Rock Mech. En.* **2013**, *32*, 1101–1111. (In Chinese) [CrossRef]
6. Xie, H.P.; Gao, F.; Ju, Y. Research and development of rock mechanics in deep ground engineering. *Chin. J. Rock Mech. En.* **2015**, *34*, 2161–2178. (In Chinese) [CrossRef]
7. He, M.C.; Xie, H.P.; Peng, S.P.; Jiang, Y.D. Research on deep mining rock mechanics and engineering disaster control. *Coal Mine Support* **2007**, *3*, 1–14. (In Chinese)
8. Yao, J.R.; Li, X.B.; Sun, L.C.; Wang, M.; Yao, H.; Wang, Q.P.; Li, W.P.; Wang, S.F. Non-explosive mechanized mining of deep hard ore by using high-frequency crushing hammer. *Met. Mine* **2021**, *50*, 27–33. (In Chinese) [CrossRef]
9. Xia, Y.M.; Zhang, K.; Liu, J.S. Design optimization of TBM disc cutters for different geological conditions. *World J. Text Eng. Technol.* **2015**, *3*, 218–231. [CrossRef]
10. Rostami, J. Performance prediction of hard rock Tunnel Boring Machines (TBMs) in difficult ground. *Tunn. Undergr. Sp. Tech.* **2016**, *57*, 173–182. [CrossRef]
11. Ciccu, R.; Grosso, B. Improvement of disc cutter performance by water jet assistance. *Rock Mech. Rock Eng.* **2014**, *47*, 733–744. [CrossRef]
12. Kocis, I.; Kristofic, T.; Gajdos, M.; Horvath, G.; Jankovic, S. Utilization of electrical plasma for hard rock drilling and casing milling. In Proceedings of the SPE/IADC Drilling Conference and Exhibition, London, UK, 17–19 March 2015. [CrossRef]
13. Shi, X.M.; Duan, Y.L.; Han, B.; Zhao, J. Enhanced rock breakage by pulsed laser induced cavitation bubbles: Preliminary experimental observations and conclusion. *Geomech. Geophys. Geo Energ. Geo Res.* **2020**, *6*, 25. [CrossRef]
14. Kahraman, S.; Canpolat, A.N.; Fener, M. The influence of microwave treatment on the compressive and tensile strength of igneous rocks. *Int. J. Rock Mec. Min. Sci.* **2020**, *129*, 104303. [CrossRef]
15. Zheng, Y.L.; Zhang, Q.B.; Zhao, J. Challenges and opportunities of using tunnel boring machines in mining. *Tunn. Undergr. Sp. Tech.* **2016**, *57*, 287–299. [CrossRef]
16. Aydin, G.; Karakurt, I.; Aydiner, K. Wear performance of saw blades in processing of granitic rocks and development of models for wear estimation. *Rock Mech. Rock Eng.* **2013**, *46*, 1559–1575. [CrossRef]
17. Lu, G.M.; Zhou, J.J.; Li, Y.H.; Zhang, X.W.; Gao, W.Y. The influence of minerals on the mechanism of microwave-induced fracturing of rocks. *J. Appl. Geophys.* **2020**, *180*, 104123. [CrossRef]
18. Zhao, Q.H.; Zhao, X.B.; Zheng, Y.L.; Li, J.C.; He, J.L.; Zou, C.J. Heating characteristics of igneous rock-forming minerals under microwave irradiation. *Int. J. Rock Mech. Min.* **2020**, *135*, 104519. [CrossRef]
19. Hu, J.J.; Xie, H.P.; Sun, Q.; Li, C.B.; Liu, G.K. Changes in the thermodynamic properties of alkaline granite after cyclic quenching following high temperature action. *Int. J. Min. Sci. Technol.* **2021**, *31*, 843–852. [CrossRef]

20. Hu, J.J.; Xie, H.P.; Gao, M.Z.; Li, C.B.; Sun, Q. Damage mechanism and heat transfer characteristics of limestone after thermal shock cycle treatments based on geothermal development. *Int. J. Rock Mech. Min. Sci.* **2022**, *160*, 105269. [CrossRef]
21. Hartlieb, P.; Leindl, M.; Kuchar, F.; Antretter, T.; Moser, P. Damage of basalt induced by microwave irradiation. *Miner. Eng.* **2012**, *31*, 82–89. [CrossRef]
22. Ge, Z.L.; Sun, Q.; Hu, J.J.; Guan, Y.H.; Liu, W.; Wang, S.F.; Geng, J.S. Fracture characteristics and thermal damage mechanism of shale under microwave radiation. *J. Therm. Anal. Calorim.* **2024**, *149*, 13147–13160. [CrossRef]
23. Ge, Z.L.; Sun, Q.; Xue, L.; Yang, T. The influence of microwave treatment on the mode I fracture toughness of granite. *Eng. Fract. Mech.* **2021**, *249*, 107768. [CrossRef]
24. Bai, G.G.; Sun, Q.; Jia, H.L.; Ge, Z.L.; Li, P.F. Variations in fracture toughness of SCB granite influenced by microwave heating. *Eng. Fract. Mech.* **2021**, *258*, 108048. [CrossRef]
25. Hassani, F.; Nekoovaght, P.M.; Gharib, N. The influence of microwave irradiation on rocks for microwave-assisted underground excavation. *J. Rock Mech. Geotech. Eng.* **2016**, *8*, 1–15. [CrossRef]
26. Zhou, Z.L.; Cai, X.; Ma, D.; Cao, W.Z.; Chen, L.; Zhou, J. Effects of water content on fracture and mechanical behavior of sandstone with a low clay mineral content. *Eng. Fract. Mech.* **2018**, *193*, 47–65. [CrossRef]
27. Lu, G.M.; Feng, X.T.; Li, Y.H.; Hassani, F.; Zhang, X.W. Experimental investigation on the effects of microwave treatment on basalt heating, mechanical strength, and fragmentation. *Rock Mech. Rock Eng.* **2019**, *52*, 2535–2549. [CrossRef]
28. Dai, J.; Du, W.P.; Wu, T. Experimental study of rock fracture after microwave irradiation and impact load. *J. Henan Polytech. Univ. Nat. Sci.* **2016**, *35*, 420–423. (In Chinese) [CrossRef]
29. Dai, J.; Pan, Y.B.; Meng, Z. Experimental study on influential factors of rock strength weakening under microwave irradiation. *J. Xi'an Univ. Sci. Technol.* **2016**, *36*, 364–368. (In Chinese) [CrossRef]
30. Hartlieb, P.; Toifl, M.; Kuchar, F.; Meisels, R.; Antretter, T. Thermo-physical properties of selected hard rocks and their relation to microwave-assisted comminution. *Miner. Eng.* **2015**, *91*, 34–41. [CrossRef]
31. Qin, L.; Dai, J.; Teng, P. Study on the effect of microwave irradiation on rock strength. *J. Eng. Sci. Technol. Rev.* **2015**, *8*, 91–96. [CrossRef]
32. Xia, K.W.; Wang, Z.; Wu, B.B.; Xu, Y.; Yue, T.L. Research progress on dynamic response of deep rocks under coupled hydraulic-mechanical loading. *J. China Coal Soc.* **2024**, *49*, 454–478. (In Chinese) [CrossRef]
33. Satish, H.; Ouellet, J.; Raghavan, V.; Radziszewski, P. Investigating microwave assisted rock breakage for possible space mining applications. *Min. Technol.* **2006**, *115*, 34–40. [CrossRef]
34. Li, X.; Wang, S.; Xu, Y.; Yao, W.; Xia, K.W.; Lu, G.M. Effect of microwave irradiation on dynamic mode-I fracture parameters of Barre granite. *Eng. Fract. Mech.* **2020**, *224*, 106748. [CrossRef]
35. Jia, H.L.; Wang, T.; Xiang, W.; Tan, L.; Shen, Y.J.; Yang, G.S. Influence of water content on the physical and mechanical behaviour of argillac-eous siltstone and some microscopic explanations. *Chin. J. Rock Mech. En.* **2018**, *37*, 1618–1628. (In Chinese) [CrossRef]
36. Zhang, E.F.; Yang, G.S.; Tang, L.Y.; Yang, Q.; Xie, Z.W. Study on influence of water content to damage and degradation laws of argillaceous siltstone. *Coal Sci. Technol.* **2019**, *47*, 14–20. (In Chinese) [CrossRef]
37. Zhao, K.; Ran, S.H.; Zeng, P.; Yang, D.X.; Teng, T.Y. Effect of moisture content on characteristic stress and acoustic emission characteristics of red sandstone. *Rock Soil Mech.* **2021**, *42*, 899–908. (In Chinese) [CrossRef]
38. Dai, J.; Wang, Y.L.; Huang, B.B.; Wang, Y.P.; Li, H. Experimental study on the effect of water on the degradation of hard rock under microwave irradiation. *Chin. J. Undergr. Sp. Eng.* **2020**, *16*, 691–696. (In Chinese)
39. Peinsitt, T.; Kuchar, F.; Hartlieb, P.; Moser, P.; Kargl, H.; Restner, U.; Sifferlinger, N. Microwave heating of dry and water saturated basalt, granite and sandstone. *Int. J. Min. Miner. Eng.* **2010**, *2*, 18–29. [CrossRef]
40. Deng, H.F.; Li, J.L.; Wang, K.W.; Liu, J.; Zhu, M.; Lu, T. Research on secondary porosity changing law of sandstone under saturation-air dry cycles. *Rock Soil Mech.* **2012**, *33*, 483–488. (In Chinese) [CrossRef]
41. Li, H.G.; Li, H.M.; Xu, G.S. Influence of water content on mechanical characteristics of weakly cemented sandstone. *J. Min. Strata Control Eng. J.* **2021**, *3*, 043029. (In Chinese) [CrossRef]
42. Wang, L. *Fundamentals of Stress Wave*; National Defense Industry Press: Beijing, China, 2005. (In Chinese)
43. Zhang, Q.B.; Zhao, J. Quasi-static and dynamic fracture behaviour of rock materials: Phenomena and mechanisms. *Int. J. Fract.* **2014**, *189*, 1–32. [CrossRef]
44. Zhou, Z.L.; Cai, X.; Ma, D.; Du, X.M.; Chen, L.; Wang, H.Q.; Zang, H.Z. Water saturation effects on dynamic fracture behavior of sandstone. *Int. J. Rock Mech. Min. Sci.* **2019**, *114*, 46–61. [CrossRef]
45. Chen, R.; Xia, K.; Dai, F.; Lu, F.; Luo, S.N. Determination of dynamic fracture parameters using a semi-circular bend technique in split Hopkinson pressure bar testing. *Eng. Fract. Mech.* **2009**, *76*, 1268–1276. [CrossRef]
46. Kuruppu, M.D.; Obara, Y.; Ayatollahi, M.R.; Chong, K.P.; Funatsu, T. ISRM-suggested method for determining the mode I static fracture toughness using semi-circular bend specimen. *Rock Mech. Rock Eng.* **2014**, *47*, 267–274. [CrossRef]
47. Zhu, Q.Q.; Ma, C.; Li, X.B.; Li, D.Y. Effect of filling on failure characteristics of diorite with double rectangular holes under coupled static-dynamic loads. *Rock Mech. Rock Eng.* **2021**, *54*, 2741–2761. [CrossRef]

48. Sutton, M.A.; Orteu, J.; Schreier, H. *Image Correlation for Shape, Motion and Deformation Measurements*; Springer: New York, NY, USA, 2009. [CrossRef]
49. Ping, Q.; Luo, X.; Ma, Q.Y.; Yuan, P. Broken energy dissipation characteristics of sandstone specimens under impact loads. *Chin. J. Rock Mech. Eng.* **2015**, *34* (Suppl. S2), 4197. (In Chinese) [CrossRef]
50. Wen, S.; Zhao, X.W.; Chang, Y.L.; Li, B. Energy dissipation of dynamic failure of mixed rock specimens subject to SHPB compression. *J. Basic Sci. Eng.* **2021**, *29*, 483–492. (In Chinese) [CrossRef]
51. Chen, P.Y.; Zhao, F.J.; Chen, B.; Tian, X.Y. Energy dissipation characteristics of prefabricated fractured rock under impact load. *Miner. Eng. Res.* **2021**, *36*, 17–23. (In Chinese) [CrossRef]
52. Hu, J.; Gong, F.Q.; Jia, H.Y. Research on mechanical and energy dissipation characteristics of red sandstone in SHPB compression test. *Gold Sci. Technol.* **2020**, *28*, 411–420. (In Chinese) [CrossRef]
53. Wang, M.X.; Wang, H.B.; Zong, Q. Experimental study on energy dissipation of mudstone in coal mine under impact loading. *J. China Coal Soc.* **2019**, *44*, 1716–1725. (In Chinese) [CrossRef]
54. Gao, F.; Yang, G.; Xiong, X.; Zhou, K.P.; Cong, L.; Li, J.L. Experimental study on the dynamic mechanical characteristics of slope rock under low-temperature conditions. *Chin. J. Eng.* **2023**, *45*, 171–181. (In Chinese) [CrossRef]
55. Ping, Q.; Ma, Q.Y.; Yuan, P. Energy dissipation analysis of stone specimens in SHPB tensile test. *J. Min. Saf. Eng.* **2013**, *30*, 401–407. (In Chinese)
56. Sun, Y.J.; Qi, C.Z.; Zhu, H.T.; Guo, Y.P.; Wang, Y.Q. Energy analysis on rock dynamic fracture process. *Chin. J. Undergr. Sp. Eng.* **2020**, *16*, 43–49. (In Chinese)
57. Ravi-Chandar, K. *Dynamic Fracture*; Elsevier: Amsterdam, The Netherlands, 2004. [CrossRef]

applied sciences

MDPI

Article

Research on Analytical Solution of Stress Fields in Adjacent Tunnel Surrounding Rock Under Blasting and Verification Analysis

Tao Luo [1,2,*], Yong Wei [1], Junbo Zhao [1], Yelong Xie [3], Yan Hu [3], Xiaoming Lou [1] and Xiaofeng Huo [1]

[1] School of Zijin Geology and Mining, Fuzhou University, Fuzhou 350100, China; wayne_jie@163.com (Y.W.); 15765801773@163.com (J.Z.); 162202223@fzu.edu.cn (X.L.); huoxiaofeng0705@gmail.com (X.H.)
[2] School of Resource Engineering, Longyan University, Longyan 364012, China
[3] Zhongchang Nuclear Engineering (Fujian) Construction and Development Co., Ltd., Nanping 353200, China; 13675094319@163.com (Y.X.); xiaoshuyi531@gmail.com (Y.H.)
* Correspondence: luotao@lyun.edu.cn

Abstract

In tunnel blasting, an analytical solution for dynamic stress in the surrounding rock of adjacent tunnels is critical for dynamic response analysis, mechanical evaluations, and crack propagation control. Previous studies on stress field analytical solutions primarily modeled rock as a linear elastic material, focusing mainly on the P-wave effects from instantaneous detonation. Based on Heelan's short cylindrical cavity model, this paper derives an analytical solution for blast-induced dynamic stresses in adjacent tunnel rock, incorporating both induced SV-waves and a rock mass damage factor through rigorous theoretical analysis. Numerical case studies and field measurements were used to analyze stress propagation during tunnel blasting, and theoretical results were compared with measured data. The key findings were as follows: Radial stress > axial stress > hoop stress. All three stresses decay with increasing distance and damage factor, following an inversely proportional relationship with distance. Radial stress decays faster than axial and hoop stresses. Stress also decays exponentially over time, with the peak occurring after the transverse wave arrival. The theoretical results show approximately 10% deviation from the existing empirical formulas, while field measurements closely match the theoretical model, showing consistent stress trends and an average error of 7.02% (radial), 7.56% (axial) and 7.05% (hoop), confirming the reliability of the proposed analytical solution.

Keywords: blasting; adjacent tunnel; stress analytical solution; stress field; field measurements

1. Introduction

Tunnel engineering, a key technology for traversing complex terrains and reducing transportation distances, is widely applied in expressways, railways, and urban transit systems. Among construction methods, drilling and blasting methods remain predominant due to their advantages in speed, cost-effectiveness, and adaptability to varying geological conditions [1]. However, the transient stress waves generated during tunnel blasting operations can significantly affect existing tunnels [2], with potential damage requiring careful evaluation. In closely spaced adjacent tunnels, dynamic loading from subsequent blasts can pose serious safety risks to neighboring tunnel structures, including lining failures. As the primary source of dynamic loading, blast-induced stress waves critically impact surrounding rock masses and support systems. Understanding their response and control has become a key technical challenge in safe tunnel construction. Therefore, obtaining analytical solutions for dynamic stresses in surrounding rocks under blasting loads is essential for understanding stress propagation, controlling blast vibrations, and ensuring the integrity of existing tunnel support structures.

Current research challenges in analytical solutions for blast-induced stress mainly fall into two areas: The first is the complexity of stress superposition from transient P- and S-wave interactions, compounded by stress propagation through heterogeneous and discontinuous materials [3,4]. The second is difficulties in validation due to limitations in in situ stress measurement embedding sensors in rock, which is technically challenging, and field data are often affected by interference from air shock waves and construction equipment during signal acquisition [5]. Analytical solutions for dynamic stresses in surrounding rocks of adjacent tunnels are crucial for understanding stress evolution, calculating dynamic responses, analyzing mechanical behavior, and optimizing blast control parameters.

As a form of transient energy excitation, blast-induced stress waves have long been a central topic in explosion theory and its applications. Based on charge geometry, blasting can be classified into spherical charges and cylindrical charges. Tunnel excavation blasting typically uses cylindrical charges, and extensive international research has focused on their associated stress field characteristics.

Theoretical developments have led to three mainstream models for calculating blast-induced stress fields: the Starfield superposition method, the ideal fluid medium model, and Heelan's short-cylinder solution. Starfield [6] proposed decomposing cylindrical charges into multiple equivalent spherical charges. Experimental studies on stress wave propagation, showed that this superposition method yields acceptable agreement with measured data in the medium-to-far fields. Based on this framework, Liu [7] derived computational formulas for stress fields generated by cylindrical charges. Neiman et al. [8] treated rock media as incompressible ideal fluids under transient blast loading, where particle velocity fields satisfy Laplace's equations. Using boundary mapping methods, they developed solutions for stress fields around cylindrical charges under various borehole boundary conditions. Heelan [9] established a classical analytical solution for cylindrical charge blasting using elastodynamic theory. This model assumes an infinite, isotropic rock mass and provides an expression for radial stress (σ_r) distribution due to single hole blasting:

$$\sigma_r = P_0(\mathrm{r}/r_0)^{\alpha} e^{-\beta(r-r_0)} \tag{1}$$

where

P_0 is the initial blast pressure (MPa);
r_0 is the charge radius (m);
α and β are rock mass attenuation coefficients.

Because the mechanical properties of the rock mass will change after blasting [10], this model accounts for rock mass damage or jointed fissures, leading to prediction errors exceeding 40% in weak surrounding rock. To address this, Henrych [11] introduced an attenuation factor and proposed a modified stress wave propagation equation for heterogeneous rock masses. The relationship between the attenuation coefficient β and the rock mass integrity coefficient K_v is expressed as follows:

$$\beta = 0.85 \times (1 - K_v) \times \ln\left(\frac{r}{r_0}\right) \tag{2}$$

This model has shown good accuracy in hard rock tunnels but lacks adaptability in fractured zones. To address this, Zhang et al. [12] introduced a damage factor to revise the stress wave equation, establishing a nonlinear attenuation model that accounts for fracture propagation. Their results showed that stress attenuation in jointed rock masses exceeds that in intact rock by over 50%. Ma et al. [13] discretized cylindrical charges into multiple linked spherical charges and derived the dynamic stress field using vector superposition methods. Lei et al. [14] developed analytical solutions for stress fields under different detonation transmission directions using an equivalent unit spherical charge superposition method, incorporating directional effects. Favreau [15], Graff [16], and Achenbach [17] analyzed stress wave fields induced by cylindrical charges based on spherical cavity expansion models. Blair [18] extended Heelan's short-cylinder solution by integrating finite detonation velocity and charge length, advancing a superposition model for long charges. Gao et al. [19] further partitioned elongated cylindrical charges into a series of instantaneously detonated short charges, deriving vibration velocity formulas for arbitrary points using time-delayed superposition of detonation wave propagation.

In experimental and field-testing research, advanced techniques such as dynamic caustics and ultra-dynamic strain measurement have been widely adopted. Yang et al. [20] systematically investigated stress–strain fields in centrally initiated cylindrical charges and the evolution of local stress fields near blast-induced crack tips using dynamic caustics. Their study revealed significant principal stress differences across orientations and pronounced end effects. Ding [21] conducted scaled model experiments to analyze stress wave superposition between adjacent boreholes. Qiu [22] used physical modeling to simulate blast-induced disturbances in underground tunnels, mapping surface strain distributions at various blast locations. Yang [23] employed a high-speed camera-based digital image correlation system in dual-borehole blast tests, showing that peak stress at the midpoint between two charges reached 2.4–2.7 times that of single-hole blasting. Liu et al. [24] developed an embedded strain bar system incorporating MEMS sensors and wireless transmission, enabling real-time monitoring of dynamic stress fields during construction of the Qingdao Jiaozhou Bay Tunnel. Their signal processing algorithm combining wavelet denoising with Kalman filtering improved the signal-to-noise ratios from 15 dB to 28 dB. Zhang [25] optimized field blast monitoring protocols to capture dynamic strain on tunnel primary lining (TSL) surfaces. Luo et al. [26–28] performed in situ dynamic strain measurements on existing tunnel surfaces, quantitatively confirming strain amplitudes on blast-facing sides.

In a numerical simulation, Gao et al. [19] used 3D dynamic finite element software to model cylindrical charge blasting in infinite rock masses. By developing a quarter-symmetry finite element model, they systematically examined the effects of initiation positions on explosive energy transmission and stress field distribution. Liang et al. [29] applied numerical simulation to analyze rock fragmentation under stress wave superposition from multi-source charge detonations, confirming that stress wave interactions from spherical charges align with theoretical rock-breaking mechanics. Xiang et al. [30] proposed a numerical analysis model for strip charge stress fields by integrating the Starfield superposition method with dynamic finite element analysis, providing a comprehensive understanding of the formation and evolution mechanisms of blast-induced stress fields under diverse initiation modes.

In summary, despite substantial progress in studying stress fields from cylindrical charge blasting, three critical limitations remain:

1. Methodological limitations: The widespread use of equivalent spherical charge models—which inherently omit S-wave generation—fails to reflect actual field conditions, where cylindrical charges produce both P- and S-waves.
2. Dynamic modeling oversights: Current superposition approaches for elongated charges, based on short-cylinder solutions often overlook two key factors: (a) axial pressure effects from instantaneous detonation of short cylindrical cavities, and (b) the evolution of rock mass damage in media treated as continuous, homogeneous materials.

3. Validation constraints: Most existing measurements focus on surface particles, lacking in situ dynamic stress data from within the rock mass, which limits the verification of theoretical formulations models.

To address these gaps, this paper builds on explosion stress wave propagation theory and Heelan's short-cylinder solution [9], incorporating the axial pressure generated by the instantaneous detonation of short cylindrical cavities (Figure 1) and the effects of rock mass damage. Based on the phase delay effect and stress wave propagation intervals, a porous superposition method is used to derive an analytical solution for the stress in the surrounding rock of adjacent tunnels under blasting. Theoretical formulas are employed to calculate stress magnitude and propagation patterns of radial particles in adjacent tunnels. These results are then validated through field stress measurements and numerical simulations. The findings provide a theoretical basis for calculating dynamic stress in surrounding rock, understanding the propagation of internal cracks and fissures, and improving surrounding rock reinforcement and blast control.

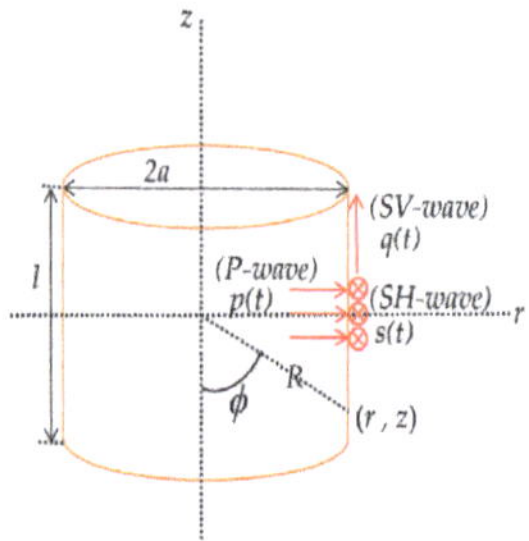

Figure 1. The instantaneous detonation pressure acts on the short-cylinder cavity: $p(t)$, $q(t)$ and $s(t)$ represent the radial, axial, and tangential pressures, respectively, which are generated by P-waves, SV waves, and SH waves (MPa); R is the distance from the charge center to the measurement point (m); ϕ is the angle from the vertical direction; $2a$ is the cylinder diameter (m); and l is the cylinder height (m).

2. Theoretical Derivation of Analytical Solutions for Blasting Stress Fields in Adjacent Tunnels

During tunnel face blasting, both surface waves (Rayleigh and Love) and body waves are generated and propagate through the surrounding rock mass. Body waves include P-waves and S-waves, with the latter divided into SV and SH components. P-waves travel faster than S-waves. The combined action of P-, SV-, and SH-waves induces three-dimensional dynamic stresses in the surrounding rock of adjacent tunnels: radial stress (σ_r), axial stress (τ_z), and hoop stress (τ_θ). The mechanical model for an adjacent tunnel's surrounding rock under coupled wave actions is shown in Figure 2.

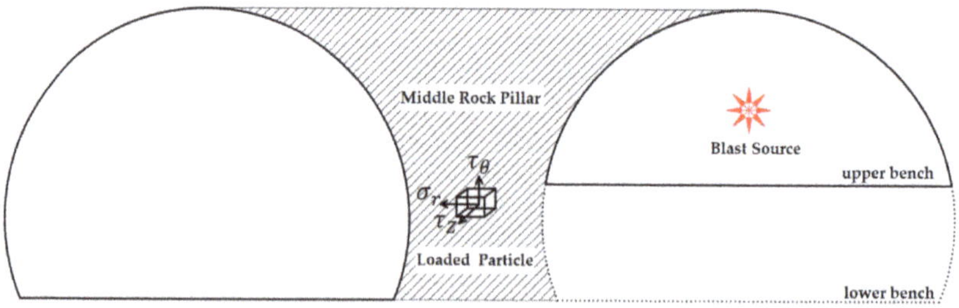

Figure 2. Schematic diagram of the stress distribution on internal material particles in surrounding rock mass of adjacent tunnels.

2.1. Blasting Stress Field of a Single-Hole Cylindrical Charge Based on Heelan's Short-Cylinder Solution

2.1.1. Stress Field of a Short-Cylinder Charge Incorporating Instantaneous Detonation Axial Pressure and Rock Mass Damage

Heelan modeled the stress induced by the instantaneous detonation of a short cylindrical cavity charge as a transient pressure applied to the boundary of a short cylindrical cavity (Figure 1), from which displacement solutions were derived. The results show that detonation generates both P-waves and S-waves, with the latter further decomposing into SH-waves (parallel to the horizontal plane) and SV-waves (perpendicular to it). Figure 3 shows the particle vibration direction as the angle between the wave propagation direction and the short cylindrical cavity wall varies. The displacement solutions for these three wave types are given by [9]:

$$\begin{bmatrix} \vec{u_P} \\ \vec{\omega_P} \end{bmatrix} = \left(\frac{F_1(\phi)}{R} \frac{\mathrm{d}}{\mathrm{d}t} \left\{ p(t - \frac{R}{C_p}) \right\} + \frac{G_1(\phi)}{R} q(t - \frac{R}{C_p}) \right) \begin{bmatrix} \sin\phi \\ -\cos\phi \end{bmatrix} \tag{3}$$

$$\begin{bmatrix} \vec{u_{SV}} \\ \vec{\omega_{SV}} \end{bmatrix} = \left(\frac{F_2(\phi)}{R} \frac{\mathrm{d}}{\mathrm{d}t} \left\{ p(t - \frac{R}{C_s}) \right\} + \frac{G_2(\phi)}{R} q(t - \frac{R}{C_s}) \right) \begin{bmatrix} \cos\phi \\ \sin\phi \end{bmatrix} \tag{4}$$

$$\vec{v_{SH}} = \frac{K(\phi)}{R} \frac{\mathrm{d}}{\mathrm{d}t} \left\{ s(t - \frac{R}{C_s}) \right\} \tag{5}$$

where $\vec{u_P}$ and $\vec{u_{SV}}$ are the horizontal displacements of the P-wave and SV-wave (m); $\vec{\omega_P}$ and $\vec{\omega_{SV}}$ are their vertical displacements (m); and $\vec{v_{SH}}$ is the tangential displacement of the SH-wave (m). In tunnel face blasting, $\vec{u}$, $\vec{w}$ and $\vec{v}$ represent the radial, axial and hoop displacements, respectively (m); C_p and C_s are the propagation velocities of the P-wave and S-wave (m/s); $p(t)$, $q(t)$, $s(t)$, ϕ and R are defined as in Figure 1; $F_1(\phi)$ and $G_1(\phi)$ are the source functions for the P-wave; $F_2(\phi)$ and $G_2(\phi)$ are for the SV-wave; and $K(\phi)$ is for the SH-wave.

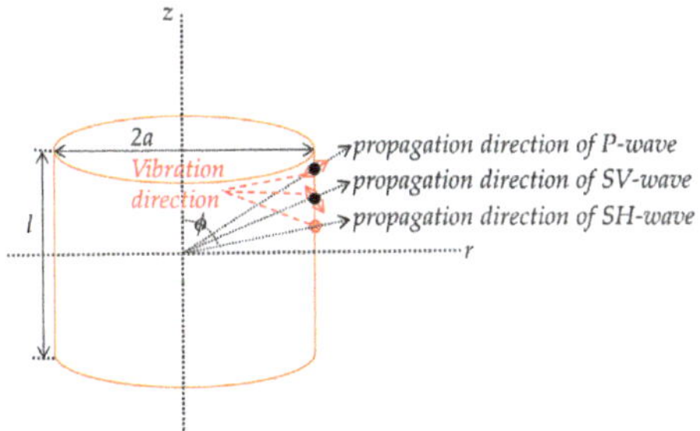

Figure 3. The propagation direction of three types of waves and the vibration direction of particles.

Source function definitions [9]:

$$\begin{cases} F_1(\phi) = \frac{\Delta}{4\pi G C_p}\left(1 - \frac{2C_s^2 cos^2\phi}{C_p^2}\right) \\ G_1(\phi) = -\frac{A}{4\pi G C_p^2} C_s^2 cos\phi \\ F_2(\phi) = \frac{\Delta}{4\pi G C_s} \sin 2\phi \\ G_2(\phi) = \frac{A}{4\pi G} \sin\phi \\ K(\phi) = \frac{\Delta}{4\pi G C_s} \sin\phi \end{cases} \tag{6}$$

where Δ is the volume of the short-cylinder cavity (m^3), A is its lateral surface area (m^2), and G is the shear modulus of the surrounding rock mass (GPa). The rock damage factor D is defined as follows [31]:

$$D = 1 - \left(\frac{v_{p,post}}{v_{p,pre}}\right)^2 \tag{7}$$

where $v_{p,pre}$ and $v_{p,post}$ are the P-wave velocities of the rock before and after blasting, respectively (m/s), measured on-site using acoustic testing. The uncertainty in the damage factor D is ± 0.01. Since the P-wave velocity in rock mass after blasting is generally lower than that before blasting, the damage factor D ranges from 0 to 1, with higher values indicating more severe damage to the surrounding rock. Based on the existing formula [32], and incorporating the damage factor, $p(t)$, $q(t)$ and $s(t)$ can be expressed as follows:

$$\begin{cases} p(t) = P_m e^{-\frac{(at+b)}{(1-D)}} \\ q(t) = \mu_d(1-\lambda)p(t) \\ s(t) = -\lambda p(t) \\ P_m = \frac{1}{8} n \rho_e D_e^2 \left(\frac{d_c}{d_b}\right)^3 \end{cases} \tag{8}$$

where P_m is the maximum pressure accounting for the combined effects of explosive stress waves and detonation product expansion (MPa); a and b are detonation decay coefficients of the explosion [33]; μ_d is the dynamic Poisson's ratio of the rock ($\mu_d = 0.8\mu$); λ is the lateral pressure coefficient, defined as $\lambda = \mu_d/(1 - \mu_d)$; n is the pressure amplification coefficient of the detonation products; ρ_e is the explosive density (kg/m^3); D_e is the explosive detonation velocity (m/s); d_c is the charge diameter (m); and d_b is the borehole diameter (m).

During cylindrical charge detonation, the rock mass within the blast volume is subjected first to the shock front and then to explosive gas pressure. This process creates "crushed" and "fractured" zones, characterized by radial cracks, circumferential cracks, and non-radial fractures [34,35]. While blasting can effectively break rock within the intended excavation zone, it may also damage surrounding rock that must be preserved, compromising its mechanical properties, bearing capacity, and stability [36]. To assess such effects, Yu and Vongpaisal developed the Blast Damage Index (BDI) for mining applications [37].

$$\mathrm{BDI} = \frac{\mathrm{IS}}{\mathrm{DR}} = \frac{\mathrm{VdC}}{\mathrm{KT}} \tag{9}$$

where IS is the induced stress; DR is the damage resistance; V is the vector sum of PPV (mm/s); d is the specific gravity of the rock mass (kg/m^3); C is the compressional wave velocity of the rock mass (mm/s); K is the site quality constant; and T is the dynamic tensile strength of the rock mass (N/m^2).

This relationship accounts for both the rock mechanics and wave propagation effects, but it depends on numerous site-specific parameters. To simplify the representation of the rock mass, this paper introduces the damage factor D ($0 < D < 1$). By incorporating D into the exponent of the stress attenuation term (Equation (8)), the model quantitatively captures the influence of rock degradation on stress wave propagation. Physically, increasing D reflects progressively damaged material, such as the formation and growth of microcracks, voids, or other defects. These features act as energy dissipation mechanisms by (i) reflecting, refracting, and scattering incident stress waves, which reduces the measured wave amplitude; (ii) converting mechanical wave energy into heat and other dissipated forms through crack friction and plastic deformation; and (iii) reducing the rock mass's overall load bearing capacity.

The expression $e^{-(at+b)/(1-D)}$ models this behavior. The term $1/(1 - D)$ amplifies the decay rate: as D increases towards 1 (severely damaged rock), $(1 - D)$ decreases, making the exponent more negative and leading to faster stress amplitude decay of the stress amplitude P_m. This matches physical observations—higher damage leads to stronger attenuation of stress waves. Thus, $1/(1-D)$ effectively quantifies how damage progressively accelerates wave energy dissipation and scattering.

During tunnel face blasting, the radial, axial, and hoop displacements of the surrounding rock are expressed as follows:

$$\begin{aligned}\vec{u} = \vec{u_P} + \vec{u_{SV}} &= \left(\tfrac{F_1(\phi)}{R}\tfrac{d}{dt}\left\{p(t-\tfrac{R}{C_p})\right\} + \tfrac{G_1(\phi)}{R}q(t-\tfrac{R}{C_p})\right)\sin\phi \\ &+\left(\tfrac{F_2(\phi)}{R}\tfrac{d}{dt}\left\{p(t-\tfrac{R}{C_s})\right\}+\tfrac{G_2(\phi)}{R}q(t-\tfrac{R}{C_s})\right)cos\phi\end{aligned} \tag{10}$$

$$\begin{aligned}\vec{\omega} = \vec{\omega_P} + \vec{\omega_{SV}} &= -cos\phi\left(\tfrac{F_1(\phi)}{R}\tfrac{d}{dt}\left\{p(t-\tfrac{R}{C_p})\right\} + \tfrac{G_1(\phi)}{R}q(t-\tfrac{R}{C_p})\right) \\ &+\left(\tfrac{F_2(\phi)}{R}\tfrac{d}{dt}\left\{p(t-\tfrac{R}{C_s})\right\} + \tfrac{G_2(\phi)}{R}q(t-\tfrac{R}{C_s})\right)\sin\phi\end{aligned} \tag{11}$$

$$\vec{v} = \vec{v_{SH}} = \frac{K(\phi)}{R}\frac{d}{dt}\left\{s(t-\frac{R}{c_s})\right\} \tag{12}$$

By differentiating the displacement field with respect to time, the vibration velocity field of the short-cylinder charge is derived as follows:

$$\begin{aligned}\vec{V_r} = \tfrac{\partial u}{\partial t} &= \left(\tfrac{F_1(\phi)}{R}\tfrac{d^2}{dt^2}\left\{p(t-\tfrac{R}{c_p})\right\} + \tfrac{G_1(\phi)}{R}\tfrac{d}{dt}\left\{q(t-\tfrac{R}{c_p})\right\}\right)\sin\phi \\ &+\left(\tfrac{F_2(\phi)}{R}\tfrac{d^2}{dt^2}\left\{p(t-\tfrac{R}{c_s})\right\} +\tfrac{G_2(\phi)}{R}\tfrac{d}{dt}\left\{q(t-\tfrac{R}{c_s})\right\}\right)cos\phi\end{aligned} \tag{13}$$

$$\begin{aligned}\vec{V_z} = \tfrac{\partial \omega}{\partial t} &= -cos\phi\left(\tfrac{F_1(\phi)}{R}\tfrac{d^2}{dt^2}\left\{p(t-\tfrac{R}{C_p})\right\} + \tfrac{G_1(\phi)}{R}\tfrac{d}{dt}\left\{q(t-\tfrac{R}{C_p})\right\}\right) \\ &+\left(\tfrac{F_2(\phi)}{R}\tfrac{d^2}{dt^2}\left\{p(t-\tfrac{R}{C_s})\right\} +\tfrac{G_2(\phi)}{R}\tfrac{d}{dt}\left\{q(t-\tfrac{R}{c_s})\right\}\right)\sin\phi\end{aligned} \tag{14}$$

$$\vec{V_\theta} = \frac{\partial v}{\partial t} = \frac{K(\phi)}{R}\frac{d^2}{dt^2}\left\{s(t-\frac{R}{C_s})\right\} \tag{15}$$

where $\vec{V_r}$, $\vec{V_z}$ and $\vec{V_\theta}$ represent the radial, axial, and hoop vibration velocities, respectively, in units of m/s.

During blasting operations, the relationships between the stresses (σ_r, τ_z, τ_θ) generated by P-waves, SV-waves, and SH-waves and the corresponding particle vibration velocities ($\vec{V_r}$, $\vec{V_z}$, $\vec{V_\theta}$) can be expressed as follows [38]:

$$\begin{cases}\sigma_r = \vec{V_r}\rho C_p \\ \tau_z = \vec{V_z}\rho C_s \\ \tau_\theta = \vec{V_\theta}\rho C_s\end{cases} \tag{16}$$

where ρ is the rock density, kg/m^3.

By synthesizing Equations (13)–(16), the stress field induced by a short-cylinder charge blasting can be derived as follows:

$$\begin{aligned}\sigma_r &= \left(\tfrac{F_1(\phi)}{R}\tfrac{d^2}{dt^2}\left\{p(t-\tfrac{R}{C_p})\right\} + \tfrac{G_1(\phi)}{R}\tfrac{d}{dt}\left\{q(t-\tfrac{R}{C_p})\right\}\right)\rho C_p\sin\phi \\ &+\left(\tfrac{F_2(\phi)}{R}\tfrac{d^2}{dt^2}\left\{p(t-\tfrac{R}{C_s})\right\} + \tfrac{G_2(\phi)}{R}\tfrac{d}{dt}\left\{q(t-\tfrac{R}{C_s})\right\}\right)\rho C_p cos\phi\end{aligned} \tag{17}$$

$$\begin{aligned}\tau_Z &= -\left(\tfrac{F_1(\phi)}{R}\tfrac{d^2}{dt^2}\left\{p(t-\tfrac{R}{C_p})\right\} + \tfrac{G_1(\phi)}{R}\tfrac{d}{dt}\left\{q(t-\tfrac{R}{C_p})\right\}\right)\rho C_s cos\phi \\ &+\left(\tfrac{F_2(\phi)}{R}\tfrac{d^2}{dt^2}\left\{p(t-\tfrac{R}{C_s})\right\}+ \tfrac{G_2(\phi)}{R}\tfrac{d}{dt}\left\{q(t-\tfrac{R}{C_s})\right\}\right)\rho C_s\sin\phi\end{aligned} \tag{18}$$

$$\tau_\theta = \frac{\rho C_s K(\phi)}{R}\frac{d^2}{dt^2}\left\{s(t-\frac{R}{C_s})\right\} \quad (19)$$

2.1.2. Superposition of Blasting Stress Fields Considering Phase Delay Effects

The detonation of an elongated cylindrical charge can be modeled as the superposition of instantaneous detonations from multiple short-cylinder segments that detonate sequentially. Accounting for time delays due to detonation wave propagation, the stress field for cylindrical charge blasting (e.g., with bottom initiation) can be derived by combining the superposition model in Figure 4 with Equations (17)–(19):

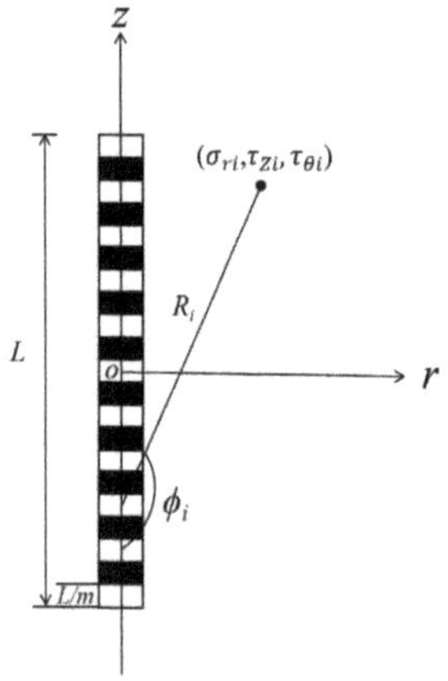

Figure 4. Calculation model for superposition of the stress field in columnar charge blasting: L is the total length of the cylindrical charge (m); and m is the number of segmented short-cylinder charges into which the elongated charge is divided. According to reference [17], explosive charges with a length-to-diameter ratio (L/D) of less than 10 are classified as short cylindrical charges.

Based on this criterion, elongated cylindrical charges can be divided and calculated. By substituting $q(t) = \mu_d\,(1-\lambda)p(t)$, and $s(t) = -\lambda p(t)$ (from Equation (8)), the stress field of the columnar charges in matrix form is as follows:

$$\sigma_i = \begin{pmatrix}\sigma_{ri}\\ \tau_{z_i}\\ \tau_{\theta_i}\end{pmatrix} = \rho C \begin{pmatrix} \left(\frac{F_2(\phi_i)}{R_i}\frac{d^2}{dt^2} + \mu_d(1-\lambda)\frac{G_2(\phi_i)}{R_i}\frac{d}{dt}\right)\left\{p\left(t-\frac{(i-1)L}{mD_e}-\frac{R_i}{C_p}\right)\right\} \\ \left(\frac{F_1(\phi_i)}{R_i}\frac{d^2}{dt^2} + \mu_d(1-\lambda)\frac{G_1(\phi_i)}{R_i}\frac{d}{dt}\right)\left\{p\left(t-\frac{(i-1)L}{mD_e}-\frac{R_i}{C_s}\right)\right\} \\ \lambda\frac{K(\phi_i)}{R_i}\frac{d^2}{dt^2}\left\{p\left(t-\frac{(i-1)L}{mD_e}-\frac{R_i}{C_s}\right)\right\} \end{pmatrix} \boldsymbol{R}_\theta \quad (20)$$

$$C = \begin{pmatrix} C_p & 0 & 0\\ 0 & -C_s & 0\\ 0 & 0 & -C_s\end{pmatrix}, \boldsymbol{F}_i = \begin{pmatrix} \left(\frac{F_2(\phi_i)}{R_i}\frac{d^2}{dt^2} + \mu_d(1-\lambda)\frac{G_2(\phi_i)}{R_i}\frac{d}{dt}\right)\left\{p\left(t-\frac{(i-1)L}{mD_e}-\frac{R_i}{C_p}\right)\right\} \\ \left(\frac{F_1(\phi_i)}{R_i}\frac{d^2}{dt^2} + \mu_d(1-\lambda)\frac{G_1(\phi_i)}{R_i}\frac{d}{dt}\right)\left\{p\left(t-\frac{(i-1)L}{mD_e}-\frac{R_i}{C_s}\right)\right\} \\ \lambda\frac{K(\phi_i)}{R_i}\frac{d^2}{dt^2}\left\{p\left(t-\frac{(i-1)L}{mD_e}-\frac{R_i}{C_s}\right)\right\} \end{pmatrix}, \boldsymbol{R}_\theta = \begin{pmatrix} \cos\theta & -\sin\theta & 0\\ \sin\theta & \cos\theta & 0\\ 0 & 0 & 1\end{pmatrix} \quad (21)$$

$$\sigma_i = \rho \boldsymbol{C}\boldsymbol{F}_i\boldsymbol{R}_\theta,\ \sigma = \sum_{i=1}^{m}\sigma_i \quad (22)$$

In existing studies, the analytical solutions for derived blasting stress fields based on short cylindrical charges are formulated as follows [19]:

$$\begin{pmatrix}\sigma_{ri}\\ \tau_i\end{pmatrix} = \rho\begin{pmatrix} C_p & 0\\ 0 & -C_s\end{pmatrix}\begin{pmatrix} sin\phi_i\\ cos\phi_i\end{pmatrix}\left(\frac{F_1(\phi_i)}{R_i}\frac{d^2}{dt^2}\left\{p(t-\frac{(i-1)L}{mD_e}-\frac{R_i}{C_p})\right\}\right) + \rho\begin{pmatrix} C_p & 0\\ 0 & C_s\end{pmatrix}\begin{pmatrix} cos\phi_i\\ sin\phi_i\end{pmatrix}\left(\frac{F_2(\phi_i)}{R_i}\frac{d^2}{dt^2}\left\{p(t-\frac{(i-1)L}{mD_e}-\frac{R_i}{C_s})\right\}\right) \quad (23)$$

$$\begin{bmatrix} \upsilon_r \\ \tau \end{bmatrix} = \sum_{i=1}^{m} \begin{bmatrix} \sigma_{ri} \\ \tau_i \end{bmatrix} \tag{24}$$

Compared to the existing research Formulas (23) and (24), the Equations (20)–(22) derived in this study introduce two key advancements: the inclusion of instantaneous detonation axial pressure $q(t)$, and the rock mass damage factor D.

2.2. Multi-Borehole Stress Field Superposition Based on Stress Wave Propagation Time Intervals

In multi-borehole blasting, stress field superposition must account for time interval effects caused by differences in stress wave propagation paths (see Figure 5). These delays result from variations in propagation distances and anisotropic wave velocities in the rock mass, necessitating precise phase corrections for accurate dynamic stress predictions.

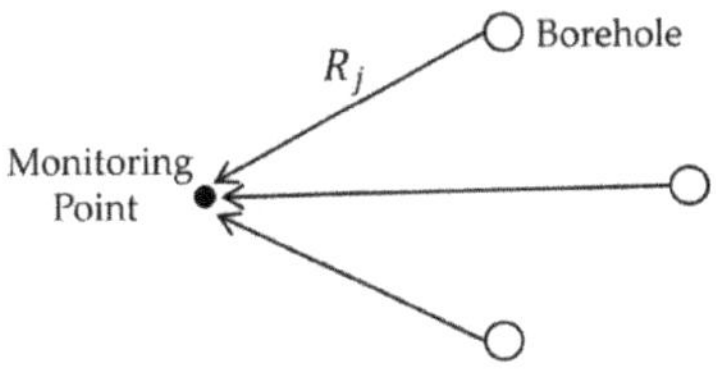

Figure 5. Schematic diagram of simultaneous explosion of multiple blastholes at the same point.

Based on the single-hole stress field derived above, the total stress field at a given observation point under multi-borehole blasting is obtained by superposition:

$$\begin{bmatrix} \sigma_r \\ \tau_Z \\ \tau_\theta \end{bmatrix} = \sum_{j=1}^{n} \begin{bmatrix} \sigma_{rj}(t - t_j) \\ \tau_{Zj}(t - t_j) \\ \tau_{\theta j}(t - t_j) \end{bmatrix} \tag{25}$$

where n is the number of boreholes; t_j is the time required for stress waves generated from the j-th borehole to reach the observation point, given by $t_j = R_j / C_{p,s}$; R_j is the distance from the j-th borehole to the observation point (m); and C is the corresponding P-wave or S-wave propagation velocity (m/s).

3. Application Examples and Comparative Analysis

3.1. Selection of Relevant Parameters and Calculation Points

A tunnel in Fujian Province is a double-track highway tunnel with a 10 m net spacing that was constructed using the upper- and lower-step method, with simultaneous construction of the left and right tunnels. The surrounding rock is weathered granite with well-developed joint fissures; its physical and mechanical properties are listed in Table 1. Engineering practice indicates that slotted holes in a single section carry a high charge, producing the greatest stress and vibration. Therefore, as an example, eight secondary slotted holes were detonated simultaneously on the upper steps. Each borehole (dot) has a diameter of 40 mm, an explosive charge diameter of 32 mm, a hole depth of 3 m, a charge of 2.4 kg per hole, and bottom initiation. The layout is shown in Figure 6. The explosion has a detonation velocity of 3200–3800 m/s, a density of 1200 kg/m^3 and a detonation product pressure coefficient (n) of 10 [31]. Due to the extensive jointing in the rock mass, the rock damage factor is taken as D = 0.2.

Table 1. Physical and mechanical parameters of tunnel surrounding rock.

Density ($kg \cdot m^{-3}$)	Compressive Strength (MPa)	Tensile Strength (MPa)	Elastic Modulus (GPa)	P-Wave Velocity (m/s)	S-Wave Velocity (m/s)	Poisson's Ratio
2650	54.50	7.50	30.70	3036	1704	0.23

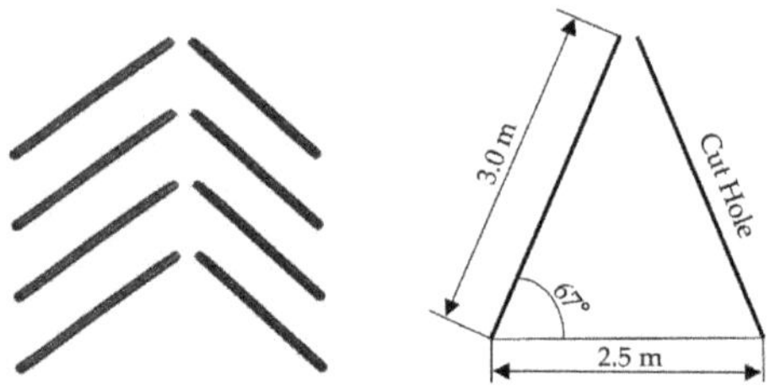

Figure 6. Schematic diagram of cutting hole.

According to existing studies [39], the blast-facing sidewall of the adjacent tunnel is a critical risk area. To evaluate the stress magnitude and distribution in the rock mass at this location, six calculation points (Points A–F) are placed at 1-m intervals along a horizontal line within the sidewall (see Figure 7). The first point (Point A) is located 11.42 m from the center of the cut holes. According to reference [17], a single cylindrical charge with an aspect ratio of 10 was divided into 15 short cylindrical charge units (m = 15), each 16 cm long with a base radius of 1.6 cm. These parameters were then used in Equations (20)–(22) and (25) for calculation.

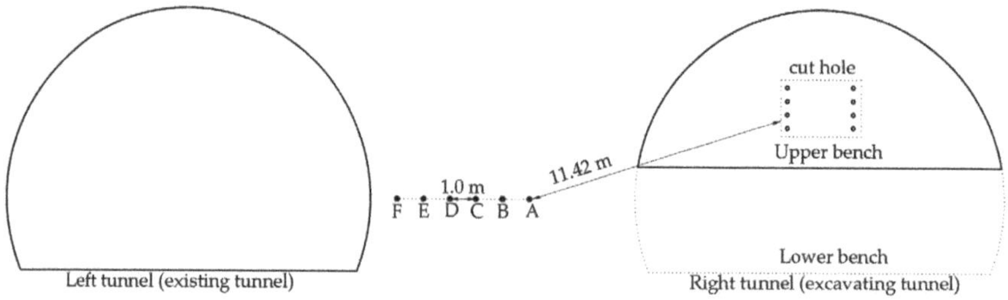

Figure 7. Schematic diagram of stress calculation points.

3.2. Calculation Results and Comparative Analysis

The peak stress points A~F are shown in Table 2. The uncertainty in the peak stress is ±0.02. The results indicate that during blasting at the excavation face, the peak stresses in the rock mass follow the order: radial stress > axial stress > hoop stress.

Table 2. Stress calculation results.

Calculation Point	Distance to the Cut Hole Center (m)	Peak Stress (MPa)		
		Radial	Axial	Hoop
A	11.42	1.60	0.93	0.76
B	12.38	1.37	0.78	0.65
C	13.35	1.18	0.68	0.56
D	14.32	1.02	0.59	0.49
E	15.29	0.90	0.52	0.43
F	16.26	0.80	0.46	0.38

To further examine stress evolution, Figure 8 shows the triaxial stress variations of radial points A–F along the tunnel axis, based on the data in Table 2. The results indicate that radial stress decays inversely with distance and attenuates more rapidly than axial stress. Hoop stress remains consistently lower than both radial and axial stresses. Based on Table 2, the time history curve of radial stress at Point A is plotted in Figure 9. This curve shows that radial stress decreases over time and exhibits a double peak. The peak occurs after the arrival of the S-wave, as the superposition of P- and S-waves causes a sudden increase in stress, indicating their combined effect on the surrounding rock particles. According to Table 2, radial, axial, and hoop stresses decay with distance following the power law $\sigma = \sigma_0 d^{\alpha}$, where σ_0 is the stress at d = 0, d is the distance from the cut hole center, and $\alpha = -2.004 \pm 0.01063$ is the common power coefficient, identical for all stresses (see Figure 7). The initial stresses are σ_{r0} = 215.427 ± 5.89566 MPa, σ_{a0} = 123.68921 ± 3.36189 MPa, and σ_{h0} = 102.10915 ± 2.85727 MPa.

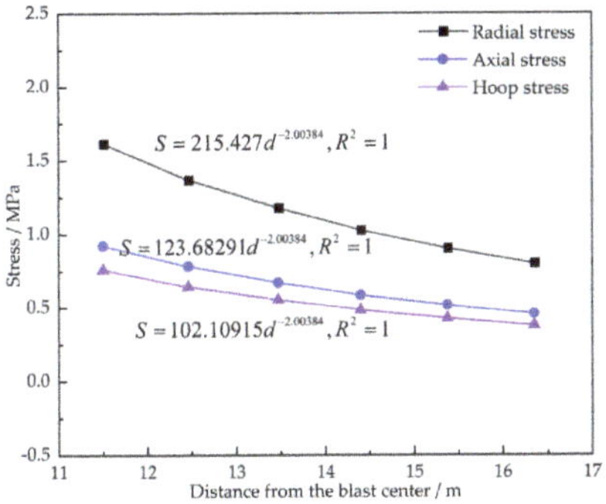

Figure 8. Stresses decrease as a power law of the distance to the cut hole center.

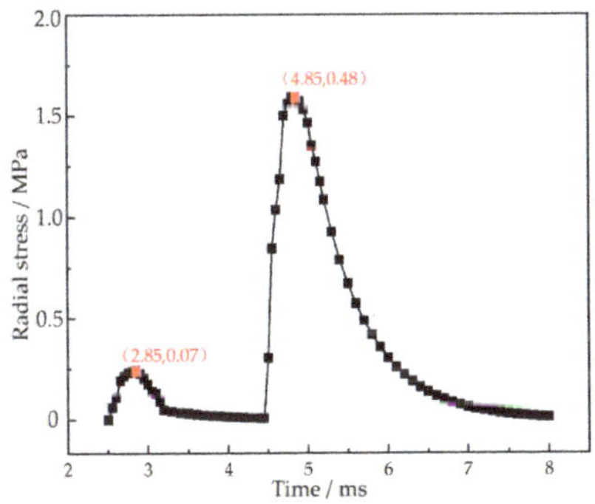

Figure 9. The radial stress–time history curve of point A.

To compare the stress results, parameters from the engineering case were applied to Equations (20)–(22) and (25) from this study—which account for the instantaneous axial pressure from short-cylinder charges and the rock mass damage factor D—and to Equations (23) and (24) from previous studies. The computed stress values are shown in Figures 10 and 11.

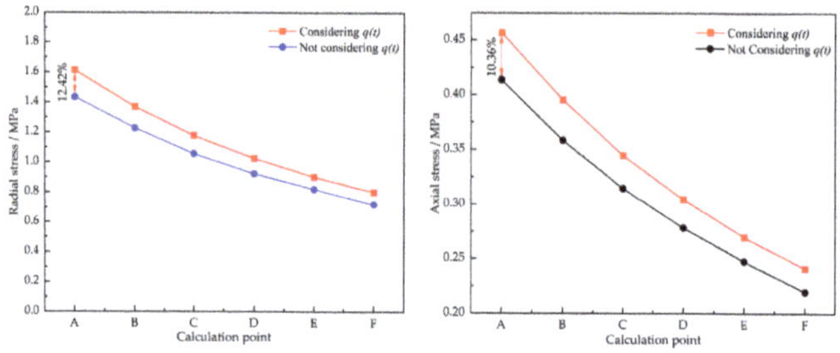

(**a**) Radial stress (**b**) Axial stress

Figure 10. Comparison of radial and axial stresses with and without considering relevant factors q(t) and D.

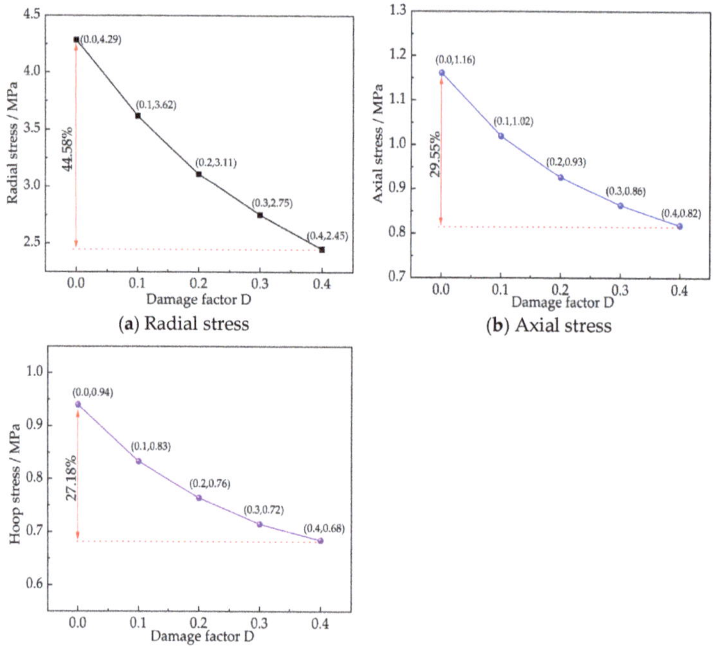

(**a**) Radial stress (**b**) Axial stress (**c**) Hoop stress

Figure 11. Variation patterns of triaxial stresses at calculation point A with different damage factors.

Comparative analysis of instantaneous detonation axial pressure ($q(t)$): As shown in Figure 10, radial stress is higher when instantaneous detonation axial pressure is considered. The maximum difference reaches 12.42%, with an average difference of 11.25% across all calculation points. Similarly, Figure 10 shows that axial stress is also higher with this factor included, with a maximum difference of 10.36%, and an average difference of 9.68%.

Comparative analysis of rock mass damage (D): As illustrated in Figure 11, incorporating the rock mass damage factor D reduces all the stress components. Stress attenuation follows a nonlinear trend as D increases. At D = 0.4, radial stress decreases by 44.58%, axial stress by 29.55%, and hoop stress by 27.18%, compared to the undamaged case (D = 0).

Based on the results in Table 2 and the comparative analysis in Figures 10 and 11, the formulas derived in this study (Equations (20)–(22) and (25))—which account for instantaneous detonation axial pressure and rock mass damage—demonstrate greater accuracy and rationality than existing models.

4. Field Monitoring and Stress Analysis

4.1. Engineering Situations

In a twin-tunnel highway project in Fujian Province, the left and right tunnels were excavated simultaneously, with the left tunnel advancing more rapidly. The upper bench excavation area of the active tunnel is approximately 77 m^2. The blasting operations use digital electronic detonators, with 40 mm diameter boreholes and No. 2 rock emulsion explosive cartridges (32 mm diameter). The initiation sequence follows a ring-by-ring (row-by-row) pattern with a 50 ms inter-row delay, achieving a 3.0 m cyclic advance. Detailed blasting parameters are provided in Table 3, and the blasting network is shown in Figure 12.

Table 3. Blasting parameters of upper bench.

Borehole Type	Number of Boreholes	Charge per Hole (kg)	Total Charge (kg)	Delay Segment
Cut Holes	6	1.5	9.0	Segment 1 (0 ms)
	8	2.4	19.2	Segment 2 (50 ms)
Auxiliary Holes	11	1.8	19.8	Segment 3 (100 ms)
	13		23.4	Segment 4 (150 ms)
	16		28.8	Segment 5 (200 ms)
	24		43.2	Segment 6 (250 ms)
Contour Holes	18	1.5	27.0	Segment 7 (300 ms)
	18		27.0	Segment 8 (350 ms)
Floor Holes	16	1.8	28.8	Segment 9 (400 ms)
Total	130		226.2	

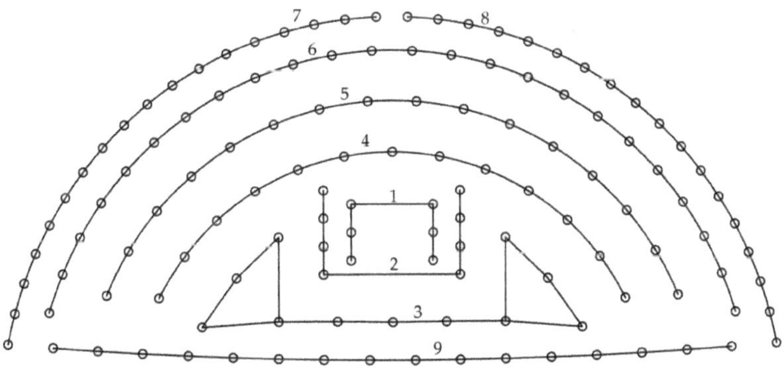

Figure 12. Blasting network diagram.

4.2. Monitoring Test Setup, Materials, and Methods

The purpose of dynamic strain monitoring is to measure stress in the surrounding rock of the adjacent tunnel during blasting, verify the accuracy of the previously derived analytical stress solutions, and analyze stress variation patterns induced by blasting. An ultra-dynamic strain measurement system (Figure 13) was used to monitor dynamic strain in the tunnel surrounding rock. The testing procedure, shown in Figure 14, operates as follows: voltage signals were collected via embedded sensors, converted into strain values using Equation (26), and dynamic stress peaks are then calculated based on Hooke's law (Equation (27)) [40].

$$\varepsilon = 4 \times U/K \times U_0 \times n \times G \tag{26}$$

$$\sigma = E\varepsilon \quad (27)$$

where U is the peak output voltage (V), K is the sensitivity coefficient of the strain gauge, U_0 is the bridge voltage of the strain amplifier (V), n is the number of effective bridge arms, G is the gain of the strain amplifier, and E is the elastic modulus of the rock mass (GPa). The dynamic strain test uses a half-bridge circuit, with the instrument parameters listed in Table 4.

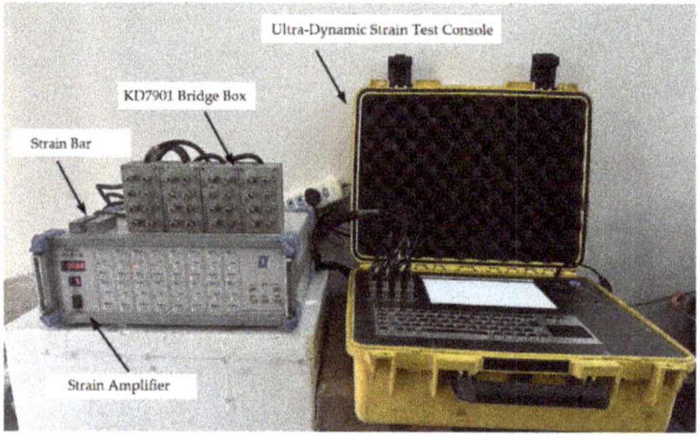

Figure 13. Ultra-dynamic strain measurement system.

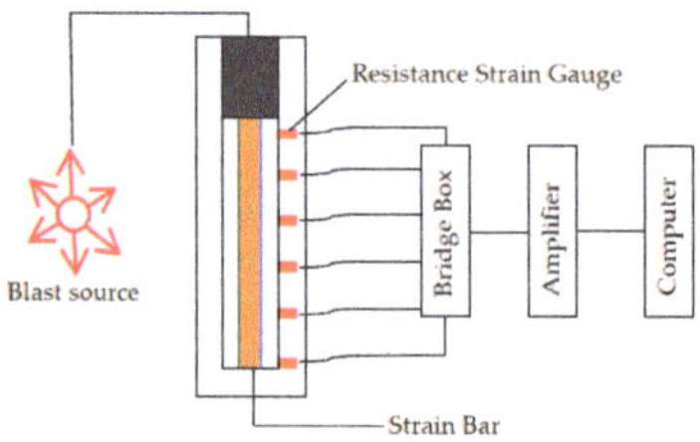

Figure 14. Flowchart of the testing system.

Table 4. Key parameters of the instrument.

K	U_0/V	n	G
2.0	4.0	1	100

The operating principle of the triggered voltage signal recording system is illustrated in Figure 15. When blast loads act on the strain bar, it undergoes minor deformation, causing the bonded strain gauges to change resistance. The resistance change produces a microvolt-level output via a Wheatstone bridge. The signal is then amplified by an ultra-dynamic strain amplifier, triggering a pulse that converts the microscopic deformation into a millivolt-level voltage signal recorded by the system.

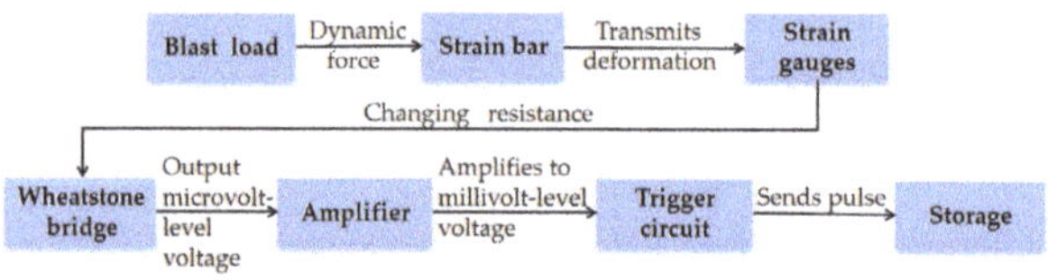

Figure 15. Schematic diagram of the voltage signal recording mechanism.

The measurement point layout is shown in Figure 16. Horizontal boreholes are drilled into the sidewall of the adjacent existing tunnel, with five strain measurement points spaced at 1 m intervals (Point 1 is 1 m from the sidewall). Prefabricated strain bars, matched to the physical and mechanical properties of the tunnel surrounding rock, are used. Triaxial resistance strain gauges are mounted on the bars to measure strain via electrical resistance (Figure 17). The instrumented bars are inserted into the boreholes and fixed with grout. A schematic diagram of the strain bar inserted into the borehole is shown in Figure 18, where the specifications of the PVC pipe in the diagram are consistent with those of the on-site drilling. To determine the rock mass damage factor D at the same locations, a YL-LCT single-transmitter dual-receiver sonic probe (Figure 19) is used to measure P-wave velocities. Sonic testing boreholes are drilled 0.5 m from the strain measurement holes (Figure 20), ensuring spatial alignment for accurate comparative analysis.

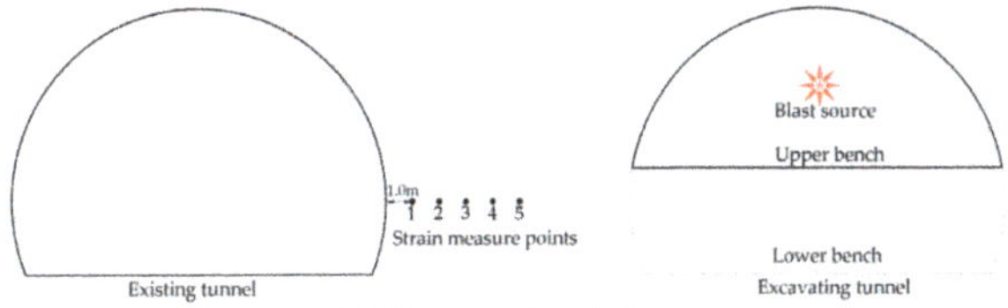

(**a**) Cross-sectional view.

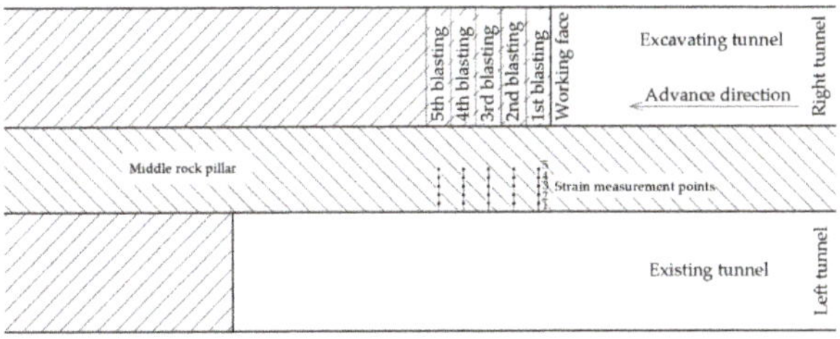

(**b**) Plan view.

Figure 16. Layout of monitoring points.

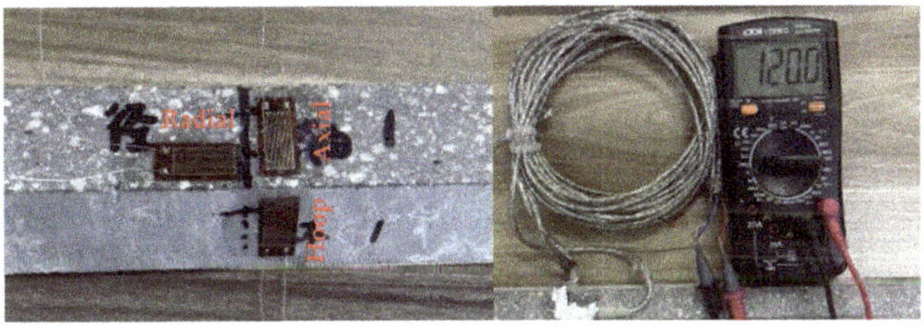

Figure 17. Strain bar, strain gauge and resistance testing.

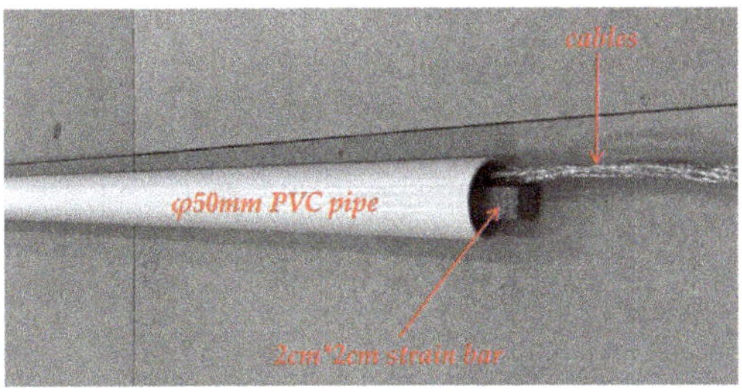

Figure 18. Schematic diagram of strain bar installation into borehole.

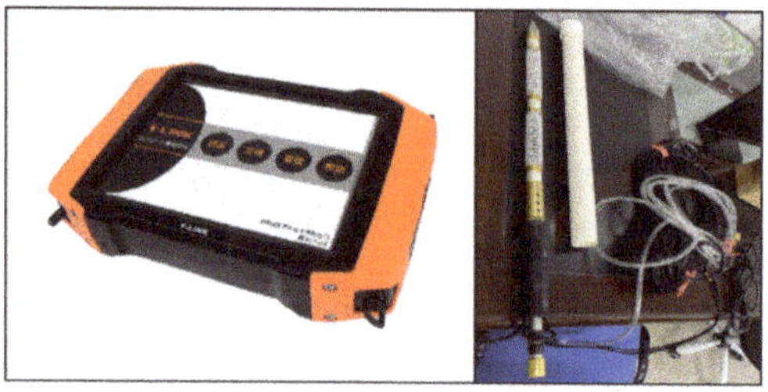

Figure 19. YL-LCT single-transmitter dual-receiver sonic probe.

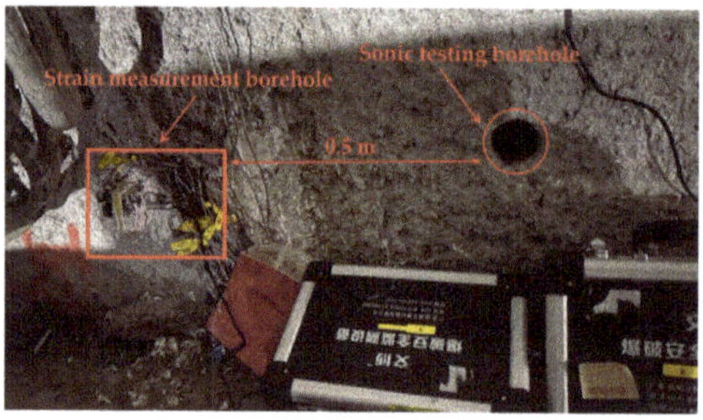

Figure 20. Layout of wave velocity testing boreholes.

During on-site dynamic strain testing, intense blasting vibrations caused collisions between data cables and surrounding structures, while mechanical equipment and welding operations introduced significant signal interference. To ensure measurement accuracy, the following measures were implemented before testing: data cables at borehole exits were wrapped with foam sleeves, and the remaining segments were enclosed in stainless steel junction boxes to prevent impact damage. Testing instruments were grounded using steel rebars to eliminate electromagnetic interference. All mechanical equipment was relocated, and welding operations were suspended during testing. Instrument connections were established and activated only after confirming the absence of interference sources. Sonic testing was completed prior to blasting. The ultra-dynamic strain amplifier was activated just before the blast and deactivated afterward. The field-testing configuration is shown in Figure 21. Each blasting round required approximately 15 min for complete monitoring, and the procedure was repeated over five consecutive rounds.

(**a**) Strain monitoring.

(**b**) Wave velocity monitoring.

Figure 21. On-site Monitoring Layout.

4.3. Monitoring Results

The wave velocity test results of the surrounding rock prior to the first blasting cycle are shown in Figure 22. The extracted results of measuring points 1–5 are shown in Table 5. Substituting the data into Equation (7), the calculated damage factors for monitoring Points 1 to 5 are 0.20, 0.18, 0.14, 0.10, and 0.09, respectively. Taking the radial strain test at monitoring Point 1 during the first blasting cycle as an example, the measured voltage signal is shown in Figure 23. The stress peak occurs at approximately 55 ms, closely aligning with the 50 ms delay associated with the highest charge, a single time-delay interval of the cut holes, confirming the reliability of the dynamic strain measurements. The wave velocities and dynamic strain data from the remaining four points were substituted into Equations (7), (26) and (27) to calculate the corresponding damage factors and dynamic stresses. The results are presented in Table 6, where "—" indicates that dynamic strain data were not obtained at certain points due field complexity and signal interference. The uncertainty in the peak stress and peak voltage is ±0.02.

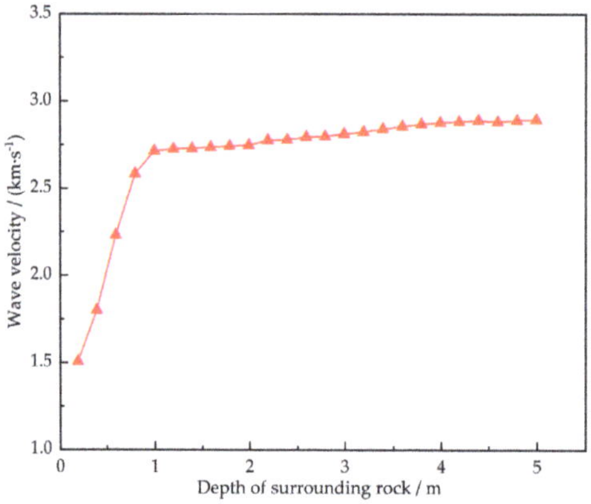

Figure 22. Rock wave velocity test results (first blasting).

Table 5. Sonic wave test results for measuring points 1–5.

	Point 1	Point 2	Point 3	Point 4	Point 5
hole depth (m)	1.0	2.0	3.0	4.0	5.0
original rock wave velocity (km/s)	3.036	3.036	3.036	3.036	3.036
wave velocity before the first blasting (km/s)	2.716	2.750	2.816	2.881	2.897
D	0.20	0.18	0.14	0.10	0.09

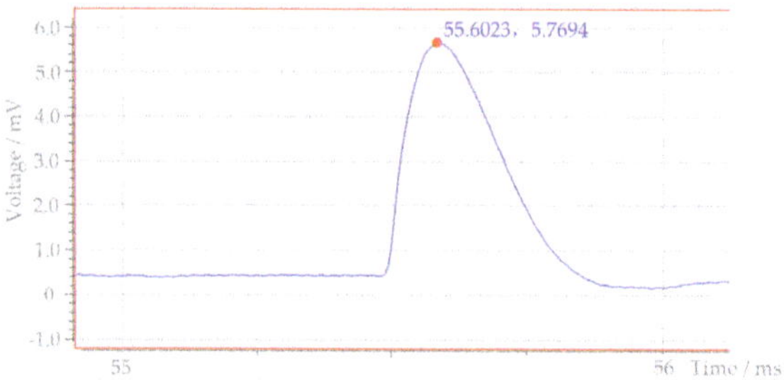

Figure 23. Measured voltage signal (radial direction at monitoring point 1 during the first blasting).

Table 6. Stress measurement results.

Blasting Times	Damage Factor D	Measure Point	Peak Voltage (mV)			Peak Stress (MPa)		
			Radial	Axial	Hoop	Radial	Axial	Hoop
1	0.20	1	5.77	3.19	2.72	0.89	0.49	0.42
	0.18	2	6.27	3.57	2.94	0.96	0.55	0.46
	0.14	3	6.76	—	3.20	1.04	—	0.50
	0.10	4	7.17	4.20	3.41	1.10	0.65	0.52
	0.09	5	—	4.79	3.89	—	0.74	0.60
2	0.22	1	5.70	3.14	2.70	0.88	0.48	0.42
	0.20	2	6.19	3.55	—	0.95	0.54	—
	0.15	3	6.73	3.76	3.18	1.03	0.58	0.49
	0.10	4	6.99	4.10	3.43	1.07	0.63	0.51
	0.10	5	7.97	4.69	3.78	1.22	0.72	0.59
3	0.23	1	5.57	3.02	2.70	0.87	0.48	0.41
	0.21	2	6.08	3.49	—	0.93	0.54	—
	0.16	3	6.68	3.74	3.17	1.03	0.58	0.49
	0.14	4	—	4.04	3.32	—	0.62	0.50
	0.10	5	7.83	4.61	3.71	1.20	0.71	0.57
4	0.23	1	5.73	3.14	2.70	0.87	0.48	0.41
	0.21	2	5.85	3.34	2.81	0.90	0.51	0.43
	0.17	3	6.53	3.65	3.09	1.00	0.56	0.47
	0.15	4	6.66	3.90	3.16	1.02	0.60	0.49
	0.12	5	7.56	4.45	3.60	1.16	0.68	0.55
5	0.25	1	5.68	3.12	2.63	0.85	0.46	0.40
	0.23	2	5.85	—	2.78	0.90	—	0.43
	0.15	3	6.62	3.69	3.12	1.02	0.57	0.48
	0.12	4	6.89	4.03	3.25	1.06	0.62	0.50
	0.10	5	7.81	4.58	—	1.20	0.70	—

As shown in Table 6, at each monitoring point, the stresses hierarchy is as follows: radial stress > axial stress > hoop stress. Additionally, all three stresses decrease with the components of increasing rock mass damage and with greater distance from the blast source.

4.4. Stress Propagation Law and Comparative Analysis

Based on the data in Table 6, the average damage factors and stress values at monitoring points 1–5 over five blasting cycles are summarized Table 7. To further examine the stress propagation behavior in the surrounding rock of the adjacent tunnel with respect to distance from the blast center, Figure 24 was generated using the data in Table 7. As shown in Figure 24, radial, axial, and hoop stresses all decrease with increasing distance, with radial stress showing the fastest attenuation rate, followed by axial and hoop stresses.

Table 7. The average value of measured damage factor and stress.

Measure Point	Damage Factor D	Average Stress (MPa)			Corresponding Theoretical Calculation Points	Theoretical Stress (MPa)		
		Radial	Axial	Hoop		Radial	Axial	Hoop
1	0.23	0.88	0.48	0.41	F	0.80	0.46	0.38
2	0.21	0.93	0.54	0.44	E	0.90	0.52	0.43
3	0.15	1.03	0.58	0.49	D	1.02	0.59	0.49
4	0.12	1.06	0.62	0.50	C	1.18	0.68	0.56
5	0.10	1.20	0.71	0.57	B	1.37	0.78	0.65

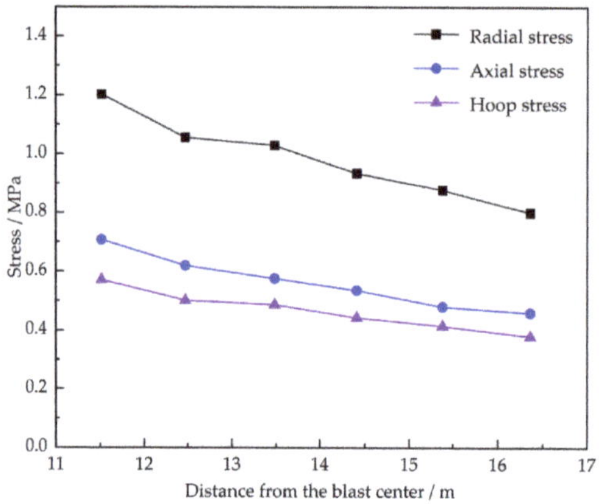

Figure 24. Propagation law of measured stress with the distance from the blast center.

The mean relative error (MRE) between measured and theoretical stress values is defined as follows:

$$\mathrm{MRE}_{(i)} = \left| \frac{\overline{\sigma}^{(i)}_{\text{measured}} - \sigma^{(i)}_{\text{theoretical}}}{\sigma^{(i)}_{\text{theoretical}}} \right| \times 100\% \tag{28}$$

$$\overline{\sigma}^{(i)}_{\text{measured}} = \frac{1}{n}\sum_{j=1}^{n} \sigma_{ij} \tag{29}$$

where $\overline{\sigma}^{(i)}_{\text{measured}}$ is the mean measured stress at the i-th monitoring point, $\sigma^{(i)}_{\text{theoretical}}$ is the theoretical stress at the i-th monitoring point, n is the number of blast events, and σ_{ij} is the measured stress at the i-th monitoring point during the j-th blast.

And the average error (AE) of all measuring points is defined as follows:

$$\mathrm{AE} = \frac{1}{k}\sum_{i=1}^{k} \mathrm{MRE}_{(i)} \tag{30}$$

where k is the number of measuring points.

A comparison between measured and theoretical stress values is presented in Figure 25. The results show that the measured stresses are slightly lower than theoretical predictions, with average errors of 7.02% for radial stress, 7.56% for axial stress, and 7.05% for hoop stress.

Analysis of discrepancies: According to Reference [41], borehole stemming can improve explosive energy utilization by approximately 7.5%. In this test, however, the cut holes were unstemmed, leading to energy loss. Additionally, signal attenuation occurred due to excessive cable lengths at the signal receivers. According to reference [40], the voltage drop (ΔU) along the cable is calculated as follows [42]:

$$\Delta U = \Delta U_0\% \times I \times L \tag{31}$$

where $\Delta U_0\%$ is the percentage voltage drop per kilometer of cable, I is the load current (A), and L is the cable length (km). Based on data from reference [42], $\Delta U_0\%$ was taken as 0.766%. With a measured load current of I = 0.52 A and cable length L = 0.1 km, the voltage drop calculated using Equation (31) is 0.3983 mV. This voltage drop reduced the peak voltage during measurements, resulting in values lower than theoretical predictions and contributing to the previously noted error, further lowering the reduced measured stress values. These factors collectively explain the observed discrepancies.

Despite these minor deviations, the measured and theoretical values align well, with errors within acceptable limits (≤8%), confirming the reliability of the analytical stress field solutions developed in this study. In tunnel engineering, the MRE between theoretical and measured stress values typically remains below 10%, a range acceptable for meeting project reliability requirements [43,44].

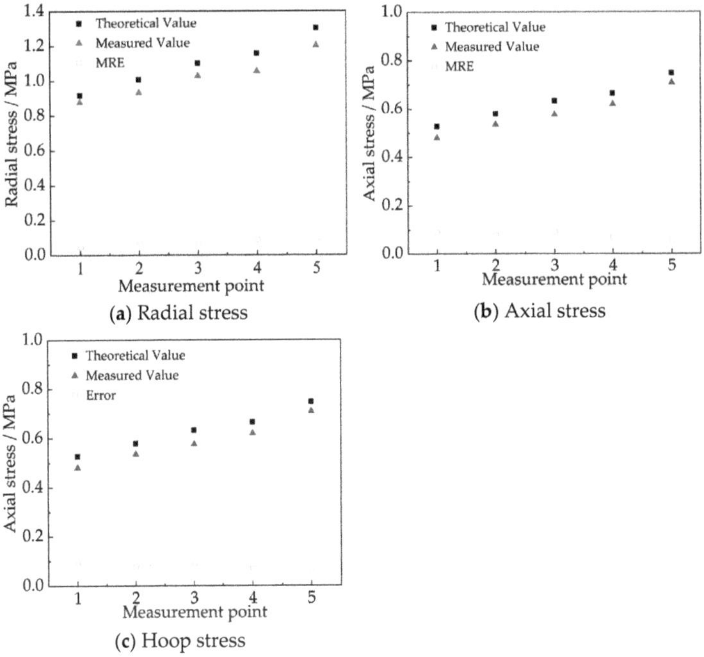

Figure 25. Comparison between measured and theoretical values of stress.

5. Conclusions

Based on Heelan's short cylindrical cavity solution, this study derives an analytical model for dynamic stresses in the surrounding rock of adjacent tunnels by incorporating the axial pressure $q(t)$ generated by instantaneous detonation and the rock mass damage factor D. A case study was conducted to investigate stress magnitudes and propagation behavior, with validation through field measurements. The theoretical stress values derived from the model show consistent trends and minor numerical deviations when compared to measured data.

(1) Stress magnitude and propagation laws: Both theoretical and measured results exhibit consistent triaxial stress trends in the following order: radial stress > axial stress > hoop stress. All three stresses decrease with increasing distance and damage factor, showing an inverse relationship with distance. Radial stress decays at the highest rates, followed by axial and hoop stresses. Stress also decays exponentially over time, with the peak occurring after the arrival of the transverse S-wave.

(2) Advantages of proposed stress model: Incorporating instantaneous detonation pressure significantly improves accuracy compared to conventional models that neglect this factor, reducing average errors by 11.25% in radial stress and 9.68% in axial stress. Including the damage factor D reveals nonlinear stress attenuation with increasing damage. At D = 0.4, compared to the intact rock condition (D = 0), stress reductions

reach 44.58% (radial), 29.55% (axial), and 27.18% (hoop), confirming the importance of accounting for detonation pressure and rock mass damage in stress prediction.
(3) Validation through field measurements: Measured stresses show small deviations from theoretical values, with average errors of 7.02% (radial), 7.56% (axial), and 7.05% (hoop). This close agreement validates the reliability of the proposed analytical solution.

Note: The primary limitations of the stress calculation model in this study are its initial assumption of heterogeneous, anisotropic rock masses as homogeneous media, with stress computed through the introduction of a damage factor. This elasticity-based model simplifies reality and does not fully capture the complex behavior of anisotropic rock. In practice, stress wave propagation in such media is nonlinear, and influenced by transmission, refraction, and reflection at rock mass discontinuities, as well as strain-softening effects in tunnel rock [45]. During multi-borehole stress superposition, the rock between the cut holes and the central pillar is treated as a continuous medium. However, auxiliary (relief) and perimeter pilot holes cause additional stress attenuation due to wave scattering at these cavities. To address these limitations, future research will focus on the following: (1) Stress wave propagation in anisotropic and fractured rock masses using a true triaxial Hopkinson bar system to simulate in situ stress conditions, and ultrasonic transmission techniques to monitor P-wave velocity evolution under varying stress paths [46]. (2) Discrete element modeling (DEM) to simulate contact mechanics across rock discontinuities. (3) Numerical simulation tools to study stress wave behavior in rock masses containing pre-existing cavities. These approaches aim to improve theoretical accuracy and better reflect real-world conditions.

Author Contributions: Methodology, T.L., X.L. and X.H.; resources, Y.H. and Y.X.; data curation, Y.W., J.Z., Y.H. and Y.X.; writing—original draft, Y.W. and J.Z.; writing—review and editing, T.L., X.L. and X.H.; funding acquisition, T.L. All authors have read and agreed to the published version of the manuscript.

Funding: This work is supported by the Xiaoming Lou: Natural Science Foundation of China (No. 52109124).

Institutional Review Board Statement: Not applicable.

Informed Consent Statement: Not applicable.

Data Availability Statement: The original contributions presented in this study are included in the article. Further inquiries can be directed to the corresponding author.

Conflicts of Interest: Authors Yelong Xie and Yan Hu were employed by the company Zhongchang Nuclear Engineering (Fujian) Construction and Development Co., Ltd. The remaining authors declare that the research was conducted in the absence of any commercial or financial relationships that could be construed as a potential conflict of interest.

References

1. Liu, F.-X. Technological advancements and prospects of drill and blast tunnel construction equipment. *Mod. Tunn. Technol.* **2024**, *61*, 190–202.
2. Shi, J.-J.; Xu, W.-X.; Zhang, H.; Ma, X.-Y.; An, H.-M. Dynamic response and failure mechanism of deep-buried tunnel with small net distance under blasting load. *Buildings* **2023**, *13*, 711. [CrossRef]
3. You, S.; Yang, R.-S.; Xiao, C.-L.; Ding, C.-X.; Li, C.-X.; Yang, Z.; Li, J. Study on the superposition effect of stress waves and crack propagation law between blastholes at different angles. *Opt. Lasers Eng.* **2024**, *178*, 108193. [CrossRef]
4. Ding, X.-H.; Yang, Y.-Q.; Zhou, W.; An, W.; Li, J.-Y.; Ebelia, M. The law of blast stress wave propagation and fracture development in soft and hard composite rock. *Sci. Rep.* **2022**, *12*, 17120. [CrossRef]
5. Luo, H.-H.; Li, X.-L.; Wang, J.-G.; Li, G.-T.; Li, Z.-L. Study on peak overpressure prediction of underground blasting shock waves on straight-through roadways. *J. Chin. Saf. Sci.* **2019**, *29*, 57–62.

6. Starfield, A.M.; Pugliese, J.M. Compression waves generated in rock by cylindrical explosive charges: A comparison between a computer model and field measurements. *J. Rock. Mech. Min. Sci.* **1968**, *1*, 65–77. [CrossRef]
7. Liu, J.-C.; Gao, W.-X.; Zhang, S.-H.; Hu, Y. Study on initial stress and damage law of hole wall in water interval blasting. *J. Arab. Geosci.* **2022**, *15*, 96. [CrossRef]
8. Neiman, I.-B. Modeling the explosion of a system of borehole charges in a scarp. *Sov. Min.* **1986**, *22*, 108–113. [CrossRef]
9. Heelan, P.-A. Radiation from a cylindrical source of finite length. *Geophysics* **1953**, *18*, 685–696. [CrossRef]
10. Yue, G.-W.; Li, M.; Wang, L.; Liang, W.-M. Optimal layout of blasting holes in structural anisotropic coal seam. *PLoS ONE* **2019**, *14*, e0218105. [CrossRef]
11. Henrych, J. The dynamics of explosion and its use. *J. Appl. Mech.* **1980**, *47*, 218. [CrossRef]
12. Zhang, M.-W.; Shimada, H.; Sasaoka, T.; Matsui, K.; Dou, L.-M. Evolution and effect of the stress concentration and rock failure in the deep multi-seam coal mining. *Environ. Earth Sci.* **2014**, *72*, 629–643. [CrossRef]
13. Ma, J.-J. Theoretical Calculation and Verification of the Range of Damage Caused by Columnar Charge Blasting. Master's Thesis, Kunming University of Science and Technology, Kunming, China, 2021.
14. Lei, T.; Kang, P.-L.; Ye, H.-W.; Li, L.; Wang, Q.-Z. Study on the direction effect of stress wave superposition and fracturedistribution in rock mass during cylindrical charge blasting. *J. Rock. Mech. Eng.* **2024**, *43*, 399–411.
15. Favreau, R.-F. Generation of strain waves in rock by an explosion in a spherical cavity. *J. Geophys. Geophys. Res.* **1969**, *74*, 4267–4280. [CrossRef]
16. Graff, K.-F. *Wave Motion in Elastic Solid*; The Ohio State University Press: Columbus, OH, USA, 1991.
17. Achenbach, J.-D. Wave propagation in elastic solids. *J. Appl. Mech.* **1980**, *41*, 544. [CrossRef]
18. Blair, D. Seismic radiation from an explosive column. *Geophysics* **2010**, *75*, 55–65. [CrossRef]
19. Gao, Q.-D.; Jin, J.; Wang, Y.-Q.; Lu, W.-B.; Leng, Z.-D.; Chen, M. Acting law of in-hole initiation position on distribution of blast vibration field. *Explos. Shock.* **2021**, *41*, 138–152.
20. Yang, R.-S.; Guo, Y.; Li, Q.; Xu, P.; Chen, C. Evolution law on explosive stress and strain field of column charges at middle detonation position. *J. Chin. Coal. Soc.* **2019**, *44*, 3423–3431.
21. Ding, C.-X.; Yang, R.-S.; Feng, C. Stress wave superposition effect and crack initiation mechanism between two adjacent boreholes. *J. Rock. Mech. Min. Sci.* **2021**, *138*, 104622. [CrossRef]
22. Qiu, J.; Li, X.; Hu, C.; Zhao, Y.; Li, D.; Liang, L. Physical Model Test on the Deformation Behavior of an Underground Tunnel Under Blasting Disturbance. *Rock. Mech. Rock. Eng.* **2020**, *54*, 91–108. [CrossRef]
23. Yang, L.-Y.; Ding, C.-X.; Yang, R.-S.; Yang, Q.-C. Experimental and Theoretical Analysis of Stress Superposition in Double-Hole Blasts. *J. Test. Eval.* **2020**, *48*, 201800932. [CrossRef]
24. Liu, G.-Y.; Liu, X.-M.; Chen, S.-H. Influence of delay time on superposition effect of surface vibration. *Eng. Blast.* **2022**, *28*, 63–70.
25. Zhang, Z.; Shi, X.; Qiu, X.; Ouyang, J. On-site tests investigating the effects of blasting vibration amplitude on the dynamic response of thin spray-on liner. *Tunn. Undergr. Space Technol.* **2024**, *145*, 105620. [CrossRef]
26. Luo, C.; Yang, X.-A.; Li, K.; Wang, B.; Xiao, C.-Y. Dynamic strain test and measurement of a tunnel in nearby blasting vibration. *Vib. Shock.* **2020**, *39*, 262–268.
27. Zhang, C.-H. Analysis of Velocity Maximum of the Linging of Existing Tunnel by the Blasting of the Tunnel Near-By. Master's Thesis, Lanzhou University, Lanzhou, China, 2009.
28. Liang, Q.-G.; Li, D.-W.; Zhu, Y.; Xie, F.-H. *Vibration Control Technology for Blasting Construction Near Tunnels*; Science Press: Beijing, China, 2015.
29. Liang, R.; Zhu, M.; Zhou, W.-H.; Huang, X.-B.; Du, K.-F.; Zhou, Y.-T. Rock-breaking Characteristics of stress wave superposition between spherical charges. *J. Yangtze. River. Sci. Res. Inst.* **2020**, *37*, 67–72.
30. Xiang, W.-F.; Shu, D.-Q.; Zhu, C.-B. Stress field analysis of strip charge blasting based on starfield superposition method. *Explos. Shock.* **2004**, *05*, 437–442.
31. Zhang, B. Damage Characteristics of Surrounding Rock for Large-section Tunnel Excavated by Bench Blasting. *Blasting* **2023**, *40*, 69–76.
32. Dai, J. *Dynamic Behaviors and Blasting Theory of Rock*; Beijing Metallurgical Industry Press: Beijing, China, 2013.
33. Lu, W.-B. Study on the Propagation and Effects of Stress Waves in Rock Blasting. Ph.D. Thesis, Wuhan University, Wuhan, China, 1994.
34. Onederra, A.-I.; Furtney, K.-J.; Sellers, E.; Iverson, S. Modelling blast induced damage from a fully coupled explosive charge. *J. Int. Rock. Mech. Min. Sci.* **2013**, *58*, 73–84. [CrossRef]
35. Song, W.-H.; Jiao, H.-C.; Wang, Y.-W. Crack closure effect during the impact coal seam with high-pressure air blasting and the influence of gas drainage efficiency. *Front. Earth Sci.* **2023**, *11*, 1131386. [CrossRef]
36. Costamagna, E.; Oggeri, C.; Segarra, P.; Castedo, R.; Navarro, J. Assessment of contour profile quality in D&B tunnelling. *Tunn. Undergr. Space Technol.* **2018**, *75*, 67–80. [CrossRef]
37. Yu, T.-R.; Vongpaisal, S. New blasting damage criteria for underground blasting. *CIM Bull.* **1996**, *89*, 139–145.

38. Cao, F.; Ling, T.-H.; Zhang, S. Safety threshold of blasting vibration velocity of highway tunnel considering influence of stress wave transmission. *Vib. Shock.* **2020**, *39*, 154–159.
39. Cao, M.-X.; Yan, S.-H.; Zheng, Y.-Q.; Shao, C.; Wang, N.-X.; Du, J.-X. Analysis of blasting vibration response of adjacent tunnel. *Eng. Blasting* **2025**, *31*, 130–138.
40. Lou, X.-M.; Lin, J.; Wang, X.-L.; Liu, H.-Y.; Wang, G. Characteristics of axial distribution of borehole wall pressure along theborehole of coupled charges with different borehole diameters under spherical detonation wave. *J. Harbin. Inst. Technol.* **2025**, *57*, 148–159.
41. Luo, Y.; Shen, Z.-W. Investigation on length of stemming material and its effect in hole-charged blasting. *Mech. Pract.* **2006**, *02*, 48–52.
42. Wang, G.-Z.; Pan, Q.-W. Voltage loss and voltage drop of low-voltage cables. *Wire. Cable.* **2014**, *1*, 17–19.
43. Wang, B.; He, C.; Wu, D.-X.; Geng, P. Inverse analysis of in-situ stress field of Cangling super-long highway tunnel. *Rock. Mech.* **2012**, *33*, 628–634.
44. Wang, T. Multi-Scale Analysis of In-Situ Stress Field of He-Jian No. 1 Tunnel Area of He-Jian Expressway in Yunnan Province. Master's Thesis, Chang'an University, Xi'an, China, 2022.
45. Cui, L.; Sheng, Q.; Zhang, J.; Dong, Y.-K.; Guo, Z.-S. Evaluation of input geological parameters and tunnel strain for strain-softening rock mass based on GSI. *Sci. Rep.* **2022**, *12*, 20575. [CrossRef]
46. Tian, X.-C.; Tao, T.-J.; Liu, X.; Jia, J.; Xie, C.-J.; Lou, Q.-X.; Chen, Q.-Z.; Zhao, Z.-H. Calculation of hole spacing and surrounding rock damage analysis under the action of in situ stress and joints. *Sci. Rep.* **2022**, *12*, 22331. [CrossRef]

applied sciences

MDPI

Article

Study on Optimization of Downward Mining Schemes of Sanshandao Gold Mine

Weijun Liu [1], Zhixiang Liu [1] and Zaiyong Li [2,*]

[1] School of Resources and Safety Engineering, Central South University, Changsha 410083, China; weijunliu@csu.edu.cn (W.L.); liulzx@csu.edu.cn (Z.L.)
[2] Guizhou Provincial Bureau of Geology and Mineral Resources Geology 113 Brigade, Liupanshui 553004, China
* Correspondence: zaiyongli2025@163.com

Abstract

To address the challenges associated with deep ground pressure control at the Sanshandao Gold Mine, a pre-controlled top-to-middle and deep-hole upper and lower-wall goaf subsequent filling mining method was proposed. Three distinct downward mining schemes were designed, the excavation procedure is systematically designed with 18 steps, and the temporal and spatial evolution characteristics of stress and displacement were analyzed using FLAC3D. The results revealed that stress concentration occurred during excavation steps 1–3. As excavation progressed to steps 4–9, the stress concentration area shifted primarily to the filling zones of partially excavated and filled sections. By steps 10–12, the stress concentration in these areas was alleviated. Upon completion of all excavation and filling steps, a small plastic zone was observed, accompanied by an alternating distribution of high and low stress within the backfill. Throughout the excavation process, vertical displacement ranged from 4.42 to 22.73 mm, while horizontal displacement ranged from 1.72 to 3.69 mm, indicating that vertical displacement had a more significant impact on stope stability than horizontal displacement. Furthermore, the fuzzy comprehensive evaluation method was applied to optimize the selection among the three schemes, with Scheme 2 identified as the optimal. Field industrial trials subsequently confirmed the technical rationality and practical applicability of Scheme 2 under actual mining conditions.

Keywords: mining schemes; numerical simulation; fuzzy comprehensive evaluation; optimization; field industrial trials

1. Introduction

As the first undersea bedrock mining metal mine in China, the Sanshandao Gold Mine features an outcrop on the seabed, with the ore body gradually extending landward as it descends [1–3]. With the ongoing deepening of the mine, the in situ stress has progressively increased, leading to more pronounced ground pressure phenomena. Stress gradient zones have formed near geological structures such as active faults and around the excavation stopes. These zones are susceptible to external dynamic disturbances, which may trigger geological disasters such as large-scale rockbursts and fault slips in the goaf [4–6]. Addressing the challenges of deep resource mining safety and ground pressure regulation at the Sanshandao Gold Mine, the implementation of a rational mining plan is crucial to ensure mining safety and effective ground pressure control. Consequently, selecting an appropriate deep mining scheme is of significant importance for both mine safety and efficient production [7,8].

Currently, numerous studies have been conducted by domestic and foreign experts on the optimization of different mining schemes, and good results have been achieved from the aspects of numerical simulation, theoretical analysis, and mathematical methods [9–13]. For example, Liu et al. [14] used the renormalization group model and numerical simulation method to analyze the retention scheme of a security pillar for seabed metal ore mining and concluded that the width of the pillar is 10 m and the spacing is 50 m. Tan et al. [15] used the DDA numerical simulation method to study the mining method of the ore body in the caving area in the southern part of Hainan Iron Mine and obtained a safe and reliable mining scheme. Wang et al. [16] used the numerical simulation method to establish three different mining schemes for the Panlong lead–zinc mine, analyzed the deformation of the strata of the three different schemes, and finally obtained the optimal scheme. Zhou et al. [17] investigated the optimal mining width of the slope of the Panzhihua #7 coal mine by the FLAC3D numerical simulation method and determined that the 6 m was the optimal mining width. Zhao et al. [18] used theoretical analysis and numerical simulation methods to analyze the effects of several downward cementation filling methods on the stability of the backfill roof and verified the rationality of the methods through experimental methods. Shi et al. [19] used numerical simulation methods to analyze three different types of support methods, including the anchor net cable support, anchor injection, and U-shaped shed support, and the results showed that the U-shaped shed support had the best effect. Liang et al. [20] used FCE and TODIM hybrid methods to optimize the optimal mining method for deep seabed gold deposits and proved the feasibility of the method through sensitivity analysis. Mijalkovski et al. [21] determined the optimal mining scheme by applying the FCE and TOPSIS methods to the mining method selection of an underground metal lead–zinc mine. Yavuz et al. [22] used the analytic hierarchy process (AHP) and Yager method to optimize the underground mining method for the Istanbul lignite mine and determined the optimal mining scheme. Bajićsanja et al. [23] used the FCE and AHP methods to optimize the mining method for the Borska Reka copper mine in Serbia to determine the optimal underground mining scheme. To solve the problem of phosphate rock mining in Dingxi mine, Hu et al. [24] used the numerical simulation method to determine the optimal mining method, applied the method to the actual field engineering, and achieved good results. Hou et al. [25] used Kunyang phosphate rock as the research object and used the entropy-weighted TOPSIS method to optimize the selection of the three mining methods for the deep phosphate ore body and determined that the open-pit mining scheme was the optimal method. Alpayi et al. [26] used the AHP and Yager methods to study the problem of underground mining method selection for different deposit shapes and ore bodies. The research methods utilized in this study primarily rely on numerical simulation and mathematical modeling. While these approaches have yielded certain insights, they are not without limitations. Although numerical simulation and mathematical methods are advantageous due to their simplicity, intuitive results, and computational efficiency, they are also prone to issues such as result variability and significant model inaccuracies. Moreover, given the inherent complexity of mining plan selection and the multitude of influencing factors, the preferred solutions identified in this study lack validation through on-site industrial trials, which are critical for assessing their reliability and practical applicability. Therefore, to optimize mining plans effectively, it is essential to integrate multiple methodologies, including numerical simulation, theoretical analysis, and mathematical modeling, complemented by field industrial trials to ensure the reliability and suitability of the proposed solution.

Therefore, to enhance the resource utilization rate of the deep ore body below −900 m in the Sanshandao Gold Mine, a pre-controlled mining method involving sub-mining and sub-filling of the upper and lower walls of deep holes in the top area is proposed based on the characteristics of deep ore body mining. Three distinct downward mining schemes were designed, and their spatial and temporal evolution characteristics of stress and displacement were analyzed using the FLAC3D method. This analysis revealed the displacement and stress evolution patterns of the three schemes, providing insights into the stress evolution laws and displacement characteristics associated with each mining approach. The fuzzy comprehensive evaluation method was then employed to optimize and identify the most suitable mining plan. Finally, the rationality of the optimal downward mining scheme was rigorously validated through field industrial trials.

2. Engineering Background

The Sanshandao Gold Mine is situated in Laizhou City, Shandong Province, and comprises three mining areas: Xinli, Xishan, and Xiling. The ore body is primarily deposited within the F1 fault zone of Sanshandao, with lithology dominated by pyrite sericite fractured rocks, pyrite sericite granitic fractured rocks, and sericite granites. The lithology of the deposit's roof and floor consists of sericite granite, sericite granitic fractured rock, and similar materials. In the alteration zone near the main section, local rocks are more fragmented, exhibiting stronger alteration, more developed fissures, increased core fracturing, and relatively lower solidity. Currently, the Xishan mine has advanced to the middle section at a depth of −960 m. During the excavation of mining roadways near the F1 fault, varying degrees of collapse have been observed. The high-temperature and high-humidity working environment significantly hampers labor efficiency, posing a pressing need for the mine to reform its existing mining methods. Enhancing the level of automation and intelligent production operations is crucial to reducing the safety risks associated with personnel operations. To address these challenges, a pre-controlled top-to-middle and deep-hole upper and lower-wall goaf subsequent filling mining method has been proposed for implementation in the Xishan ore mine.

The deep ore body in the Xishan mining area of the Sanshandao Gold Mine has an average dip angle of 44° and a horizontal thickness of 40 m. The F1 fault is exposed in the middle of the ore body, with its occurrence nearly identical to that of the ore body. The ore rock mass within the F1 fault is highly fragmented, making it a typical inclined thick and large broken ore body.

The ore-controlling structure in the mining area is the Sanshandao Fault, with the primary fault in the zone being F1. The overall strike of the F1 fault is 40°, trending southeast, with a dip angle ranging from 35° to 45°. Both the strike and dip exhibit gentle undulations, indicative of compressional shear characteristics. The F1 fault lies within the interior of the ore body, significantly compromising the integrity of the mining field and complicating the extraction of the ore body on the fault's upper plate, as illustrated in Figure 1. Figure 2 depicts the panel of the 945 m stope, while Figure 3 provides a schematic of the mining methods employed in the Xishan mining area.

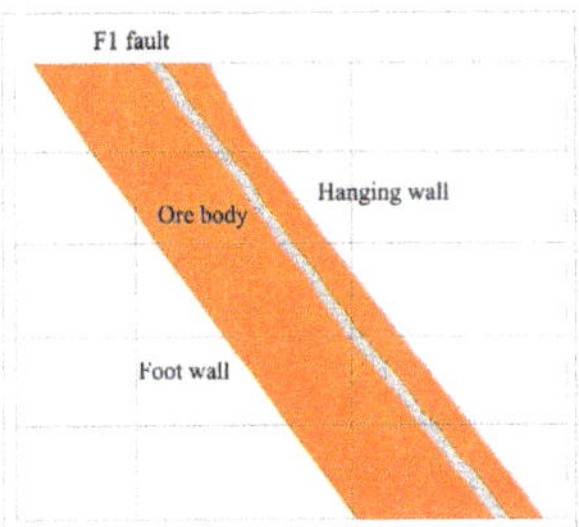

Figure 1. Schematic diagram of the F1 fault.

Figure 2. The panel of 945 m stope.

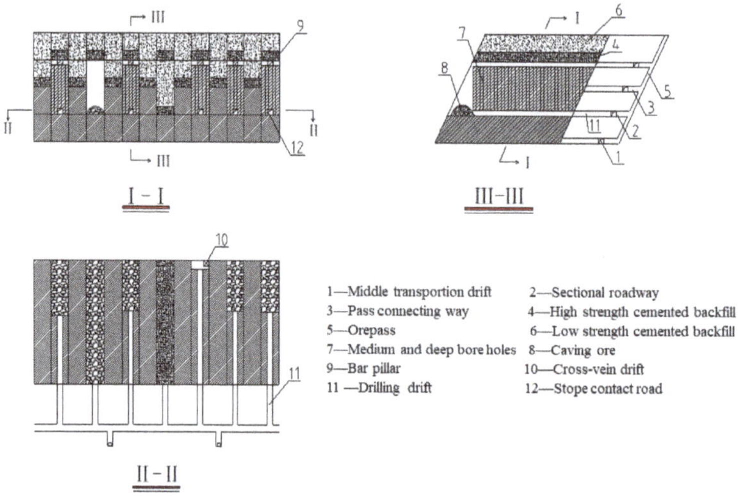

Figure 3. Diagram of the mining methods used in the Xishan mining area.

Through on-site sampling of deep rock specimens from the Sanshandao Gold Mine, a series of static load mechanical tests were performed on the rock samples. The basic physical and mechanical parameters of the rocks obtained are shown in Table 1.

Table 1. Rock mass mechanical parameters.

Rock Mass	Elastic Modulus (Pa)	Shear Modulus (Pa)	Density (g/cm^3)	Internal Friction Angle (°)	Cohesion (MPa)	Tensile Strength (MPa)
Wall rock	1.096×10^{10}	6.895×10^{9}	2.7	36.94	42.78	8.54
Ore body	32.18×10^{9}	2.024×10^{9}	2.7	32.60	21.45	4.91
1:4 backfill body	2.894×10^{7}	2.449×10^{7}	2.2	43.50	3.10	1.51
1:8 backfill body	1.234×10^{7}	9.71×10^{6}	2.1	38.70	1.71	0.42
Non-cemented backfills	6.32×10^{6}	3.68×10^{6}	1.6	32.00	0.17	0.03

3. Numerical Simulation of Downward Mining Schemes

3.1. Computational Model Establishment

The mine is mined by the pre-controlled top-to-middle and deep-hole upper and lower-wall goaf subsequent filling mining method, the ore body is divided into mine houses and pillars along the strike of the ore body and arranged at intervals, and the production is organized into 14 groups of ore house pillars. The height of the mining stage is 75 m, the mining section is set up in the mining stage, the mining section is 15 m or 30 m high, the segmented height used in the separated mine houses is staggered, and the trackless mining method outside the lower-wall vein is used to mine from top to bottom. For the 15 m high mine house, the cement–sand ratio is 1:4, the backfill is 5 m, the 1:4 backfill forms a preliminary condensation, and the 1:8 backfill is 10 m. For the 30 m high mine house, the cement–sand ratio is 1:4, and the backfill is 5 m; then, after the 1:4 backfill forms preliminary condensation, the 1:8 backfill is filled 12 m; then, after the 1:8 backfill is preliminarily coagulated, the non-cemented backfills are filled for 13 m.

The calculation model focuses on the deep downward mining area at an elevation of −900 to −1000 m, with a plane range spanning from lines 1700 to 1720. The model encompasses 14 stopes arranged along the strike direction of the ore body, each with a width of 8 m. The ore body is set to a thickness of 55 m and a height of 87 m, with top and bottom pillars of 6 m each. The overall model extends vertically along the strike of the ore body, centered on the ore body, with a total vertical extension of 174 m. The coordinate system is defined as follows: Z along the strike of the ore body, X perpendicular to the strike, and Y vertically upward. The model comprises six components: the upper bedrock layer, lower bedrock layer, upper-wall rock mass, lower-wall rock mass, ore body, and an artificial false roof. The mechanical parameters of the rock mass are provided in Table 1.

Additionally, the ore body model is divided into 130 groups according to the area of the mine house and the area with different filling ratios, including one group of top pillar and one group of bottom pillar, and a total of 128 groups of mine area are divided. The ore body model is meshed, and 2 m is divided into a grid. The three-dimensional numerical model is illustrated in Figure 4.

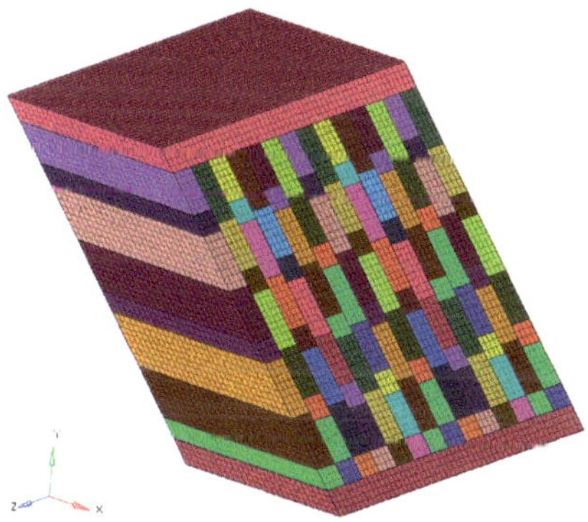

Figure 4. The constructed three-dimensional numerical model of the stope.

3.2. Determination of Rock Mass In Situ Stress

The determination of in situ stress in rock mass is a critical aspect of numerical simulation. Based on the test results of deep in situ stress in the Sanshandao Gold Mine, as well as past engineering experience and theoretical analysis, the vertical principal stress exhibits a linear increase with depth. The regression equations for the maximum horizontal principal stress, minimum horizontal principal stress, and vertical principal stress values as functions of depth are as follows:

$$\sigma_{\max} = 0.11 + 0.0539H \text{ (MPa)}$$

$$\sigma_{\min} = 0.13 + 0.0181H \text{ (MPa)}$$

$$\sigma_{z} = 0.08 + 0.0315H \text{ (MPa)}$$

where σ_{max}, σ_{min}, and σ_z are the maximum horizontal principal stress, the minimum horizontal principal stress, and the vertical stress, respectively. H is the buried depth of the measurement point, and the unit is m.

Therefore, the model employs the Mohr–Coulomb criterion, with constraints applied to all four sides and the bottom of the model. Uniform loads are applied to the top of the orebody, and gravity loads are applied to the entire model. The model depth is set at 960 m, and the results of the in situ stress tests are converted into stresses in the X and Y directions, which are then input into the model for calculation.

3.3. Design of Downward Mining Schemes

The scheme design is primarily based on the criterion of the optimal mining method. Considering that the downward mining scheme must minimize the influence of in situ stress, the ore body is divided into stopes and pillars along the strike direction using the pre-controlled top-to-middle and deep-hole upper and lower-wall goaf subsequent filling mining method at intervals. Production is organized in units of 14 groups of stopes and pillars, with the ore body divided into three sections from top to bottom. To ensure the continuity of the ore body during the mining process, mining is conducted in a pattern of five stopes followed by one pillar. To select the best downward mining scheme, three mining schemes were established based on different mining sequences. Among them, Scheme 1 is a staggered downward mining mode from one horizontal wing to the other wing, with six vertical and two oblique intervals, and the mining sequence is shown in Figure 5. Scheme 2 is a staggered downward mining mode from the horizontal center to the two wings, with five mining and one vertical and oblique separation, and the mining sequence is shown in Figure 6. Scheme 3 is a staggered downward mining mode from the horizontal two wings to the center with five mining one, vertical interval six, and oblique separation two, and the mining sequence is shown in Figure 7.

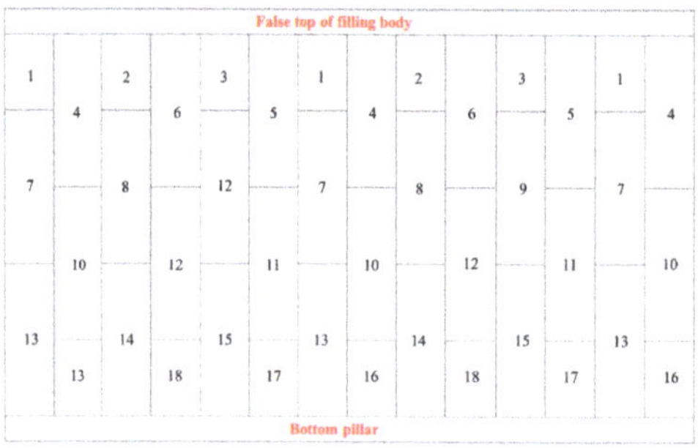

Figure 5. Scheme 1. Note: Steps 1–18 represent the excavation steps.

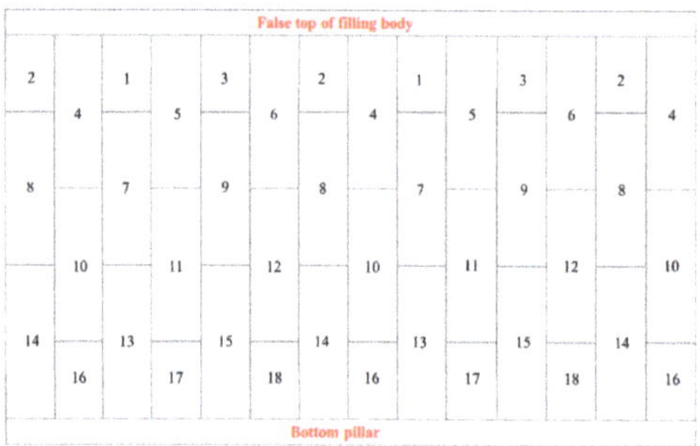

Figure 6. Scheme 2. Note: Steps 1–18 represent the excavation steps.

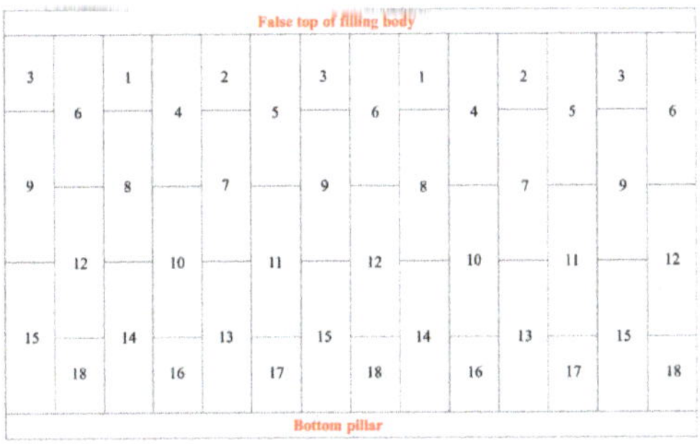

Figure 7. Scheme 3. Note: Steps 1–18 represent the excavation steps.

4. Analysis of Numerical Simulation Results

4.1. Comparative Analysis of Vertical Stress Changes of Different Mining Schemes

In the stress analysis, the maximum stress concentration depends on the strength parameters. During mining steps 1–3, the stress concentration in the top and bottom pillars is not pronounced initially, but the stress concentration on both sides of the stope progressively increases. In mining steps 4–6, the stress concentration area in Scheme 2 is smaller compared to Schemes 1 and 3, although the maximum vertical stress value in Scheme 2 is significantly higher than that in Scheme 3. In mining steps 7–9, the stress concentration phenomenon in the surrounding rock on both sides of the stope becomes more pronounced, and the stress concentration area continues to expand. The stress concentration area in Scheme 2 remains smaller than in Schemes 1 and 3, but the maximum vertical stress value in Scheme 2 is slightly higher than in Scheme 1. Specifically, the maximum vertical stress in the surrounding rock of Scheme 2 is 47.0 MPa, while the maximum stress concentration values in Schemes 1 and 3 are 46.1 MPa and 48.1 MPa, respectively.

In mining steps 10–12, the stress concentration value in the surrounding rock of Scheme 2 is relatively small. The maximum stress value in Scheme 2 remains relatively low, while in mining steps 13–15, the maximum stress value exceeds 48 MPa, with the stress concentration area in Scheme 2 being smaller than in Schemes 1 and 3. In mining steps 16–18, the maximum stress concentration value in the surrounding rock on both sides of the ore body in Scheme 3 is significantly higher than in Schemes 1 and 2, reaching 55.8 MPa. The vertical stress changes for each scheme during the mining steps are illustrated in Figure 8, while the vertical stress changes for mining steps 7–9 in the three schemes are shown in Figure 9.

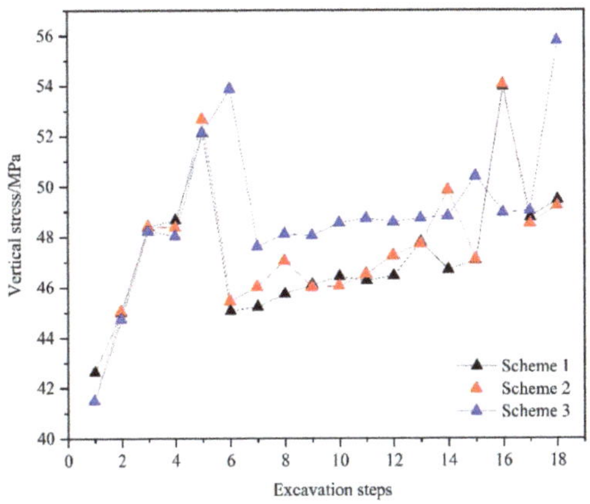

Figure 8. Variation of vertical stress in the mining steps of schemes.

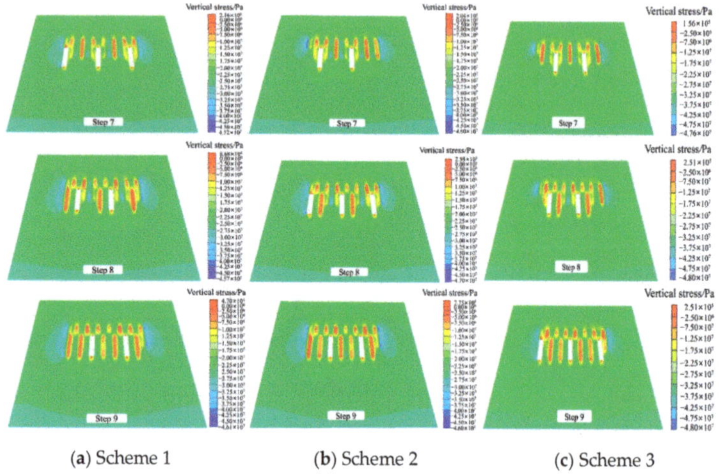

Figure 9. Vertical stress in excavation steps 7–9.

4.2. Comparative Analysis of Horizontal Stress Changes of Different Mining Schemes

During the initial mining steps 1–3, no significant stress concentration is observed at the top and bottom of the columns. However, the stress concentration on both sides of the stope progressively increases. In mining steps 4–6, the stress concentration area in Scheme 2 is smaller compared to Schemes 1 and 3, although the maximum vertical stress value in Scheme 2 is significantly higher than in Scheme 3. In mining steps 7–9, the stress concentration phenomenon in the surrounding rock on both sides of the stope becomes more pronounced, and the stress concentration area continues to expand. While the stress concentration area in Scheme 2 is less extensive than in Schemes 1 and 3, the maximum vertical stress value in Scheme 2 is slightly higher than in Scheme 1. Specifically, the maximum vertical stress in the surrounding rock of Scheme 2 is 47.0 MPa, compared to 46.1 MPa in Scheme 1 and 48.0 MPa in Scheme 3.

Similarly, the maximum stress value in Scheme 2 during mining steps 13–15 is also relatively low. However, in mining steps 16–18, the stress concentration area in Scheme 3 is significantly larger than in Schemes 1 and 2. The vertical stress changes for each scheme during the mining steps are illustrated in Figure 10, while the vertical stress changes for mining steps 7–9 in the three schemes are shown in Figure 11.

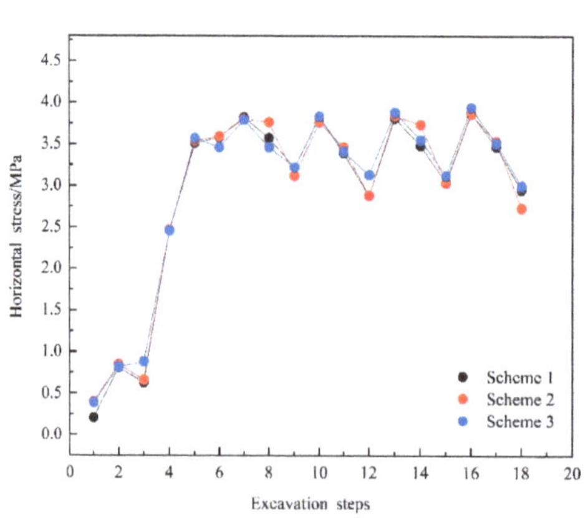

Figure 10. Horizontal stress variation of the mining steps of schemes.

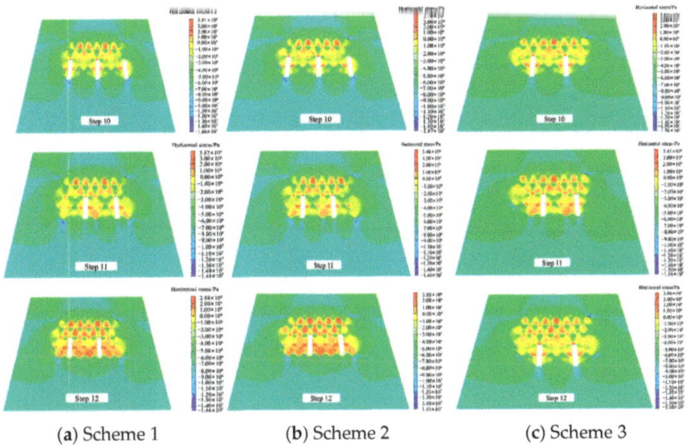

Figure 11. Horizontal stress in excavation steps 10–12.

4.3. Analysis and Comparison of Vertical Displacement Changes of Different Mining Schemes

As excavation progresses from steps 1–3, the vertical displacement of the monitoring point increases in alignment with the calculated number of steps. This growth rate is initially rapid but subsequently slows. During excavation steps 4–6, the area exhibiting significant vertical displacement expands, and vertical displacement becomes more pronounced on both sides of the lower portion of the stope. In excavation steps 7–9, the vertical displacement in the upper half of the stope's sides is minimal, while the displacement in the lower half is more substantial. During excavation steps 10–12, a notable increase in vertical displacement is observed, with the largest displacements concentrated in the roof, the area to be excavated, and the bottom plate.

In Scheme 1, the vertical displacement of the roof increases from 17.17 cm to 18.58 cm, and the vertical displacement of the bottom plate increases from 11.15 cm to 11.80 cm. In Scheme 2, the vertical displacement of the roof increases from 15.54 cm to 16.21 cm, while in Scheme 3, it increases from 18.96 cm to 20.67 cm. Scheme 2 demonstrates relatively effective control over vertical displacement. In subsequent mining steps 13–15, as the excavation range expands, the vertical displacement in all three schemes increases significantly. During mining steps 16–18, the vertical displacement contours of the three schemes show discernible but gradual changes. Notably, the vertical displacement of the roof in Scheme 2 remains largely unchanged. Figure 12 illustrates the vertical displacement changes associated with the mining steps for each scheme, while Figure 13 shows the vertical displacement changes during excavation steps 10–12 for the three schemes.

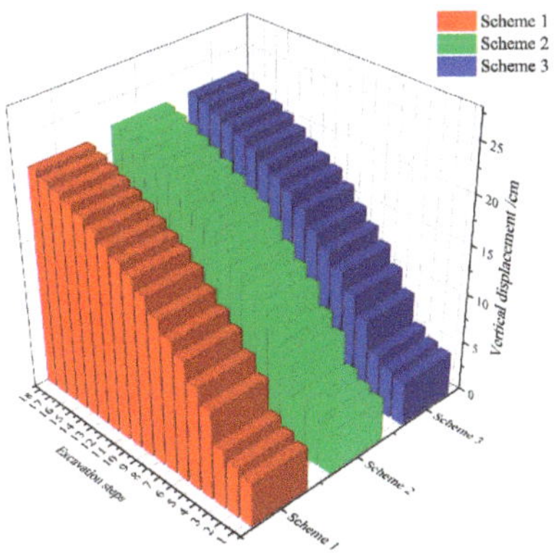

Figure 12. Vertical displacement changes of mining steps in schemes.

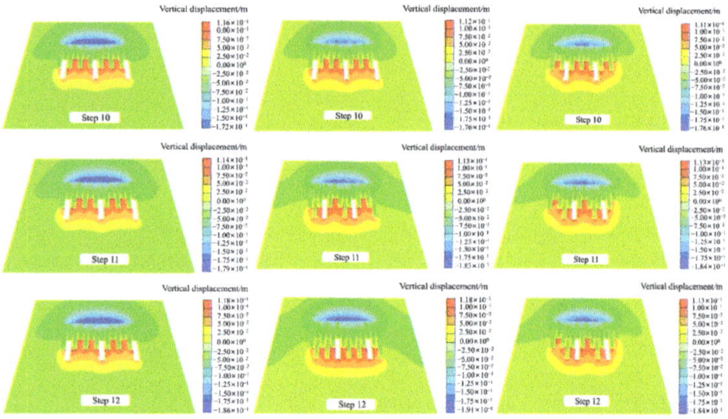

(**a**) Scheme 1 (**b**) Scheme 2 (**c**) Scheme 3

Figure 13. Vertical displacement changes in the excavation steps 10–12 of schemes.

4.4. Analysis and Comparison of Horizontal Displacement Changes of Different Mining Schemes

During excavation steps 1–3, as the mining range expands, the horizontal stress within the model gradually increases, and horizontal deformation develops from both sides of the stope toward the excavation area. In excavation steps 4–6, the maximum horizontal displacement increases to over 2.5 cm, indicating that rock deformation becomes more pronounced with the expansion of the mining range, and energy release further intensifies. During excavation steps 7–9, the horizontal displacement of the surrounding rock exceeds 3 cm, and minor displacement also occurs in the backfill of the areas where filling has been completed. By excavation steps 10–12, the maximum horizontal displacement reaches approximately 3 cm, but the rate of increase slows, suggesting that the backfill begins to effectively inhibit the deformation of the surrounding rock. In mining steps 13–15, the horizontal displacement of the stopes on both sides increases to some extent across all three schemes, but the rate of growth slows significantly. The mechanical effect of the backfill in restraining the deformation of the surrounding rock becomes more pronounced. The maximum horizontal displacement of the surrounding rock on the left side increases from 3.42 cm, 3.36 cm, and 3.64 cm in step 13 to 3.72 cm, 3.95 cm, and 3.93 cm in step 15 for Schemes 1, 2, and 3, respectively. In mining steps 16–18, the maximum horizontal displacement of the left surrounding rock is 3.95 cm, 3.41 cm, and 4.38 cm for Schemes 1, 2, and 3, respectively. Figure 14 illustrates the horizontal displacement changes at the mining steps for each scheme, while Figure 15 shows the horizontal displacement changes during excavation steps 13–15 for the three schemes.

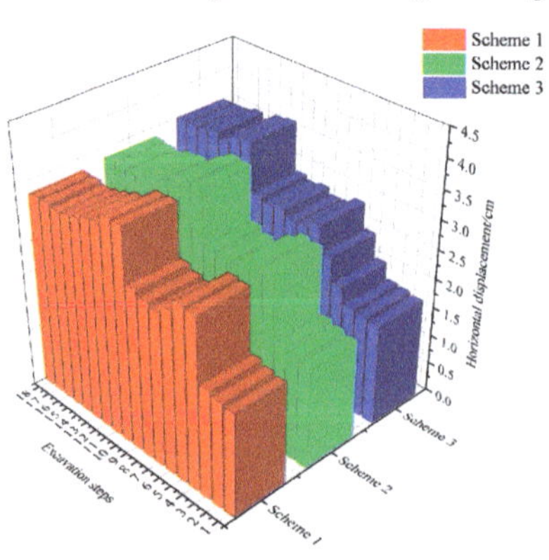

Figure 14. Horizontal displacement of the mining steps for schemes.

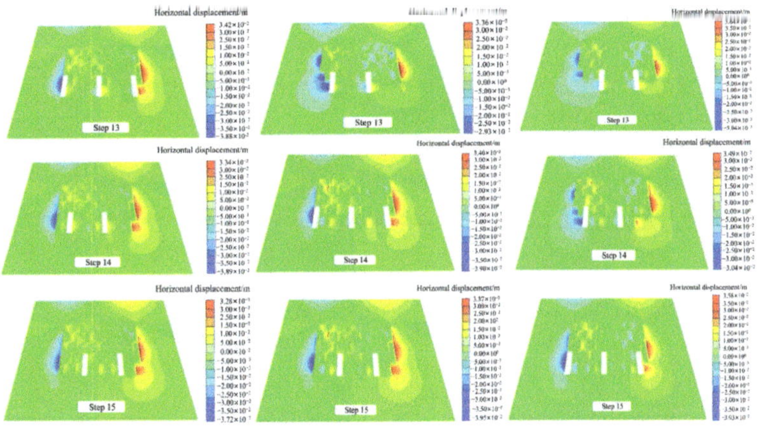

(**a**) Scheme 1 (**b**) Scheme 2 (**c**) Scheme 3

Figure 15. Horizontal displacement changes at the excavation steps 13–15 of schemes.

4.5. Evolution Analysis of Maximum and Minimum Principal Stress in Stope

During the initial excavation steps 1–3, the maximum principal stress concentration around the mining stope is not significant, with magnitudes considerably smaller than those observed in the unmined ore body and surrounding rock areas. In excavation steps 4–6, significant tensile stress concentration areas emerge on both sides of the mining stope. As excavation progresses to steps 7–12, the area of tensile stress concentration between mining chambers expands. During excavation steps 13–15, a larger area of tensile stress concentration develops in part of the backfill within the backfilling area, where tensile and compressive stresses are distributed in an alternating pattern. By excavation steps 16–18, the tensile stress concentration area reaches its maximum extent, with interconnected tensile stress zones and separated compressive stress zones in the middle. The maximum principal stresses for Schemes 1 to 3, observed during excavation steps 1 to 18, range from 44.52 to 56.77 MPa, 43.29 to 56.87 MPa, and 43.29 to 58.8 MPa, respectively. During excavation steps 1–3, compressive stress concentration becomes evident in specific filling areas, with noticeable variations in spacing between regions of low and high compressive stress. In excavation steps 4–6, certain areas of the backfill experience increased compressive stress, leading to a separation between regions of higher and lower compressive stress.

As excavation progresses to steps 7–12, the compressive stress concentration area becomes primarily localized in small sections of the surrounding rock on both sides. During excavation steps 13–15, high and low compressive stress zones alternate within the filling area, and a significant compressive stress concentration area develops in the surrounding rock on both sides of the ore body. By excavation steps 16–18, the compressive stress concentration area remains concentrated, maintaining its localized distribution. Figure 16 illustrates the relationship between the maximum principal stress and the excavation step, while Figures 17–19 depict the changes in maximum and minimum principal stresses during excavation steps 16–18 for Schemes 1, 2, and 3, respectively. The numerical simulation results reveal that, following the final three stages of excavation, the stress concentration area is distributed across the stope roof and also appears within the backfill body. This is evident from the maximum principal stress contour map, which indicates that the mechanical role of the backfill body is further enhanced, playing a significant part in regulating in situ stress and ground pressure within the mining area. From the perspective of compressive stress distribution, the maximum compressive stresses for Schemes 1, 2, and 3 are 50.06 MPa, 42.56 MPa, and 58.64 MPa, respectively. A comparison of tensile stress values shows that the maximum tensile stresses for Schemes 1, 2, and 3 are 3.71 MPa, 3.43 MPa, and 4.17 MPa, respectively. The research and comparative analysis demonstrate that Scheme 2 has the most favorable effect on ground pressure regulation among the three schemes.

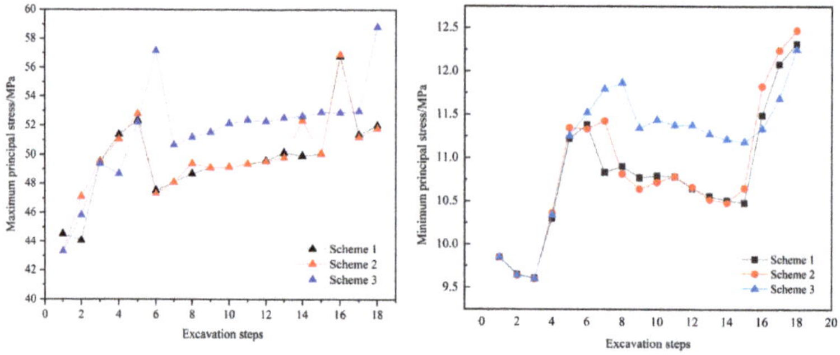

(**a**) Maximum principal stress (**b**) Minimum principal stress

Figure 16. Stress variation of the mining steps of schemes.

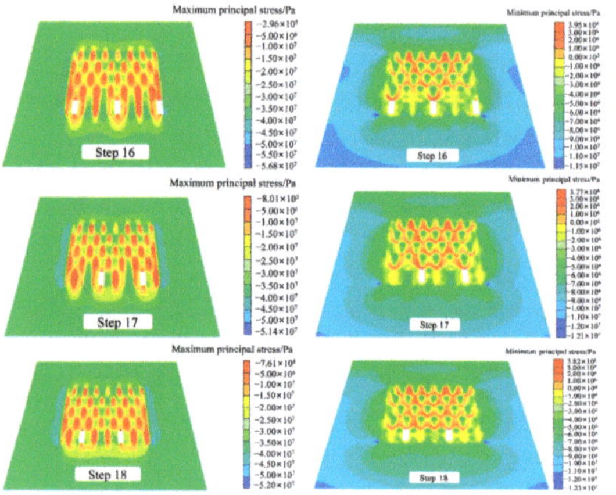

(**a**) Maximum principal stress (**b**) Minimum principal stress

Figure 17. The principal stress in excavation steps 16–18 of Scheme 1.

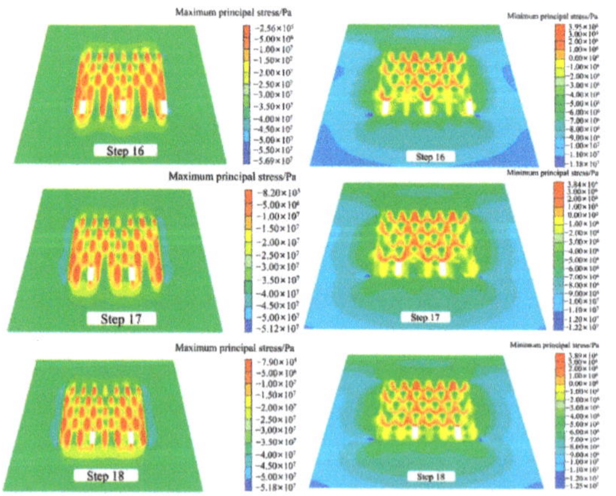

(**a**) Maximum principal stress (**b**) Minimum principal stress

Figure 18. The principal stress in excavation steps 16–18 of Scheme 2.

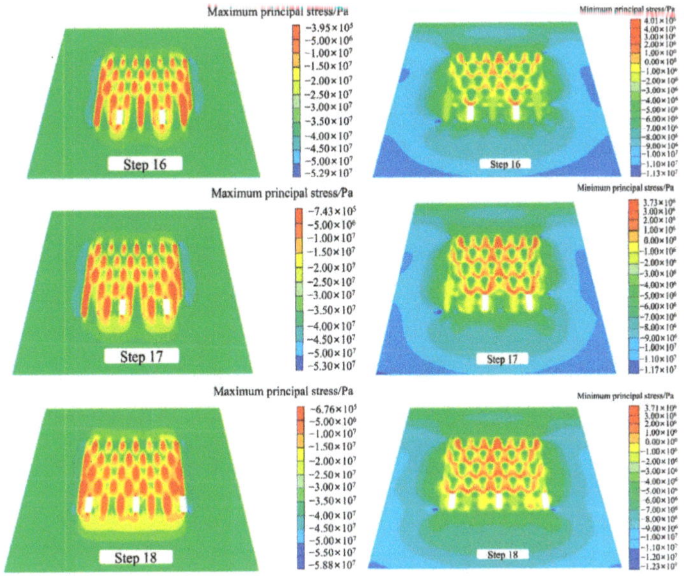

(**a**) Maximum principal stress (**b**) Minimum principal stress

Figure 19. The principal stress in excavation steps 16–18 of Scheme 3.

4.6. Stress Analysis After All Stopes Are Mined and Filled

The vertical stress distribution contour map of the stope after complete mining and filling is shown in Figure 20a. Figure 20a demonstrates that the surrounding rock in the edge area of the model is predominantly subjected to compressive stress. In contrast, within the ore body model area, the presence of the mined and filled ore body leads to the generation of both compressive and tensile stresses at different locations within the backfill. Overall, the backfill body experiences a combination of compressive and tensile stresses. The tensile stress value in the backfill of Scheme 2 is the lowest, and the maximum principal stress value in the excavation area and its vicinity is less than that in the surrounding rock at the model's edge. After the stope is fully mined and filled, the minimum principal stress distribution map, as illustrated in Figure 20b, reveals that the surrounding rock on both sides of the filling area is subjected to significant compressive stress. Additionally, the interval distribution of high and low stress areas within the filling area exhibits distinctive characteristics. The vertical stress distribution contour map of the stope after the completion of mining and filling operations is shown in Figure 21a. Figure 21a indicates that the vertical stress in Schemes 1 and 3 is predominantly compressive, while Scheme 2 exhibits a tensile stress area, albeit with a relatively low tensile stress value of 0.027 MPa. The distribution of high and low stress intervals is evident in the filling areas of all three schemes. It is observed that regions subjected to greater compressive stress correspond to low stress areas. Additionally, a significant concentration of vertical stress is evident in the surrounding rock on both sides of the filling area, with local stress in some mining areas exceeding 50 MPa. Figure 21b presents the horizontal stress distribution contour map following the full excavation and backfilling of the stope. Analysis of the Figure 21b also reveals that the stress orientation in the majority of the backfill regions aligns with the Z-axis, with stress magnitudes ranging approximately from 2.5 to 2.6 MPa.

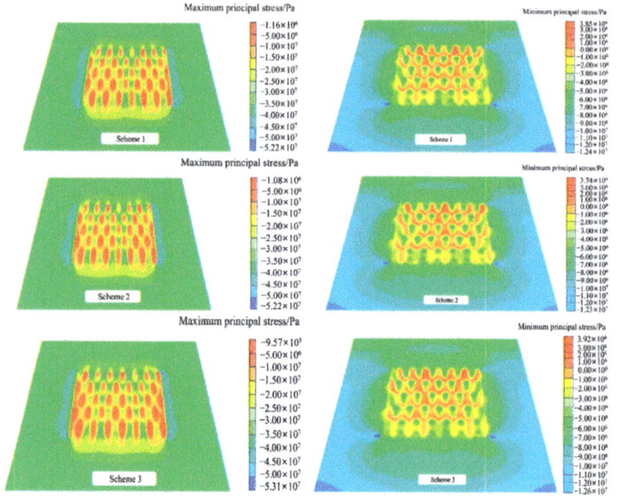

(**a**) Maximum principal stress (**b**) Minimum principal stress

Figure 20. The principal stress after all full excavation and filling.

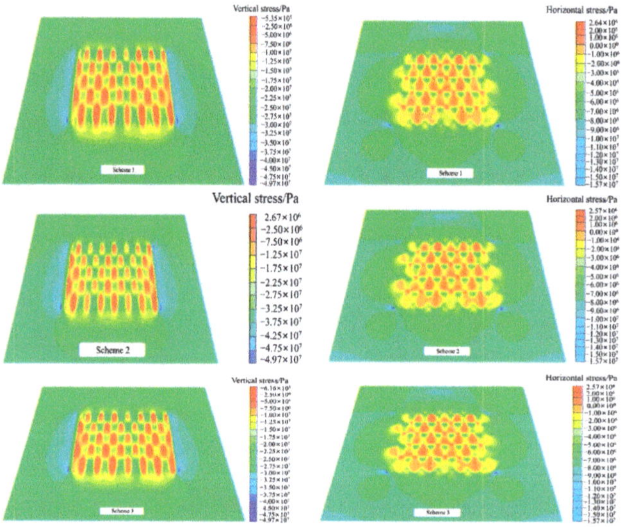

(**a**) Vertical stress (**b**) Horizontal stress

Figure 21. The stress after all full excavation and filling.

4.7. The Plastic Zone Analysis After All Stopes Are Mined and Filled

The plastic zone of Scheme 1, Scheme 2, and Scheme 3 after all stopes are mined and filled are shown in Figure 22. As can be seen from the Figure 22, the plastic zone area after the mining of the three schemes is small, independent of each other and not connected. It is mainly distributed in the top and bottom pillars and the edges of the backfill in each mine.

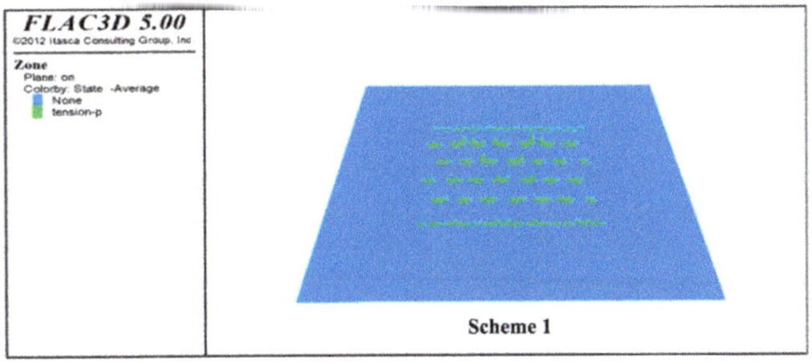

(a) Scheme 1

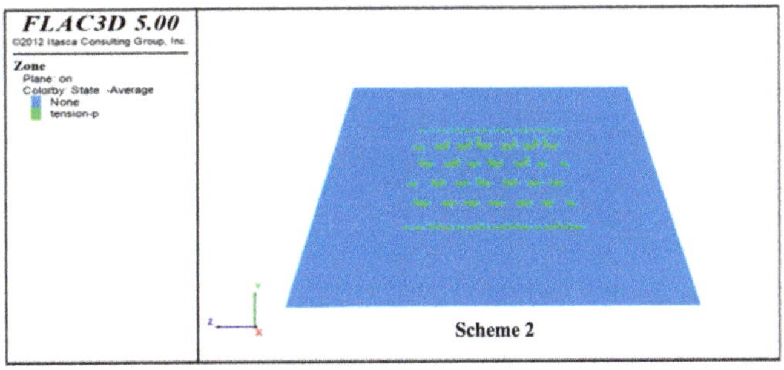

(b) Scheme 2

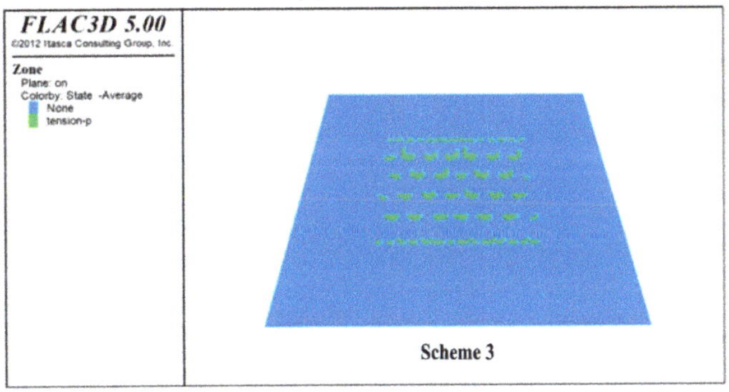

(c) Scheme 3

Figure 22. The plastic zone after full excavation and filling.

5. Optimization of Schemes for Deep Mining of Sanshandao Gold Mine

According to the analysis of the numerical simulation results of the previous three downward mining schemes, the three mining schemes have their advantages and disadvantages, and the index data are of different sizes, which is difficult to judge subjectively. Therefore, it is necessary to quantitatively analyze the evaluation indicators of each scheme by combining the AHP and FCE and optimize the three schemes, respectively [27–33]. In conclusion, Scheme 1, Scheme 2, and Scheme 3 are now preferred. Firstly, a comprehensive evaluation index system is established for the three mining schemes. The AHP then is used to objectively determine the weight of each factor, after which a fuzzy comprehensive evaluation can be established through FCE theory. Finally, the best downward mining scheme can be determined.

5.1. Comprehensive Evaluation Indicators

Based on the actual situation of the ore body after mining and filling, six indexes were selected from the simulation calculation results of the three mining schemes. These included the maximum compressive stress of the surrounding rock, the average vertical displacement of the top column monitoring point, the average vertical stress of the top column monitoring point, the maximum compressive stress of the backfill body, the maximum tensile stress of the backfill body, and the plastic failure degree. The comprehensive evaluation index factors for mining schemes are presented in Table 2.

Table 2. Comprehensive evaluation index factors.

Index Factor	Scheme 1	Scheme 2	Scheme 3
Maximum compressive stress of surrounding rock/MPa	52.16	52.21	53.06
Average vertical displacement of top column monitoring point/cm	19.10	19.10	19.07
Average vertical stress at top column monitoring point/MPa	11.32	11.27	11.12
Maximum compressive stress of backfill/MPa	39.73	39.78	41.03
Maximum tensile stress of backfill/MPa	3.85	3.74	3.92
Plastic failure degree/%	12.35	12.40	11.83

5.2. Membership Degree of Quantitative Indicators

The selected quantitative indexes are as follows: the maximum compressive stress of the surrounding rock; the average vertical displacement of the top column monitoring point; the average vertical stress of the top column monitoring point; the maximum compressive stress of the backfill body; the maximum tensile stress of the backfill body; and the index of the degree of plastic failure of each scheme. The degree of membership of the quantitative indicators is determined by the membership function method, while that of the non-quantitative indicators is determined by the relative binary comparison method [34–36].

The target eigenvalue matrix composed of m indexes of n schemes is presented below:

$$Y = \begin{bmatrix} y_{11} & y_{12} & \cdots & y_{1n} \\ y_{21} & y_{22} & \cdots & y_{2n} \\ \cdots & \cdots & \cdots & \cdots \\ y_{m1} & y_{m2} & \cdots & y_{mn} \end{bmatrix} = (y_{ij}) \tag{1}$$

where $i = 1, 2, \cdots, m; j = 1, 2, \cdots, n$.

The following steps are to be followed in order to calculate target relative dominance [37,38]:

For larger and better indicators, the following equation is used for specification [39]:

$$r_{ij} = \frac{y_{ij}}{\max y_{ij}} \tag{2}$$

For smaller and better indicators, the following equation is used for normalization:

$$r_{ij} = \frac{\min y_{ij}}{y_{ij}} \tag{3}$$

The relative superiority matrix of the target is obtained:

$$R = \begin{bmatrix} r_{11} & r_{12} & \cdots & r_{1n} \\ r_{21} & r_{22} & \cdots & r_{2n} \\ \cdots & \cdots & \cdots & \cdots \\ r_{m1} & r_{m2} & \cdots & r_{mn} \end{bmatrix} = (r_{ij}) \tag{4}$$

where $i = 1, 2, \cdots, m; j = 1, 2, \cdots, n$.

According to Table 2 of the comprehensive evaluation index factors, the eigenvector matrix can be obtained as follows:

$$R_{1\text{-}6} = \begin{bmatrix} 52.16 & 52.21 & 53.06 \\ 19.10 & 19.10 & 19.07 \\ 11.32 & 11.27 & 11.12 \\ 39.73 & 39.78 & 41.03 \\ 3.85 & 3.74 & 3.92 \\ 12.35 & 12.40 & 11.83 \end{bmatrix} \tag{5}$$

Therefore, by using Formulas (2) and (3) to normalize the above matrix, the comprehensive membership matrix is finally obtained:

$$R_{1\text{-}6} = \begin{bmatrix} 1 & 0.999 & 0.983 \\ 1 & 1 & 0.998 \\ 0.982 & 0.987 & 1 \\ 1 & 0.999 & 0.968 \\ 0.971 & 1 & 0.954 \\ 0.996 & 1 & 0.954 \end{bmatrix} \tag{6}$$

5.3. *Analytic Hierarchy Process*

The AHP is used to determine the weighting of indicators at each level by using a multi-level analytical evaluation model [40,41].

The weight matrix of six indicators can be obtained by using AHP:

$$\begin{bmatrix} 1 & 1/3 & 1 & 1 & 1/9 & 1/7 \\ 3 & 1 & 3 & 3 & 1/3 & 1/2 \\ 1 & 1/3 & 1 & 1 & 1/9 & 1/7 \\ 1 & 1/3 & 1 & 1 & 1/9 & 1/7 \\ 9 & 3 & 9 & 9 & 1 & 3 \\ 7 & 2 & 7 & 7 & 1/3 & 1 \end{bmatrix} \tag{7}$$

This is a selection step based on the analytic hierarchy process as follows:

Firstly, the judgment matrix is constructed by performing pairwise comparisons among the six criteria with respect to the target layer. This matrix forms the basis for determining the relative weight of each criterion layer.

Subsequently, for each row within the judgment matrix, the geometric mean is computed. Specifically, the product of the row elements is calculated and then raised to the power of 1/n. The m-dimensional vector is obtained:

$$\varpi_i = \sqrt[m]{\prod_{j=1}^{m} a_{ij}} \tag{8}$$

where $i = 1, 2, \cdots, m; j = 1, 2, \cdots, n$.

The vector is then normalized to the weight vector, and the weight vector is obtained:

$$\omega_i = \frac{\varpi_i}{\sum_{j=1}^{m} \varpi_j} \tag{9}$$

where $i = 1, 2, \cdots, m; j = 1, 2, \cdots, n$.

Finally, through calculation and comparison, the selection weights of mining schemes are obtained.

Finding the maximum eigenvalue $\lambda_{\max} = 6.0774$, $CI = (\lambda_{\max} - n)/(n-1) = 0.0155$. According to Table 3, $CR = CI/RI = 0.0125 < 0.1$ meets the consistency test.

Table 3. Average random consistency indicators.

n	1	2	3	4	5	6	7	8	9	10	11	12	13	14
RI	0	0	0.52	0.89	1.12	1.24	1.36	1.41	1.46	1.49	1.52	1.54	1.56	1.58

Therefore, the weight vectors that affect the selection of mining sequence can be obtained by using the square root method as $\omega = [0.0447, 0.1376, 0.0447, 0.0447, 0.4634, 0.2649]$.

5.4. Optimization of Mining Schemes in Sanshandao Gold Mine

In the deep mining process of Sanshandao gold mine, the weighted average model was used to comprehensively evaluate the above three schemes from six indicators: maximum vertical displacement of the top column, maximum vertical displacement of the bottom column, maximum compressive stress of the backfilling body, maximum tensile stress of the filling body, and the degree of damage in the plastic zone.

Based on the above calculation results, the scheme membership vector can be obtained as follows:

$$S = \omega \cdot R_{1\text{–}6} = (0.9848, 0.9993, 0.9642) \tag{10}$$

Therefore, the order priority of the three schemes is Scheme 2 > Scheme 1 > Scheme 3.

In conclusion, Scheme 2 is finally determined as the best scheme for the deep deposit of the Sanshandao gold mine.

6. Engineering Application and Limitations

The preferred Scheme 2 was implemented in an industrial trial at the mine site, with the test area selected between the 1600 and 1800 lines at the −945 m to −975 m level. Key parameters of the test area included cut-off grade: 1.00 g/t, total ore volume: 181,642 t, average grade: 2.5 g/t, and metal content: 454.105 kg. For the medium-deep hole high-section stopes in the lower wall between −945 m and −975 m, the downward drilling method was employed to extract Stopes 1#, 2#, and 3#, with subsequent analysis of economic performance metrics. The average daily production capacity in the test panel (medium-deep hole mining): 364.3 t. The total ore extracted from the three test stopes was 33,452 t, the ore recovery rate was 92.38%, and the mining dilution rate was 6.18%. Figure 23 shows the blasting effect of the 2# stope cutting patio, and Figure 24 shows the stope blasting effect. The trial achieved outstanding technical and economic performance, confirming the efficacy of the methodology. Notably, the results validated Scheme 2 as the optimal mining strategy for the given geological and operational conditions.

While the study provides valuable insights, certain limitations should be acknowledged. The optimization of mining schemes was based on numerical simulations using FLAC3D. However, numerical models inherently involve simplifications and assumptions that may not fully capture actual complexities. Future research should incorporate high-performance numerical software with enhanced computational capabilities to improve simulation accuracy and account for multi-physics interactions. Simultaneously, the fuzzy comprehensive evaluation method, while effective, introduces inherent randomness in decision-making. To strengthen the robustness of scheme selection, multi-method validation is recommended, such as theoretical analysis, physical similarity experiments, integration of machine learning algorithms, etc. Furthermore, a combination of holistic approach would better address uncertainties in geological variability, rock mass behavior, and operational constraints.

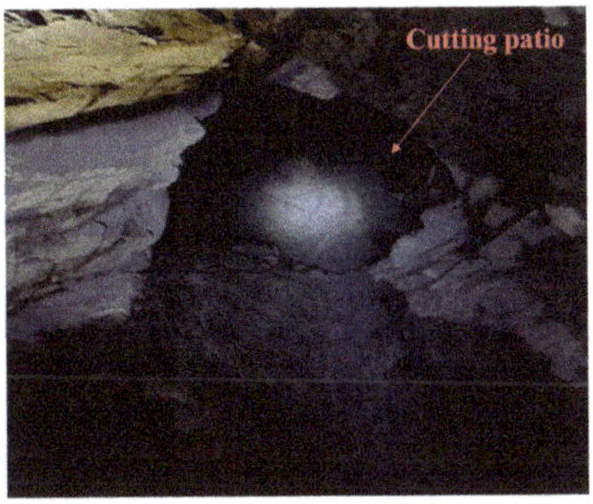

Figure 23. Blasting effect of the 2# stope cutting patio.

Figure 24. Stope blasting effect.

7. Conclusions

(1) According to the characteristics of the mining method of the deep ore body of the Sanshandao gold mine, three different downward mining schemes were designed, namely, the staggered downward mining from the horizontal wing to the other wing, the vertical interval six, and the oblique interval two; the horizontal center to the two wings separated by five mining one, the vertical interval six, and the oblique interval two; and the horizontal two wings to the center of the staggered downward mining scheme.

(2) A numerical simulation analysis was conducted on three distinct mining phases, with the objective of elucidating the evolution rules governing the maximum principal stress, minimum principal stress, equivalent strain, vertical displacement, horizontal displacement, and the distribution of the plastic zone in different mining schemes across varying mining phases. The change in vertical displacement is between 4.42 and 22.73 mm, while the change in horizontal displacement is between 1.72 and 3.69 mm. These values demonstrate that vertical displacement has a more significant impact on the stability of the stope than horizontal displacement.

(3) As the mining range is extended, the mechanical effect of the filling body is becoming increasingly pronounced. The downward mining scheme is conducive to the effective management of deep ground pressure at the Sanshandao gold mine. Scheme 2 is conducive to the control of stress concentration, the reduction in tensile stress in

the mining area, the prevention of large plastic zones and large displacements in the mining area, and the realization of ground pressure control at the Sanshandao gold mine.

(4) Based on the characteristics of three mining technology schemes, the evolution laws of stress-strain distribution and other data for three different schemes were analyzed. The AHP and FCE method was used to optimize different mining schemes in the deep deposit of Sanshandao gold mine. Six indicators (maximum compressive stress of surrounding rock, average vertical displacement of top pillar monitoring points, average vertical stress of top pillar monitoring points, maximum compressive stress in the mining area, maximum tensile stress in the mining area, and degree of plastic failure) were selected to establish an evaluation index system, and Scheme 2 was determined as the optimal scheme. Finally, the rationality and reliability of Scheme 2 were validated through on-site industrial trials.

Author Contributions: Conceptualization, W.L.; methodology, W.L. and Z.L. (Zaiyong Li); formal analysis, W.L. and Z.L. (Zaiyong Li); writing—original draft, W.L.; investigation, Z.L. (Zhixiang Liu); writing—review, Z.L. (Zhixiang Liu); supervision, Z.L. (Zhixiang Liu); resources, Z.L. (Zhixiang Liu); project administration, Z.L. (Zhixiang Liu); funding acquisition, Z.L. (Zhixiang Liu); software, Z.L. (Zaiyong Li); data curation, Z.L. (Zaiyong Li); validation, Z.L. (Zaiyong Li); visualization, Z.L. (Zaiyong Li). All authors have read and agreed to the published version of the manuscript.

Funding: This work was supported by the Fundamental Research Funds for the Central Universities of Central South University (No. 2024zzts0421) and Liupanshui Innovation Technology Center for Disaster Prevention and Resilience Enhancement in Ecologically Fragile Areas (52020-2024-PT-06).

Institutional Review Board Statement: Not applicable.

Informed Consent Statement: Not applicable.

Data Availability Statement: The data will be made available on request.

Acknowledgments: The authors would like to acknowledge the reviewers for improving the quality of the manuscript.

Conflicts of Interest: The authors declare no conflicts of interest.

References

1. Liu, Z.X.; Dang, W.G.; He, X.G. Undersea safety mining of the large gold deposit in Xinli District of Sanshandao Gold Mine. *Int. J. Miner. Metall. Mater.* **2012**, *19*, 574–583. [CrossRef]
2. Peng, K.; Li, X.B.; Wan, C.C.; Peng, S.Q.; Zhao, G.Y. Safe mining technology of undersea metal mine. *Trans. Nonferrous Met. Soc. China* **2012**, *22*, 740–746. [CrossRef]
3. Peng, K.; Liu, Z.P.; Zhang, Y.L.; Fan, X.; Chen, Q.F. Determination of isolation layer thickness for undersea mine based on differential cubature solution to irregular Mindlin plate. *J. Cent. South Univ.* **2017**, *24*, 708–719. [CrossRef]
4. Chen, Y.; Zhao, G.Y.; Wang, S.F.; Li, X. Investigations of the height of fractured zones in overburden induced by undersea mining. *Arab. J. Geosci.* **2019**, *12*, 618. [CrossRef]
5. Chen, Y.; Zhao, G.Y.; Wang, S.F.; Wu, H.; Wang, S.W. A case study on the height of a water-flow fracture zone above undersea mining: Sanshandao Gold Mine, China. *Environ. Earth Sci.* **2019**, *78*, 122. [CrossRef]
6. Li, D.; Zhang, J.F.; Wang, C.W.; Chen, Y.; Ge, D.C. Assessing Rockburst Hazards Using a Self-Developed Real-Time Microseismic Monitoring System in a Deep-Sea Goldmine. *IEEE Access* **2019**, *7*, 134360–134371. [CrossRef]
7. Liu, Z.X.; Han, K.W.; Yang, S.; Liu, Y.X. Fractal evolution mechanism of rock fracture in undersea metal mining. *J. Cent. South Univ.* **2020**, *27*, 1320–1333. [CrossRef]
8. Zhang, C.; Zhao, Y.X.; Han, P.H.; Bai, Q.S. Coal pillar failure analysis and instability evaluation methods: A short review and prospect. *Eng. Fail. Anal.* **2022**, *138*, 106344. [CrossRef]
9. Chen, X.H.; Yang, S.H.; Hu, D.B.; Li, X.H. Sustainable mining method selection by a multi-stakeholder collaborative multi-attribute group decision-making method. *Resour. Policy* **2024**, *92*, 105043. [CrossRef]

10. Jahanbani, Z.; Mortazavi, A.; Ataee-pour, M. A Causal Analysis of the Influential Criteria in Underground Mining Method Selection. *Rock Mech. Rock Eng.* **2024**, *57*, 8581–8603. [CrossRef]
11. Liu, B.; Wang, Y.J.; Zhang, J.; Hou, S.L.; He, M.C.; Wu, X.; Li, H.C. Comparative Analysis on the Safety Characteristics of N00 Mining Method with Conventional 121 Mining Method. *Geotech. Geol. Eng.* **2024**, *42*, 321–338. [CrossRef]
12. Xia, W.H.; Li, Y.; Song, W.D.; Fu, J.X.; Zhang, C.; Yu, L. Stability Analysis of Secondary Mining of Submarine Mine Under the Clamping Control of Fault and Fracture Zone. *Min. Metall. Explor.* **2024**, *41*, 909–924. [CrossRef]
13. Li, S.; Yu, L.F.; Dan, Z.Y.; Yin, T.B.; Chen, J.Y. The Recent Progress China Has Made in Mining Method Transformation, Part I: Shrinkage Method Transformed into Backfilling Method. *Appl. Sci.* **2024**, *14*, 10033. [CrossRef]
14. Liu, Z.-x.; Luo, T.; Li, X.; Li, X.-b.; Huai, Z.; Wang, S.-f. Construction of reasonable pillar group for undersea mining in metal mine. *Trans. Nonferrous Met. Soc. China* **2018**, *28*, 757–765. [CrossRef]
15. Tan, B.H.; Ren, F.Y.; Ning, Y.J.; He, R.X.; Zhu, Q. A New Mining Scheme for Hanging-Wall Ore-Body during the Transition from Open Pit to Underground Mining: A Numerical Study. *Adv. Civ. Eng.* **2018**, *2018*, 1465672. [CrossRef]
16. Wang, F.F.; Ren, Q.Y.; Chen, B.; Zou, P.; Peng, Z.J.; Hu, W.J.; Ma, Z. Numerical Investigation on Safe Mining of Residual Pillar in Goaf: A Case Study of Panlong Lead-Zinc Mine. *Geotech. Geol. Eng.* **2020**, *38*, 4269–4287. [CrossRef]
17. Zhou, Z.; Zhou, J.L.; Lai, L.; Xu, M.T.; Xu, Y.L. Determination of optimal mining width for coal mining under the slope by of using numerical simulation. *Sci. Rep.* **2024**, *14*, 1124. [CrossRef]
18. Zhao, K.; Liang, N.; Zeng, P.; Wang, W.; Gong, C.; Xiong, L.; Liu, H. Research on Optimization and Numerical Simulation of Layout Scheme of Mining Approach in Downward Slicing and Filling Method. *Appl. Sci.* **2023**, *13*, 8688. [CrossRef]
19. Shi, J.; Feng, J. Simulate on Support Technology of Deep Mining. *Geotech. Geol. Eng.* **2021**, *39*, 4663–4668. [CrossRef]
20. Liang, W.-z.; Zhao, G.-y.; Wu, H.; Chen, Y. Optimization of mining method in subsea deep gold mines: A case study. *Trans. Nonferrous Met. Soc. China* **2019**, *29*, 2160–2169. [CrossRef]
21. Mijalkovski, S.; Peltechki, D.; Despodov, Z.; Mirakovski, D.; Adjiski, V.; Doneva, N. Application of the FUZZY TOPSIS method for selecting an underground mining method. *Acta Montan. Slovaca* **2023**, *28*, 465–478. [CrossRef]
22. Yavuz, M. The application of the analytic hierarchy process (AHP) and Yager's method in underground mining method selection problem. *Int. J. Min. Reclam. Environ.* **2015**, *29*, 453–475. [CrossRef]
23. Bajić, S.; Bajić, D.; Gluščević, B.; Ristić Vakanjac, V. Application of Fuzzy Analytic Hierarchy Process to Underground Mining Method Selection. *Symmetry* **2020**, *12*, 192. [CrossRef]
24. Hu, B.Y.; Zhang, Q.L.; Li, S.; Yu, H.X.; Wang, X.M.; Wang, H. Application of Numerical Simulation Methods in Solving Complex Mining Engineering Problems in Dingxi Mine, China. *Minerals* **2022**, *12*, 123. [CrossRef]
25. Hou, D.; Xu, M.; Li, X.; Wang, J.; Wang, M.; Li, S. Optimization of mining methods for deep orebody of large phosphate mines. *Front. Built Environ.* **2023**, *9*, 1282684. [CrossRef]
26. Alpay, S.; Yavuz, M. Underground mining method selection by decision making tools. *Tunn. Undergr. Space Technol.* **2009**, *24*, 173–184. [CrossRef]
27. Sadiq, R.; Husain, T.; Veitch, B.; Bose, N. Risk-based decision-making for drilling waste discharges using a fuzzy synthetic evaluation technique. *Ocean Eng.* **2004**, *31*, 1929–1953. [CrossRef]
28. Ataei, M.; Jamshidi, M.; Sereshki, F.; Jalali, S.M.E. Mining method selection by AHP approach. *J. South. Afr. Inst. Min. Metall.* **2008**, *108*, 743–751.
29. Wen, K.-L. A Matlab toolbox for grey clustering and fuzzy comprehensive evaluation. *Adv. Eng. Softw.* **2008**, *39*, 137–145. [CrossRef]
30. Lin, Q.; Zhang, S.; Lin, H.; Zhang, K.; Fan, W.; Huang, C.; Shao, Z.; Lan, M. Failure behavior of jointed rock masses containing a circular hole under compressive-shear load: Insights from DIC technique. *Theor. Appl. Fract. Mech.* **2025**, *139*, 105089. [CrossRef]
31. Liu, W.J.; Liu, Z.X.; Wang, M.; Zhang, S.X. A comparative performance study on the development of hybrid extreme gradient boosting models for predicting rock layer subsidence in subsea gold mine. *Geomech. Geophys. Geo-Energy Geo-Resour.* **2025**, *11*, 39. [CrossRef]
32. Liu, W.J.; Liu, Z.X.; Xiong, S.; Wang, M. Comparative prediction performance of the strength of a new type of Ti tailings cemented backfilling body using PSO-RF, SSA-RF, and WOA-RF models. *Case Stud. Constr. Mater.* **2024**, *20*, e02766. [CrossRef]
33. Li, C. Application of Fuzzy Comprehensive Evaluation in Selecting Large Sports Venues. *Agro Food Ind. Hi-Tech* **2017**, *28*, 22–26.
34. Ma, W.; Wang, Y. Fuzzy comprehensive evaluation model of interuniversity collaborative learning based on network. *Open Phys.* **2017**, *15*, 427–432. [CrossRef]
35. He, R.X.; Liu, H.; Ren, F.Y.; Li, G.H.; Zhang, J.; Zhou, Y.J. Comprehensive Evaluation and Decision for Goaf Based on Fuzzy Theory in Underground Metal Mine. *Adv. Civ. Eng.* **2022**, *2022*, 3104961. [CrossRef]
36. Fu, Z.G.; Wu, X.L.; Liao, H.C.; Herrera, F. Underground Mining Method Selection With the Hesitant Fuzzy Linguistic Gained and Lost Dominance Score Method. *IEEE Access* **2018**, *6*, 66442–66458. [CrossRef]
37. Lin, Y.; Lin, C.; Qiu, X. Fuzzy Comprehensive Evaluation Method of Masonry Structure Safety Based on Grey Clustering Theory. *Math. Probl. Eng.* **2018**, *2018*, 8710192. [CrossRef]

38. Wang, Z. Fuzzy comprehensive evaluation of physical education based on high dimensional data mining. *J. Intell. Fuzzy Syst.* **2018**, *35*, 3065–3076. [CrossRef]
39. Wu, Z.; Li, Q.; Kong, D.; Chen, G.; Luo, D. The ANP-Fuzzy-TOPSIS model for the optimization of the scheme of large-section blasting. *Arab. J. Geosci.* **2020**, *13*, 47. [CrossRef]
40. Ali, M.A.M.; Kim, J.G. Selection mining methods via multiple criteria decision analysis using TOPSIS and modification of the UBC method. *J. Sustain. Min.* **2021**, *20*, 49–55. [CrossRef]
41. Zhu, L. Research and application of AHP-fuzzy comprehensive evaluation model. *Evol. Intell.* **2022**, *15*, 2403–2409. [CrossRef]

applied sciences

MDPI

Article

Numerical Investigation of the Pull-Out and Shear Mechanical Characteristics and Support Effectiveness of Yielding Bolt in a Soft Rock Tunnel

Yan Zhu [1], Mingbo Chi [1], Yanyan Tan [2], Ersheng Zha [1,*] and Yuwei Zhang [3]

[1] China Academy of Safety Science and Technology, No. 32 Beiyuan Road, Beijing 100012, China; zhuyantyy@163.com (Y.Z.); chimingbo1028@163.com (M.C.)
[2] Chinese Association of Automation, No. 95 Zhongguancun East Road, Beijing 100190, China; tanya1125@163.com
[3] School of Mechanics and Civil Engineering, China University of Mining and Technology at Beijing, D11 Xueyuan Road, Beijing 100083, China; zhyw1020@163.com
* Correspondence: zhaersheng@foxmail.com; Tel.: +86-15198167918

Abstract: Conventional bolts frequently fail under large deformations due to stress concentration in soft rock tunnels. In contrast, yielding bolts incorporate energy-absorbing mechanisms to sustain controlled plastic deformation. This study employed FLAC3D to numerically investigate the pull-out, shear, and bending behaviors of yielding bolts, evaluating their support effectiveness in soft rock tunnels. Three-dimensional finite difference models incorporating nonlinear coupling springs and the Mohr–Coulomb criterion were developed to simulate bolt–rock interactions under multifactorial loading. Validation against experimental pull-out tests and field measurements confirmed the model accuracy. Under pull-out loading, the axial forces in yielding bolts decreased more rapidly along the bolt length, reducing stress concentration at the head. The central position of the maximum load-bearing capacity in conventional bolts fractured under tension, resulting in an hourglass-shaped axial force distribution. Conversely, the yielding bolts maintained yield strength for an extended period after reaching it, exhibiting a spindle-shaped axial force distribution. Parametric analyses reveal that bolt spacing exerts a greater influence on support effectiveness than length. This study bridges critical gaps in understanding yielding bolt behavior under combined loading and provides a validated framework for optimizing energy-absorbing support systems in soft rock tunnels.

Keywords: yielding bolt; numerical simulation; pull-out; shear performance; support effect; soft rock tunnel

1. Introduction

The stability of tunnels in soft rock masses, characterized by low strength, high deformability, and time-dependent creep behavior, remains a critical concern in geotechnical engineering, particularly in deep mining and transportation infrastructure projects [1,2]. These geological conditions often lead to large-scale convergence, stress redistribution, and progressive failure during and after excavation, necessitating support systems capable of accommodating significant deformations while maintaining structural integrity [3–5]. Among reinforcement techniques, rock bolting has long been a cornerstone strategy for stabilizing underground openings [6–8]. However, conventional bolts exhibit limited adaptability under high-strain conditions, frequently suffering brittle failure induced by stress concentration in squeezing grounds [9]. This limitation has driven the development of yielding bolts, which integrate energy-absorbing mechanisms to sustain controlled plastic deformation without losing load-bearing capacity [10]. Despite their growing application, the mechanical behavior of yielding bolts under combined loading regimes—specifically pull-out, shear, and bending—and their interaction with soft rock masses remain insufficiently characterized [11,12]. Addressing these knowledge gaps is pivotal for optimizing yielding bolt design and validating their efficacy in real-world tunneling scenarios, particularly in geotechnically challenging environments.

While experimental studies on yielding bolts have demonstrated their superior elongation capacity compared to conventional systems [13], most of the research has focused on simplified laboratory pull-out tests, leaving critical questions unanswered. For instance, the shear resistance of yielding bolts has received limited attention [14,15]. Similarly, the bending behavior of these bolts under lateral loads, such as those induced by asymmetrical ground pressures, remains poorly understood [16]. Furthermore, existing studies often neglect the holistic performance of yielding bolts in full-scale tunnel models, where complex stress paths, rock–bolt coupling effects, and time-dependent deformation interplay [17]. These limitations underscore the need for advanced numerical frameworks capable of simulating multifactorial loading conditions and providing insights into bolt–rock interactions at both local and global scales.

Numerical modeling has emerged as a powerful tool for analyzing the nonlinear behavior of support systems in soft rock tunnels [18–21]. The finite difference method, implemented in software such as FLAC3D, is particularly suited for this purpose due to its ability to handle large deformations, interface slip, and elastoplastic material responses [22–24]. Prior studies by Jiang et al. [25] and Bahrani and Hadjigeorgiou [26] validated the capability of FLAC3D to replicate the pull-out and shear tests of conventional bolts, establishing its credibility for simulating bolt–rock interactions. Saadat et al. [27] employed numerical simulation methods to analyze the influence of boundary conditions on the shear behavior of rock joints. Zhang et al. [28] used Flac3D to study the development law of interface for geotechnical prestressed anchorage bolt. Zhang et al. [29] revealed the shear characteristics of the bolt–grout interface under cyclic shear loading. However, these efforts largely ignored yielding mechanisms, which introduce additional complexities, such as strain-dependent stiffness degradation, frictional energy dissipation, and post-yield load redistribution [10,30]. Recent advancements in constitutive modeling—including the integration of nonlinear coupling springs, strain-softening criteria, and anisotropic damage laws—have enhanced the fidelity of bolt simulations [31,32]. For example, Tahmasebinia et al. [9] simulated yielding bolt performance under cyclic loading, highlighting the influence of bolt geometry on energy absorption. Despite these strides, comprehensive numerical investigations integrating pull-out, shear, and bending analyses for yielding bolts remain scarce [33,34]. This gap hinders the optimization of critical design parameters, such as bolt length, spacing, and yielding element configuration, which directly govern support effectiveness and economic viability [35–37].

To address these issues, the pull-out tests of conventional bolts and previously developed yielding bolts were simulated using FLAC3D. The detailed axial force distribution and its dynamic evolution under pull-out conditions for both bolt types were investigated. Additionally, a comprehensive analysis of the shear resistance of the bolts was conducted. Furthermore, a comparative study was undertaken to evaluate the support efficacy of conventional bolts against yielding bolts within the context of soft rock tunnel environments. The findings elucidate the mechanical behavior of yielding bolts under combined loading regimes, and this work also bridges critical gaps in the understanding of energy-absorbing support systems.

A comprehensive numerical simulation of the pull-out, shear, and bonding analysis of yielding bolts was performed. A distinctive spindle-shaped axial force distribution in yielding bolts during plastic deformation was revealed, contrasting with the hourglass-shaped pattern in conventional bolts. This mechanistic insight explains the superior elongation capacity of yielding bolts.

2. Methodology

2.1. Simulation Theory

Numerical simulation of the bolt mechanical properties was conducted using the finite difference software FLAC3D 3.00. Rock bolt elements were employed to simulate the bolts, and these were defined by geometric parameters, material properties, and coupled spring parameters. A line segment between two structural nodes is represented as a single element component, with the components between two nodes sharing the same symmetric cross-section parameters. The rock bolt element effectively combines the functions of beam elements and bolt elements, making it suitable for simulating bolts subjected to both normal and axial frictional forces [22].

Each structural element component possesses its own local coordinate system, and it is used to specify inertial moments, distributed loads, and defined force and moment signs. As shown in Figure 1, the local coordinate system of the components of the simulated bolts was defined by the positions of its two nodes (1 and 2) and the vector Y with the following rules: the central axis coincides with the X axis, the X axis direction is from Node 1 to Node 2, and the Y axis lies in the cross-sectional plane.

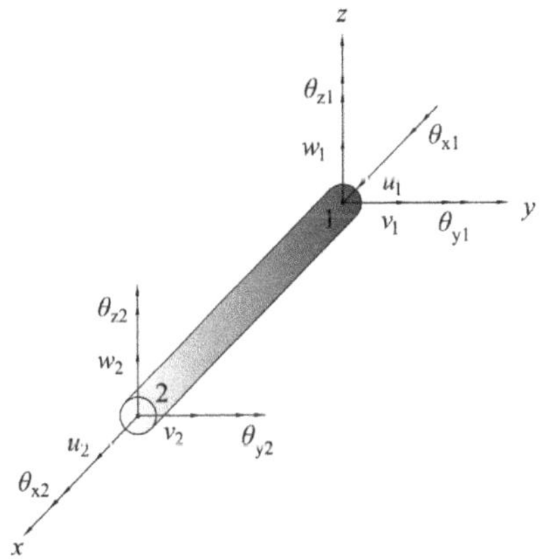

Figure 1. The local coordinate system and 12-degree-freedom component. For Node 1, u_1, v_1, and w_1 are translational displacements along the X axis, Y axis, and Z axis, respectively; and θ_{x1}, θ_{y1}, and θ_{z1} are rotation angles around the X axis, Y axis, and Z axis, respectively. The symbols of the degrees of freedom of Node 2 are consistent with that of Node 1.

The interaction between the bolt rod and the rock mass was simulated using coupling springs. These nonlinear, sliding coupling springs transmit forces and moments between structural element nodes and solid elements. The sign conventions for forces and moments on components are illustrated in Figure 2.

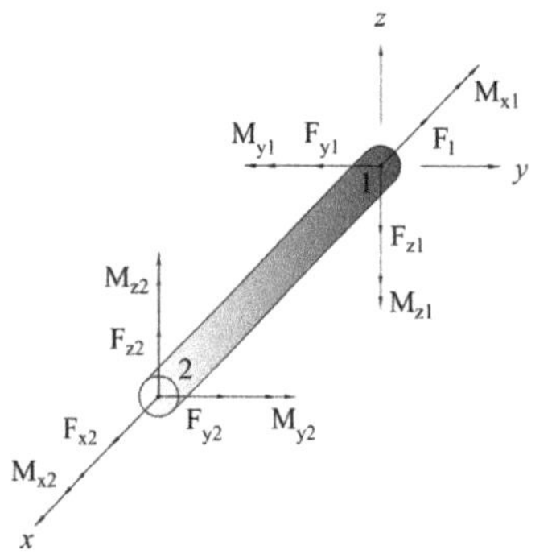

Figure 2. The symbol protocol of the endpoint force and moment of the component. For Node 1, F_{x1}, F_{y1}, and F_{z1} are forces along the X axis, Y axis, and Z axis, respectively; and M_{x1}, M_{y1}, and M_{z1} are moments along the X axis, Y axis, and Z axis, respectively. The symbols of Node 2 are consistent with that of Node 1.

During the process of bolt pullout, the axial behavior of the bolt can be described using a one-dimensional constitutive model, as expressed in the following equation:

$$K = \frac{AE}{L}, \tag{1}$$

where K represents the axial stiffness of the bolt, A denotes the cross-sectional area of the bolt, E stands for the elastic modulus, and L signifies the length of the bolt. The tensile yield strength F_t and compressive yield strength F_c of the bolt were determined, and these two limits must not be exceeded during the utilization of the bolt.

When subjected to tensile loading, the bolt's interaction with the surrounding rock is simulated through tangential coupling springs. The function of these springs is analogous to the tangential action mechanism of grouted bolts that is depicted in Figure 3; therefore, the properties of the tangential coupling springs can represent the grouting properties. The shear stress on the contact surface between the bolt and the rock mass primarily considers its cohesive force and frictional force. The influence of tangential coupling springs around the bolt is mainly reflected through parameters such as the stiffness, cohesive force, internal friction angle, outer boundary radius of the bolt, and effective stress around the bolt. The relationship between them can be expressed by the following equation [26]:

$$F_s/L = K_s(u_p - u_m), \tag{2}$$

where F_s is the shear force, K_s is the shear stiffness of the coupling spring, and u_p and u_m are the axial displacements of the rock bolt element and the surrounding medium, respectively.

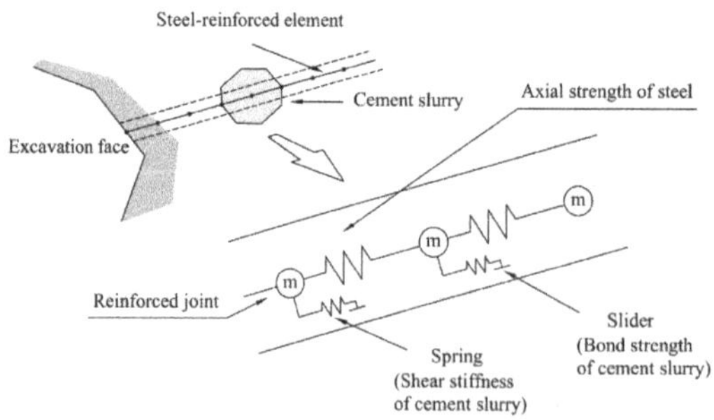

Figure 3. Mechanical model of the bolt element.

Furthermore, we have

$$F_s^{\max}/L = C_s + \sigma_m^{l\tan\phi_s}, \tag{3}$$

where C_s is the cohesive force per unit length of the shear coupling spring, ϕ_s is the friction angle of the shear coupling spring, σ_m is the pressure on the rock bolt, and l is the outer circumference of the rock bolt.

When the bolt body is subjected to shear due to the sliding of the surrounding rock, it needs to resist bending moments. When there are cracks on the contact surface, sliding due to shear deformation can easily occur, where rock bolts serve to reinforce the structure. Sliding along the crack surfaces can lead to shear deformation or even failure of the bolt. By specifying the bolt's yield strength, axial yielding can be achieved. Bolt fracture is simulated based on the tensile failure strain. The strain at various bolt positions includes the axial strain and bending plastic strain, where the axial strain is the average plastic strain along the bolt. The formula [22] for calculating plastic tensile strain is given by

$$\varepsilon_{pl} = \sum \varepsilon_{pl}^{ax} + \sum \frac{d\theta_{pl}}{2L}, \tag{4}$$

where d is the bolt diameter, L is the bolt length, and θ is the average twist angle of the bolt. If the strain exceeds a specified value, the forces and moments on the bolt components in the simulation will be reduced to zero, indicating the failure of the bolt.

2.2. Pull-Out Simulation of the Yielding Bolt

Jiang et al. utilized a three-dimensional finite difference method to establish a numerical simulation model for bolt pull-out tests [25]. Their research indicated that their numerical results aligned with the field test results, validating the feasibility of numerically simulating bolt pull-out behavior. In this paper, the finite difference software FLAC3D was employed to simulate the yielding bolt pull-out test and the shear test of the bolt body, aiming to demonstrate the feasibility of simulating the mechanical characteristics of the yielding bolt and the shear performance of the bolt through FLAC3D simulation.

To simulate the stress and deformation characteristics under the yielding bolt pull-out action, a computational model was established, as shown in Figure 4. The computational model of the anchoring body was a 10 m × 10 m × 10 m (X × Y × Z) cubic block. The bolt was 6 m in length, with the bolt head (that is, near the opening) positioned at (5, 5, 10) and bolt end at (5, 5, 4). The top face of the model shown in Figure 4 is a free boundary, the four sides were subject to normal constraints, and the bottom face was fully fixed. A constant velocity V was applied at the head of the bolt along the axis of the bolt. During pull-out simulation, the displacements perpendicular to the Z direction on two free surfaces were constrained, and the bolt was fixed with zero velocity in the X and Y directions.

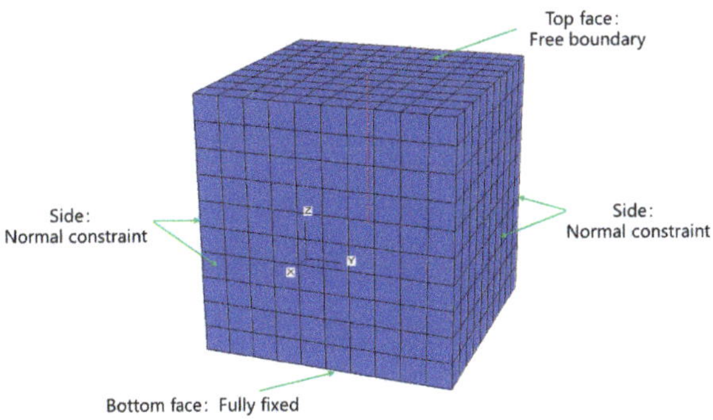

Figure 4. Calculation model. The red line refers to the bolt.

The constitutive model of the model adopted the Mohr–Coulomb model. The material parameters of the surrounding rock were selected from the fourth category, according to the standard for engineering classification of rock masses, of surrounding rocks focused on in this study [38]. The proposed rock parameters are listed in Table 1, and the bolt calculation parameters are shown in Table 2.

Table 1. The physical and mechanical parameters of the surrounding rock.

Material	Elasticity Modulus (GPa)	Poisson Ratio	Unit Weight (kN/m^3)	Cohesive Force (MPa)	Internal Friction Angle (°)
Surrounding rock	4.00	0.325	21.5	0.45	33

Table 2. The physical and mechanical parameters of the yielding bolt.

Yielding Bolt	Length (m)	Cross Sectional Area (m^3)	Elasticity Modulus (GPa)	Poisson Ratio	Tensile Failure Strain
Regular part	5.5	5×10^{-4}	200	0.25	0.1
Yielding element	0.5	5×10^{-4}	20	0.25	0.2

2.3. *Shear Performance Simulation of the Yielding Bolt*

Considering the shear performance of the yielding bolt, a model—as shown in Figure 5—was established for simulation verification. The computational model of the anchoring body was a 10 m × 10 m × 10 m (X × Y × Z) cubic block, and it was divided into two large rock blocks above and below, with an interface function between them to allow for rock sliding. The mesh division used 8-node hexahedral elements, totaling 1000 elements and 1452 nodes. The model shown in Figure 5 was divided into two parts. The boundary conditions of the upper part were as follows: the top and bottom faces were subject to normal constraints, and all of the sides were constrained along the Z-direction. The boundary conditions of the lower part were that all faces are fully fixed. A 6 m long yielding bolt was driven vertically into the center of the model at the interface, with the bolt head positioned at (5, 5, 8) and the bolt end at (5, 5, 2). The parameters of the surrounding rock and the bolt are presented in Table 1 and Table 3, respectively.

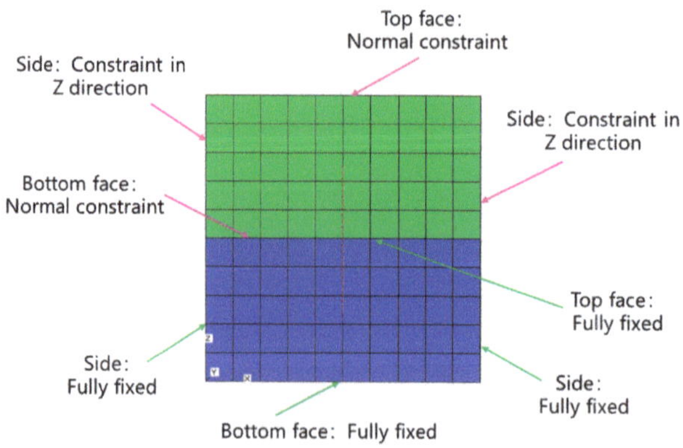

Figure 5. Shear-resistant model. The red line refers to the bolt.

Table 3. The physical and mechanical parameters of the conventional bolt

Material	Length (m)	Cross Sectional Area (m^3)	Elasticity Modulus (GPa)	Poisson Ratio	Tensile Failure Strain
Conventional bolt	6	5×10^{-4}	200	0.25	0.1

Bending tests also reflect the bolt body's shear performance. To further demonstrate the feasibility of simulating the shear performance, a bolt bending test was conducted. Part of the bolt was anchored in the rock, and a force perpendicular to the bolt body was applied at the exposed head.

A model, as depicted in Figure 6, was established, with the computational model of the anchoring body being a 10 m × 10 m × 10 m (X × Y × Z) cubic block. The mesh division used 8-node hexahedral elements, totaling 1000 elements and 1331 nodes. The boundary conditions of the model were that all faces are fully fixed. The bolt length was 6 m, with the bolt end at (5, 5, 7), where half of the bolt was anchored inside the rock and the head was outside. Just like the model shown in Figure 5, the parameters of the surrounding rock and the bolt are presented in Table 1 and Table 3, respectively.

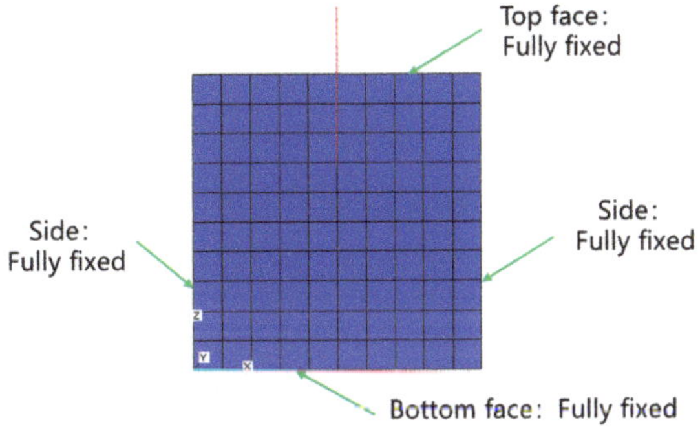

Figure 6. Stretch bending model. The red line refers to the bolt.

2.4. Simulation of the Soft Rock Tunnel Supported by Bolt

As shown in Figure 7, a model of 100 m × 100 m (X × Z) was established, with the Y axis representing unit length. The excavation of the tunnel was circular with a diameter of 10 m. The tunnel axis corresponded to the Y axis, with the vertical upward direction defined as the positive Z axis. The mesh utilized 8-node hexahedral elements, resulting in a total of 1584 elements and 3170 nodes. The model is a two-dimensional model. The boundary conditions of the model were as follows: the top face was a free boundary, the two sides were subject to normal constraints, and the bottom face was fully fixed. The vertical stress was set at 10.75 MPa and the horizontal stress along the tunnel axis and the horizontal stress perpendicular to the tunnel axis were 12.90 MPa, with a lateral pressure coefficient of 1.2. Considering the focus on soft rock as the main research subject, the surrounding rock parameters were selected from the fourth category of surrounding rocks based on the standard for engineering classification of rock masses [38], as detailed in Table 1. The constitutive model and yielding criteria for the surrounding rock adopted the Mohr–Coulomb model. In this study, the term "yielding bolt" refers to the bolt used for anchoring purposes.

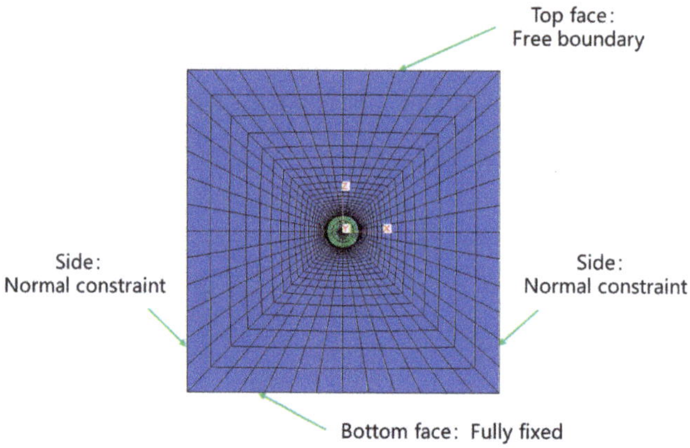

Figure 7. Chart of the plane strain model. The two red lines refer to X axis and Z axis, respectively.

To investigate the support effects of the two bolt types in soft rock tunnel excavation, a planar strain model of a tunnel was established. The rock mass was first brought to the initial equilibrium under the original rock stress state, with initial displacements set to zero. The excavated portion was replaced by a null model, and the support was then implemented. For comparison, simulations were conducted using three approaches: Approach One involved no bolt application, Approach Two involved the use of standard bolts, and Approach Three involved the use of yielding bolts. (Refer to Tables 2 and 3 for the parameters.).

To more intuitively and precisely reflect the influence of the bolts on the deformation distribution of the surrounding rock, reinforcement using bolts alone was implemented after partial stress release following tunnel excavation, and this was conducted without other combined support methods. The bolt support following tunnel excavation was radially and uniformly arranged, with radial spacing along the tunnel wall approximately 1 m apart, as illustrated in Figure 8. To provide a more intuitive representation of the surrounding rock deformation within the tunnel, 16 monitoring points were evenly distributed at positions such as the arch crown, arch shoulder, sidewalls, arch foot, and arch base during simulation, as depicted in Figure 9 and Table 4.

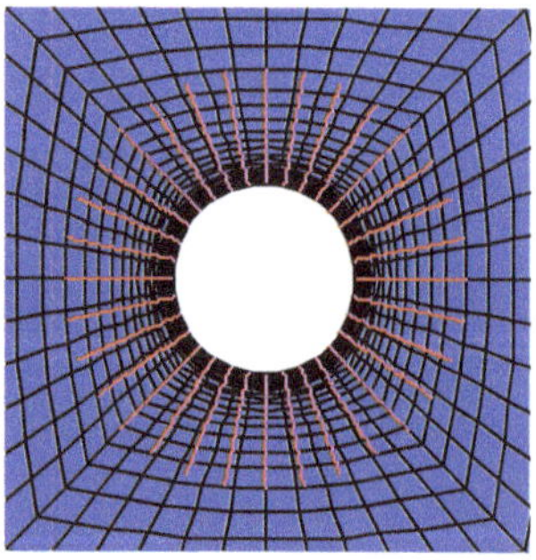

Figure 8. Diagram of bolt support. The red lines refer to the bolts.

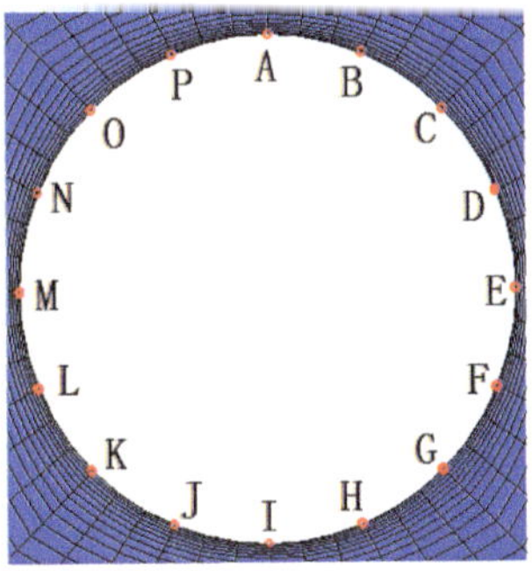

Figure 9. Diagram of the monitoring point locations on the surrounding rock.

Table 4. The monitoring point locations on the surrounding rock.

Monitoring Point	Location
A	Arch crown
B	Arch shoulder (67.5° to horizontal)
C	Arch shoulder (45° to horizontal)
D	Arch shoulder (22.5° to horizontal)
E	Sidewall (left)
F	Arch foot (22.5° to horizontal)
G	Arch foot (45° to horizontal)
H	Arch foot (67.5° to horizontal)
I	Arch base
J	Arch foot (67.5°to horizontal)
K	Arch foot (45° to horizontal)
L	Arch foot (22.5° to horizontal)
M	Sidewall (right)
N	Arch shoulder (22.5° to horizontal)
O	Arch shoulder (45° to horizontal)
P	Arch shoulder (67.5° to horizontal)

3. Results and Discussion

3.1. Feasibility Study of Pull-Out Simulations of Yielding Bolt

Guo et al. conducted indoor experimental studies using three sets of yielding bolts, with essentially identical results [13]. The bolt initially deforms elastically under loading, and when the pressure reaches 150 kN, the yielding device begins to function, causing displacement. Experimental results demonstrate that the loading remains relatively stable. A comparison between one set of indoor experiments and simulation results is shown in Figure 10. After the bolt is subjected to tension, elastic deformation occurs. In the simulation, due to the differences in the parameters of the yielding bolt, the pressure required to reach the yield deformation was approximately 160 kN. As the pull-out test progressed, the loading increased, reaching the yield strength, after which the bolt entered a stage of plastic deformation and could be sustained for a period.

Furthermore, Guo et al. conducted pull-out tests on three sets of 35 m long bolts at an engineering site, with yielding components positioned at 29 m, 32 m, and 35 m from the bolt head. As shown in Figure 11, a comparison of the numerical simulation tests under the action of yielding bolt pull-out, indoor experiments, and field tests [13] revealed that, although there are slight differences in the yield strength and yielding amount of the bolt among the three research methods, the loading characteristics, deformation trend, and support effect were consistent. Hence, it is feasible to use the finite difference software

FLAC3D to simulate pull-out tests and support effects of yielding bolts, and the simulation results align with the operating mechanism and loading characteristics of yielding bolts.

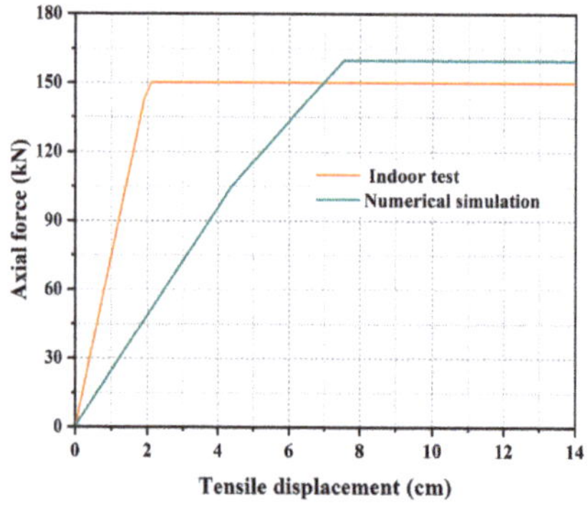

Figure 10. Comparison of the indoor pull-out test and numerical simulation results of the yielding bolt [13].

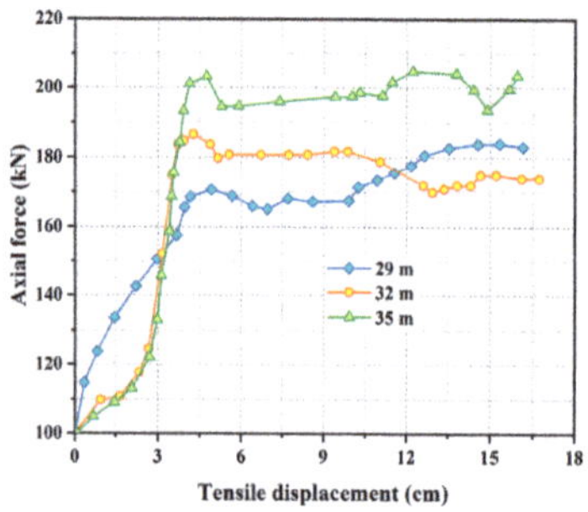

Figure 11. Results of the pull-out tests in the engineering field [13].

3.2. Comparative Analysis of the Bolts' Mechanical Behavior During Pull-Out Simulation

To analyze and compare the support effects and load characteristics of the yielding bolts and conventional bolts, models, as depicted in Figure 4, were established for the pull-out simulations. The axial forces experienced at various locations along the bolt reflected the propagation of tensile forces within the bolt, providing a direct response to the loading conditions of the bolt. The distribution of the axial forces at different positions along the two types of bolts is illustrated in Figure 12.

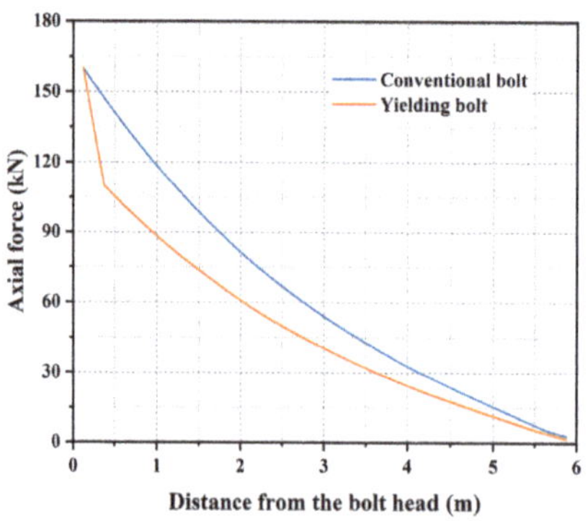

Figure 12. Axial force distribution of the different positions of the bolt.

It is evident that, under the same pull-out force, the tension borne by the bolt at different locations varies, with the axial forces experienced by the bolt gradually decreasing with increasing depth. The curve exhibits a concave shape, indicating a decreasing rate of axial force reduction along the bolt, thus approaching a plateau. As the depth increased, the yielding bolts experienced a faster decrease in axial forces compared to conventional bolts, particularly at the bolt head where higher axial forces are applied and are closer to reaching the yield strength, indicating a stronger adaptability of yielding bolts to resist larger pull-out forces. Furthermore, under the same pull-out load, the yielding bolts exhibited lower axial forces than conventional bolts at equivalent positions, suggesting that yielding bolts can withstand greater tensile forces.

The simulated pull-out involved applying a constant velocity at the bolt head, causing the pull-out force to increase incrementally until yielding. Figure 13 illustrates the variation of axial forces at different positions along the conventional and yielding bolts as the axial strain increased during the simulation process. Positions at 0.125 m, 1.125 m, 2.125 m, 3.125 m, 4.125 m, and 5.125 m from the bolt head were chosen for clarity.

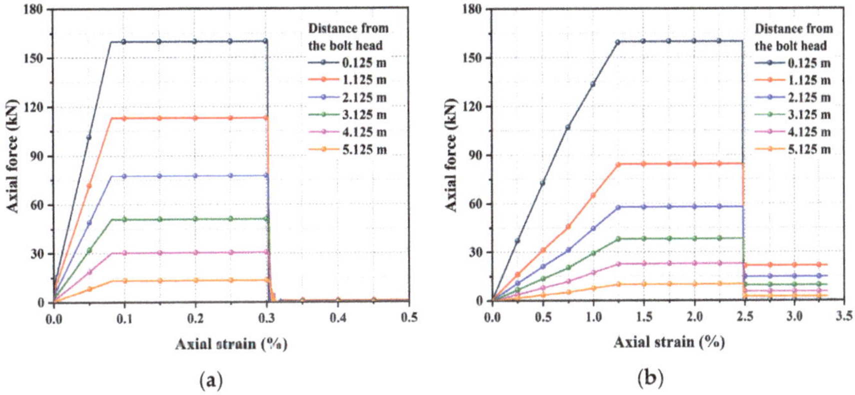

Figure 13. The axial force change process of the bolts: (**a**) conventional bolt; (**b**) yielding bolt. The six curves correspond, respectively, to the distances of 0.125 m, 1.125 m, 2.125 m, 3.125 m, 4.125 m, and 5.125 m from the bolt head.

Figure 13 shows that—before reaching yield, as the pull-out force and axial strain increase—the axial tensile forces increased at all sections, with the greatest increase being recorded near the head and there was also decreasing toward the inner end. Once the maximum axial force reached yield strength, its forces ceased, and this increased with further strain. Continued loading beyond yield led to brittle failure, analogous to necking in steel, and this caused rapid anchorage force decline and eventual failure.

Figure 14 depicts the axial forces at 0.125 m (head), 3.125 m (middle), and 5.125 m (end) for both bolt types. Prior to reaching the yield strength of the bolt, the bolt underwent elastic deformation, with the curve showing a significantly higher rate of axial force increase for conventional bolts compared to yielding bolts, highlighting the high elongation capacity of yielding bolts. According to the axial strain variation laws at the three positions of the two types of bolts, as shown in Figure 14, it is obvious that the elastic strains at which the conventional and yielding bolts reach yield strength were 0.08% and 1.26%, respectively, with the corresponding allowable elastic deformations of 4.92 mm and 75.46 mm, indicating elastic yielding distance of 70.54 mm for the yielding bolts.

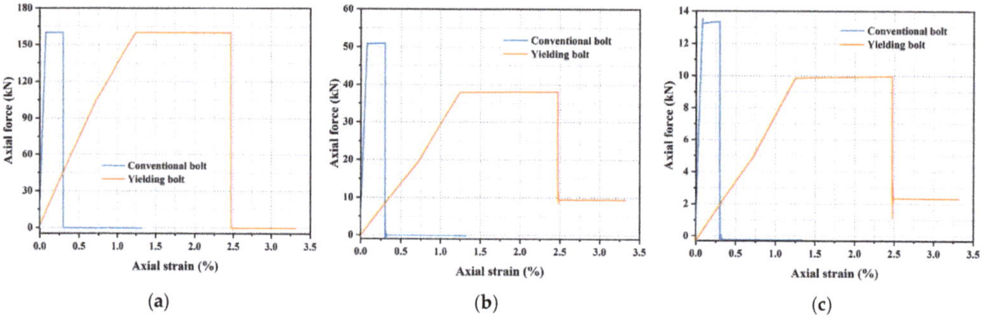

Figure 14. Comparison of the axial force distribution at different positions for the conventional bolt and yielding bolt: (**a**) bolt head; (**b**) bolt middle; and (**c**) bolt end.

After reaching yield strength, the bolt underwent plastic deformation. The plastic phase duration was longer for yielding bolts, again demonstrating their high elongation capacity. The plastic strains at which the conventional and yielding bolts reached ultimate strength were 0.3025% and 2.48%, respectively, corresponding to plastic deformations of 148.8 mm and 18.15 mm. This resulted in a plastic yielding distance of 130.65 mm for the yielding bolt.

At a distance of 0.125 m from the bolt head, the axial force acting on the yielding bolt was equivalent to that of a conventional bolt, while, at the 3.125 m and 5.125 m distances, the axial forces were notably lower for the yielding bolt compared to the conventional bolt. Under similar material conditions, this characteristic of yielding bolts in non-head sections determines their superior support effectiveness.

3.3. Feasibility Study of Shear Performance Simulations of Bolt

3.3.1. Shear Performance Simulation of Bolt

The upper rock block in Figure 5 was pushed to slide while the lower block remained fixed. As relative sliding progressed, the bolt's shear resistance was demonstrated. As illustrated in Figure 15, the bending moments at several positions along the bolt varied with calculation steps. With increasing relative sliding distance, the bending moment near the interface position gradually increased. This confirms that FLAC3D's rock–bolt element can capture the bolt's shear resistance, enabling precise analysis of its support effect. Since the yielding device minimally influences shear resistance, the shear resistance of conventional and yielding bolts was essentially identical and is not discussed further.

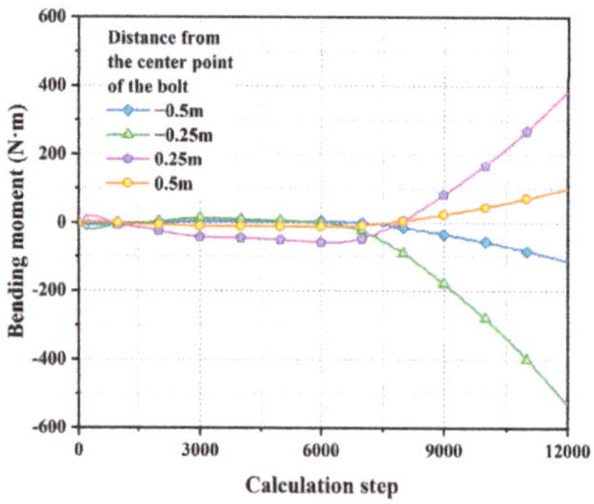

Figure 15. Variation in the bending moments at different positions along the bolt. Positions: 0.5 m and 0.25 m below the bolt center, and 0.25 m and 0.5 m above the bolt center (distances below center are negative, above are positive).

3.3.2. Bending Simulation of Bolt

A force perpendicular to the bolt body was applied at a constant velocity to the exposed bolt head, as shown in Figure 6, and this was equivalent to incrementally increasing the force during calculation. The bending moments at several positions of the bolt body varied with the calculation steps, as shown in Figure 16. With increasing force magnitude and duration, the moment in the bolt section exposed outside the rock gradually increased. This illustrated the tension bending under head loading, demonstrating the bolt body's shear resistance.

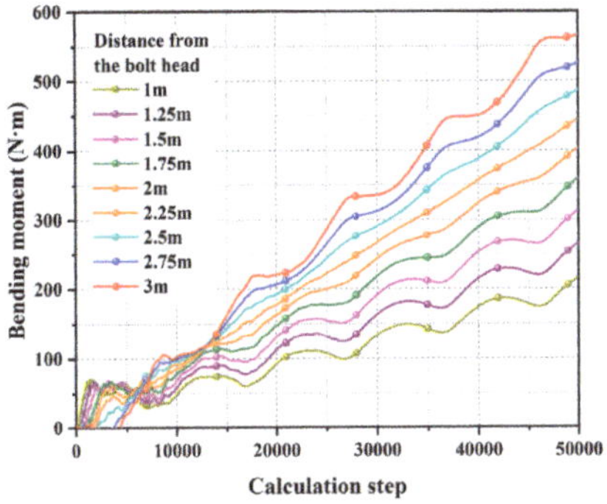

Figure 16. The bending moments varied at several bolt positions. These positions refer to the distances from the measuring points to the bolt head.

3.4. Study on the Support Effect of Bolts in Soft Rock Tunnels

3.4.1. Comparative Analysis of Surrounding Rock Deformation

To investigate the support effects of the two types of bolts after the excavation of a soft rock tunnel, a plane strain model of the tunnel, as shown in Figure 7, was established for excavation and support simulation. For comparative analysis, three scenarios were simulated: no support, conventional bolt support, and yielding bolt support. In all three scenarios, the surrounding rock mass deformed toward the tunnel, and the displacement cloud maps in the vertical and horizontal directions, along with the resultant displacement vectors, are depicted in Figures 17–19.

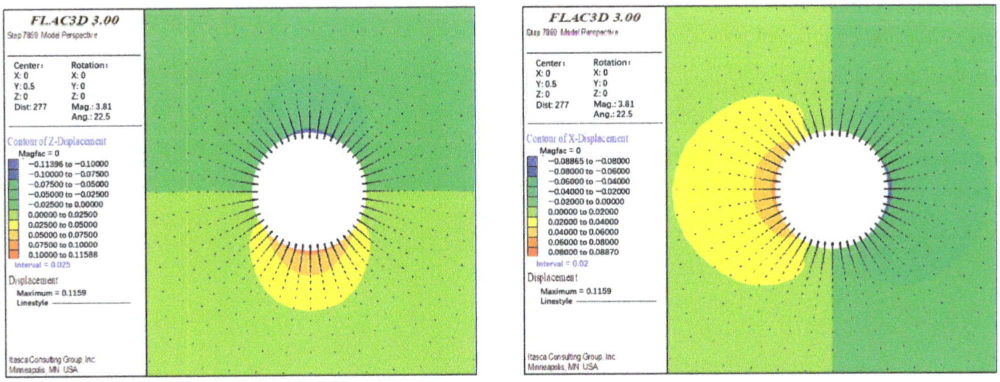

Figure 17. Displacement nephogram and resultant displacement vector diagram of the surrounding rock without supporting after excavation of the tunnel: (**a**) vertical direction; (**b**) horizontal direction.

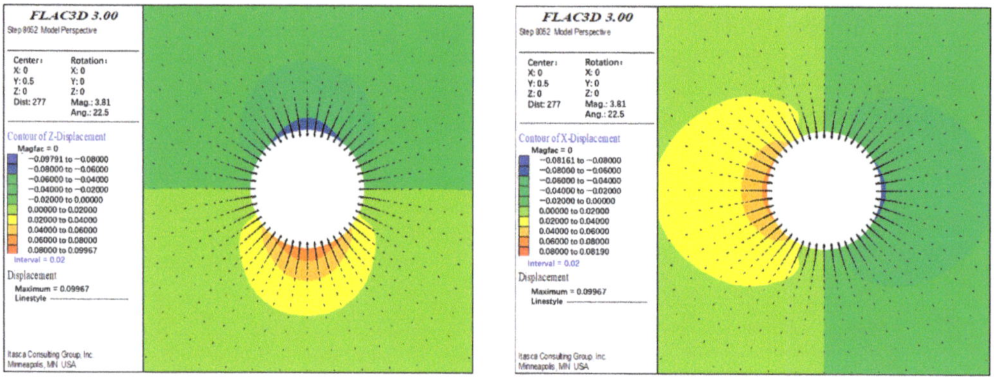

(**a**) (**b**)

Figure 18. Displacement nephogram and resultant displacement vector diagram of the surrounding rock with the conventional bolt supporting after excavation of the tunnel: (**a**) vertical direction; (**b**) horizontal direction.

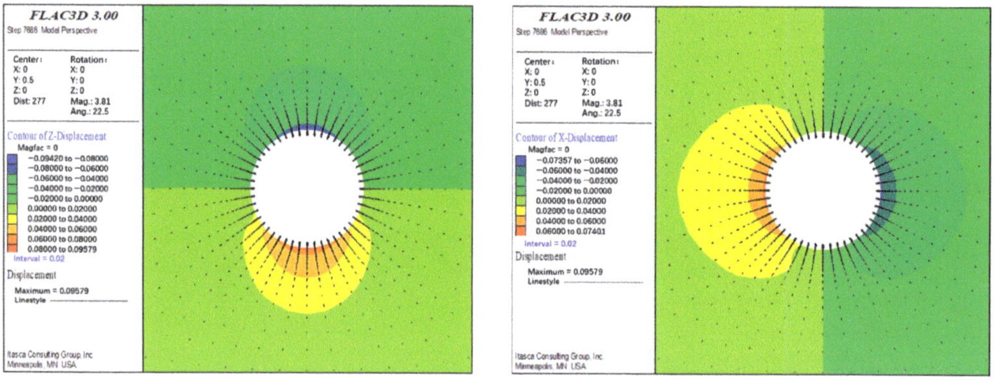

(**a**) (**b**)

Figure 19. Displacement nephogram and resultant displacement vector diagram of the surrounding rock with yielding bolt supporting after excavation of the tunnel: (**a**) vertical direction; (**b**) horizontal direction.

After excavation, maximum deformation occurred at the arch crown. In Scenario One, which was without any form of support, the stabilized settlement of the arch crown reached 11.40 cm, with a deformation of 8.89 cm on the right sidewall. In Scenario Two, which was with conventional bolt support, the stabilized settlement of the arch crown was 9.79 cm, and the sidewall deformation was 8.16 cm. For Scenario Three, which was with yielding bolt support, the arch crown settlement was 9.42 cm, and the sidewall deformation amounted to 7.40 cm. Compared to no support, the yielding bolt support increased the control of crown settlement and sidewall deformation by 17.37% and 16.76%, respectively.

Resultant displacements at the monitoring points are shown in Figure 20, clearly reflecting the rock surface deformation under the three scenarios. It can be inferred that employing the bolt support after tunnel excavation effectively constrained the deformation of the surrounding rock mass. Furthermore, when yielding bolt support was used, the deformation of the rock mass was less compared to using conventional bolt support, indicating that yielding bolts are more effective in controlling rock mass deformation during tunnel excavation.

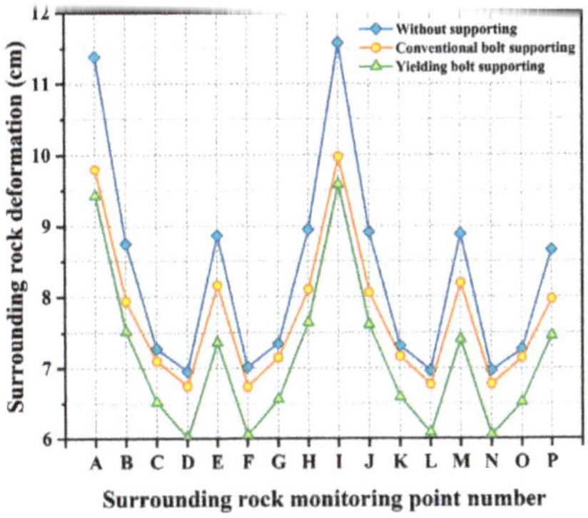

Figure 20. Displacement of the monitoring points on surrounding rock surface of the tunnel.

As shown in Figure 21, it can be seen that, regardless of the support scenario, the positions where the surrounding rock deformation was relatively large were the arch crown and arch base. It can be known from the curve with the error bar that the standard deviations of the surrounding rock deformation at the arch crown and arch base were larger, which indicates that the supporting effect of the yielding bolts was the most obvious at these two locations. This indicates that, within a certain deformation range, the greater the deformation of the surrounding rock, the more suitable it is to adopt yielding bolt support. On the other hand, even after the use of yielding bolt support, the deformation of the surrounding rock at the arch crown and arch base was still greater than that at the other monitoring positions without support. This indicates that reducing or avoiding the deformation of the surrounding rock in soft rock tunnels cannot rely solely on bolt support, it is also necessary to combine it with other technologies.

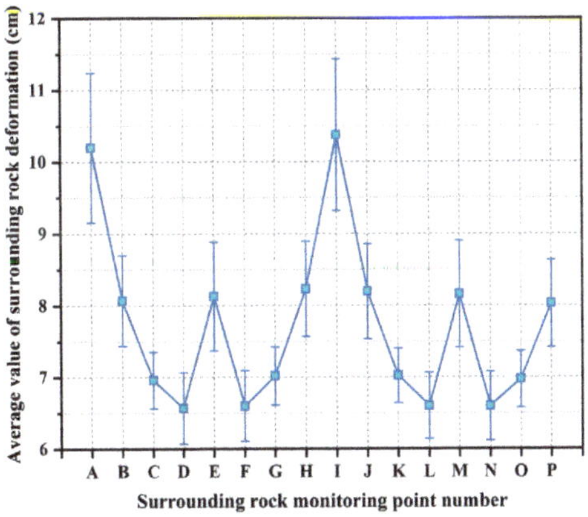

Figure 21. The average displacement of the monitoring points on the surrounding rock surface of a tunnel under three support scenarios.

3.4.2. Comparative Analysis of the Mechanical Characteristics of Bolt

After the stabilization of the tunnel, the axial force distributions of the conventional and yielding bolts were recorded, as shown in Figure 22. The thickness of the black shadow in the figure represents the magnitude of the axial force. The thicker the black shadow is, the greater the axial force at this position of the bolt. Compared to other locations in the tunnel, the deformation of the surrounding rock mass was greater at the arch crown, and the stress characteristics of the bolts at different positions at the arch crown are depicted in

Figure 22. It can be observed from the stress characteristics of the bolts shown in Figure 23 that the position of maximum load-bearing capacity of the bolt was 2.63 m away from the bolt head, and the variation of axial force at this position is illustrated in Figure 24.

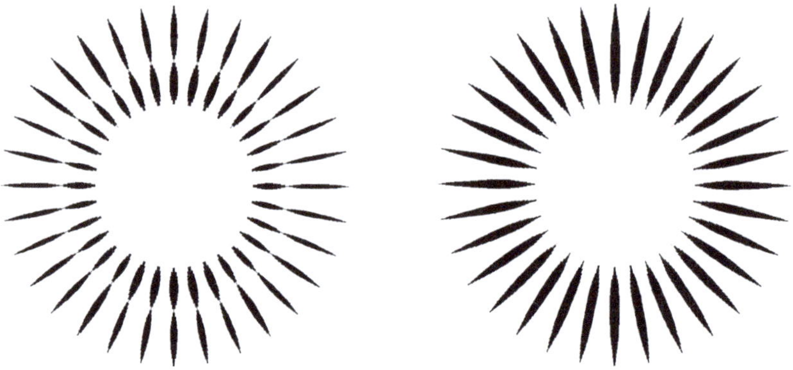

(**a**) (**b**)

Figure 22. The axial force distribution law of bolts: (**a**) conventional bolt; (**b**) yielding bolt.

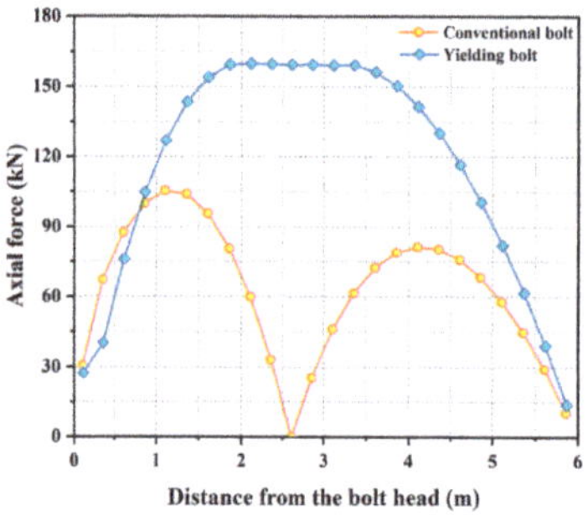

Figure 23. Axial force distribution of the bolt at the tunnel arch crown.

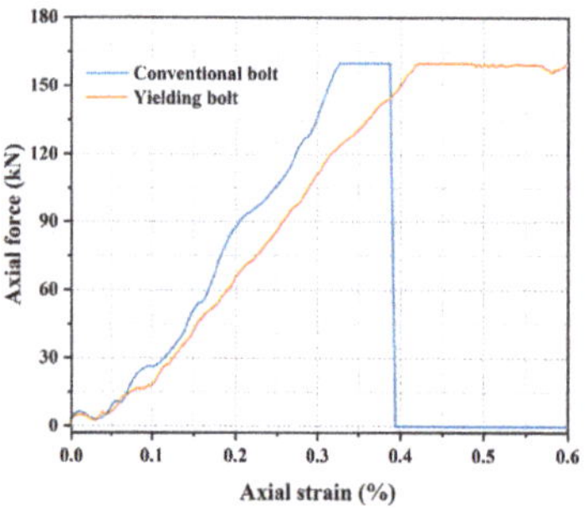

Figure 24. Axial force changes in the maximum bearing position of the bolt at the tunnel arch crown.

During support, the midsection of the bolt bears significantly more load than the two ends, and this is attributed to the stress relaxation that occurs in the rock mass at the tunnel edge, leading to a slightly inward stress concentration. Both bolt types reached yield strength during support. The conventional bolts reached yield strength faster than the yielding bolts. Furthermore, the conventional bolts maintained yield strength for a shorter duration, with the central position of the maximum load-bearing capacity quickly breaking under tension, resulting in an hourglass-shaped axial force distribution. In contrast, as shown in Figure 24, after reaching yield strength, the yielding bolts, due to the good extension effect of the yielding component, were able to sustain yield strength for a longer period, allowing for a greater strain, with the axial force exhibiting a spindle-shaped distribution.

Quantitatively comparing the speed of reaching the yield strength and the strain maintained at yield validates the yielding effect. As shown in Figures 18 and 19, in the initial stages of support, yielding bolts effectively release the deformation energy of the surrounding rock mass and improve the stress distribution in the support structure, fully embodying the support concept of the New Austrian Tunneling Method [39,40]. When the surrounding rock mass undergoes significant deformation, as shown in Figure 22, conventional bolts are prone to failure by breaking under tension, while yielding bolts accommodate larger deformations, continuing to reinforce the rock mass.

3.5. The Influence of the Yielding Bolt Length on Its Support Effect

To investigate the influence of the length of yielding bolts on the support effectiveness, eight different lengths of bolts, namely 2 m, 3 m, 4 m, 5 m, 6 m, 7 m, 8 m, and 9 m, were selected. A plane strain model of the tunnel, as shown in Figure 7, was employed for excavation and support simulation. The proportion of the yielding component of the bolt remained constant, relative to the total length of the bolt. During the simulation process, monitoring was conducted at the 16 positions shown in Figure 9. After the simulation calculations, the deformation of the surrounding rock at the four key locations, such as the arch crown, sidewalls, and arch base, were recorded and are presented in Table 5, and the total deformation of the surrounding rock at the 16 monitoring points under different lengths of bolt support is illustrated in Figure 25.

Table 5. Key point deformation of the surrounding rock with different lengths of bolt support.

Bolt Length (m)	Deformation (cm)			
	Arch Crown	Arch Base	Left Sidewall	Right Sidewall
2	9.99	10.23	7.67	7.64
3	9.68	9.83	7.49	7.48
4	9.45	9.65	7.42	7.38
5	9.39	9.58	7.39	7.33
6	9.42	9.58	7.40	7.36
7	9.41	9.60	7.42	7.37
8	9.43	9.63	7.44	7.43
9	9.49	9.68	7.48	7.47

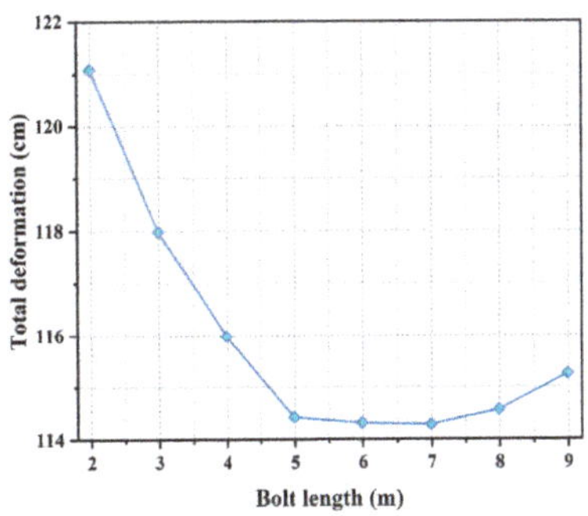

Figure 25. The total deformation of all the monitoring points of the surrounding rock.

With increasing the yielding bolts' length, the deformation at the 16 key positions of the surrounding rock and the total deformation at all points exhibited a trend of initially decreasing, then stabilizing, and then slightly increasing. As the length of the yielding bolts increased from 2 m to 5 m, the deformation of the surrounding rock decreased rapidly. For every additional meter in bolt length, the total deformation at the arch crown, left and right sidewalls, and arch base decreased by 6.13 mm. During this stage, the impact of the bolt length on support effectiveness is significant. As the length of the yielding bolts increased from 5 m to 7 m, the deformation remained relatively stable with minimal changes. Further increasing the length from 7 m to 9 m resulted in a slight increase in the deformation of the surrounding rock, but the magnitude of increase was minimal. Therefore, it can be inferred that the yielding bolts with lengths of 5 m, 6 m, and 7 m exhibited better support effectiveness. Considering utilization and cost, a preliminary determination suggests that 5 m is the optimal length for yielding bolts under these conditions.

3.6. The Influence of the Yielding Bolt Spacing on Its Support Effect

To study the impact of the spacing of yielding bolts on support effectiveness, the optimal length of 5 m was selected. Utilizing the same plane strain model of the tunnel, as shown in Figure 7, excavation and support simulations were conducted. Along the circumferential direction of the tunnel, the spacing of the yielding bolt support was set at 0.75 m, 1 m, 1.25 m, and 1.5 m. After calculations, the deformation of the surrounding rock at the arch crown, sidewalls, and arch base, as well as the total displacement change at the location of maximum deformation in the tunnel under different spacing of bolt support, were obtained, as shown in Table 6. The variation pattern of the resultant displacement at the location of maximum deformation in the tunnel is illustrated in Figure 26.

Table 6. Key point deformation of the surrounding rock with different spacings of bolt support.

Bolt Spacing (m)	Deformation (cm)			
	Arch Crown	Arch Base	Left Sidewall	Right Sidewall
0.75	9.34	9.53	7.51	7.49
1	9.39	9.57	7.39	7.33
1.25	9.94	10.16	7.79	7.77
1.5	9.92	10.13	7.77	7.74

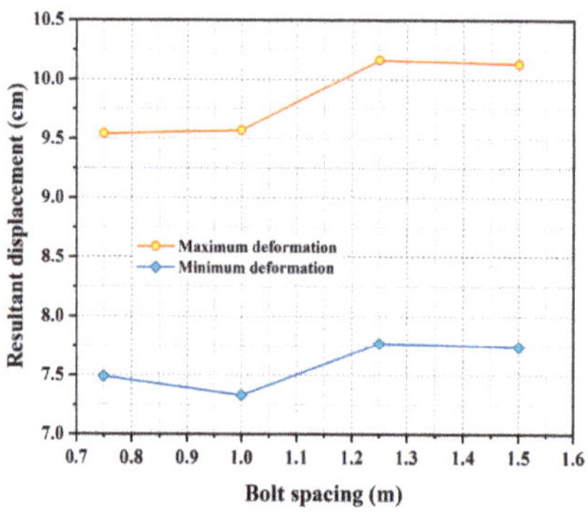

Figure 26. The resultant displacement of the maximum and minimum deformation of the surrounding rock with different spacing support.

As the bolt spacing decreases, the deformation of the surrounding rock at the arch crown, left and right sidewalls, and the arch base followed a pattern of initially stabilizing, then decreasing, and then finally stabilizing again. This pattern was similarly reflected in the resultant displacement at the location of the maximum deformation in the surrounding rock. When the bolt spacing decreased from 1.5 m to 1.25 m, the deformation of the surrounding rock did not decrease but slightly increased. As the bolt spacing decreased from 1.25 m to 1 m, there was a significant reduction in the deformation of the surrounding rock, with the resultant displacement at the location of the maximum deformation decreasing from 10.16 cm to 9.57 cm and the resultant displacement at the location of minimum deformation decreasing from 7.77 cm to 7.33 cm. Further decreasing the bolt spacing from 1 m to 0.75 m resulted in a minimal reduction in the deformation of the surrounding rock, with the resultant displacement at the location of maximum deformation decreasing from 9.57 cm to 9.54 cm. Therefore, bolt spacings of 0.75 m and 1 m demonstrated better support effects, and the difference in support effects between the two was extremely small. Under the same support area, a spacing of 1 m will save a great deal of yielding bolts compared to a spacing of 0.75 m. Considering utilization and cost, a preliminary determination suggests that 1 m is the optimal spacing for yielding bolts under these conditions.

Comparing Tables 5 and 6 and focusing on the two groups of data with the greatest change in arch crown settlement, for yielding bolt lengths of 2 m and 5 m, the difference in deformation at the arch crown was 0.69 cm, and the difference in the arch base uplift was 0.65 cm. With support spacings of 1 m and 1.25 m, the difference in arch crown deformation was 0.55 cm, and the difference in the arch base uplift was 0.59 cm. This indicates that support spacing has a greater impact on the support effectiveness than bolt length.

4. Conclusions

This study employed FLAC3D to meticulously simulate the pull-out tests of conventional bolts and previously developed yielding bolts. The axial force distribution and its dynamic evolution under pull-out conditions for both bolt types was investigated. Additionally, a comprehensive analysis of the shear resistance of the bolts was conducted. A comparative study, within the context of a soft rock tunnel environment, was undertaken to evaluate the support efficacy of conventional bolts against yielding bolts. The findings elucidate the mechanical behavior of yielding bolts under combined loading regimes. In addition, this work bridges critical gaps in the understanding of energy-absorbing support systems. Furthermore, the developed FLAC3D modeling approach provides a versatile tool for future research on advanced reinforcement strategies in geotechnically complex environments. The key conclusions are the following:

(1) The feasibility of simulating the pull-out tests, shear tests, and support effects of yielding bolts using FLAC3D was verified. The simulation results align with the working mechanism and stress characteristics of yielding bolts, establishing a basis for subsequent numerical simulation research.

(2) When subjected to pull-out forces, the maximum axial force at the bolt head decreased gradually toward the inner end of the bolt, with the reduction in the yielding bolts being greater than that of the conventional bolts. As the pull-out force increased, the axial force of the bolt correspondingly increased, with the rate of increase in yielding bolts significantly slower than that of conventional bolts, confirming the high extension rate of yielding bolts.

(3) In the support of a soft rock tunnel, compared to conventional bolts, yielding bolts exhibit a slower rate of reaching yield strength and a longer duration of maintaining yield strength. The central position of the maximum load-bearing capacity of the conventional bolts breaks under tension, resulting in an hourglass-shaped distribution of axial force, whereas yielding bolts can sustain yield strength for an extended period after reaching yield strength, with the axial force displaying a spindle-shaped distri-

bution. The elastic and plastic yielding distance of the yielding bolts are 70.54 mm and 130.65 mm, respectively. The axial force distribution law and yielding distances can provide a theoretical reference basis for engineering applications.

(4) Bolt spacing exerts a greater influence on support effectiveness in soft rock tunnels than bolt length. It was determined that a yielding bolt length of 5 m and a spacing of 1 m × 1 m are the optimal support configuration under the geological conditions discussed in this paper. This optimal support configuration and its determination method can provide a theoretical basis for more complex engineering applications.

Overall, this study addresses critical knowledge gaps regarding yielding bolt behavior under combined loading and provides a validated framework for optimizing energy-absorbing support systems in soft rock tunnels.

Author Contributions: Conceptualization, Y.Z. (Yan Zhu) and E.Z.; Methodology, M.C. and E.Z.; Software, E.Z. and Y.Z. (Yuwei Zhang); Validation, M.C.; Formal Analysis, Y.Z. (Yan Zhu); Investigation, Y.Z. (Yan Zhu) and Y.T.; Resources, M.C.; Data Curation, Y.T. and Y.Z. (Yuwei Zhang); Writing—Original Draft, Y.Z. (Yan Zhu); Writing—Review and Editing, M.C.; Visualization, Y.T. and E.Z.; Funding Acquisition, Y.Z. (Yan Zhu). All authors have read and agreed to the published version of the manuscript.

Funding: This research was funded by the Deep Earth Probe and Mineral Resources Exploration-National Science and Technology Major Project [Grant No. 2024ZD1004502], the National Natural Science Foundation of China [Grant No. 52404101], and Fundamental Research Funds for China Academy of Safety Science and Technology [Grant No. 2024JBKY14].

Institutional Review Board Statement: Not applicable.

Informed Consent Statement: Not applicable.

Data Availability Statement: The original contributions of the study are included in the article, and further inquiries can be directed to the corresponding author.

Conflicts of Interest: The authors declare no conflicts of interest.

References

1. Park, D. Roof Stability Analysis of Cylindrical Tunnels in Hard Soil/Soft Rock with Reduced Tension Strength. *Comput. Geotech.* **2023**, *164*, 105838. [CrossRef]
2. Lin, Q.; Cao, P.; Mao, S.; Ou, C.; Cao, R. Fatigue Behaviour and Constitutive Model of Yellow Sandstone Containing Pre-Existing Surface Crack under Uniaxial Cyclic Loading. *Theor. Appl. Fract. Mech.* **2020**, *109*, 102776. [CrossRef]
3. Zhao, D.; Xia, Y.; Zhang, C.; Tang, C.; Zhou, H.; Liu, N.; Singh, H.K.; Zhao, Z.; Chen, J.; Mu, C. Failure Modes and Excavation Stability of Large-Scale Columnar Jointed Rock Masses Containing Interlayer Shear Weakness Zones. *Int. J. Rock Mech. Min. Sci.* **2022**, *159*, 105222. [CrossRef]
4. Zhao, C.; Liu, J.; Lyu, C.; Chen, W.; Li, X.; Li, Z. Experimental Study on Mechanical Properties, Permeability and Energy Characteristics of Limestone from through-Coal Seam (TCS) Tunnel. *Eng. Geol.* **2022**, *303*, 106673. [CrossRef]
5. Lin, Q.; Cao, P.; Meng, J.; Cao, R.; Zhao, Z. Strength and Failure Characteristics of Jointed Rock Mass with Double Circular Holes under Uniaxial Compression: Insights from Discrete Element Method Modelling. *Theor. Appl. Fract. Mech.* **2020**, *109*, 102692. [CrossRef]
6. Moore, K.S.; Vlachopoulos, N. Effects of Rib Spacing and Grout Annulus on Grouted Rock Bolt Performance Utilizing Fiber Optic Strain Sensing. *Appl. Sci.* **2024**, *14*, 2136. [CrossRef]
7. Lou, J.; Gao, F.; Li, J.; Yuan, G.; Sharifzadeh, M. Effect of Dynamic Loading Conditions on the Dynamic Performance of MP1 Energy-Absorbing Rockbolts: Insight from Laboratory Drop Test. *Int. J. Min. Sci. Technol.* **2023**, *33*, 215–231. [CrossRef]
8. Sharifzadeh, M.; Lou, J.; Crompton, B. Dynamic Performance of Energy-Absorbing Rockbolts Based on Laboratory Test Results. Part I: Evolution, Deformation Mechanisms, Dynamic Performance and Classification. *Tunn. Undergr. Space Technol.* **2020**, *105*, 103510. [CrossRef]
9. Tahmasebinia, F.; Zhang, C.; Canbulat, I.; Vardar, O.; Saydam, S. A Numerical Study of the Behaviour of Fully Grouted Cable Bolts under Static and Dynamic Loading. In Proceedings of the ISRM International Symposium-10th Asian Rock Mechanics Symposium, Singapore, 29 October–3 November 2018.

10. Tai, Y.; Xia, H.; Huang, S.; Meng, J.; Li, W. The Influence Law of Eccentric Load on the Performance of Yielding Bolt. *R. Soc. Open Sci.* **2020**, *7*, 200227. [CrossRef]
11. Pinazzi, P.C.; Spearing, A.J.S.; Jessu, K.V.; Singh, P.; Hawker, R. Mechanical Performance of Rock Bolts under Combined Load Conditions. *Int. J. Min. Sci. Technol.* **2020**, *30*, 167–177. [CrossRef]
12. Kang, H.; Yuan, G.; Si, L.; Gao, F.; Lou, J.; Yang, J.; Dong, S. Mechanical Behavior and Failure Mechanisms of Rock Bolts Subjected to Static-Dynamic Loads. *Int. J. Min. Sci. Technol.* **2024**, *34*, 281–288. [CrossRef]
13. Guo, Y.; Jiang, F.; Zhou, H.; Xiang, X. The Mechanism of the New-Type Yielding Bolt. *Mech. Eng.* **2015**, *37*, 193–196+237. (In Chinese) [CrossRef]
14. Li, Y.; Tannant, D.D.; Pang, J.; Su, G. Experimental and Analytical Investigation of the Shear Resistance of a Rock Joint Held by a Fully-Grouted Bolt and Subject to Large Deformations. *Transp. Geotech.* **2021**, *31*, 100671. [CrossRef]
15. Sun, C.; Liu, C.; Yuan, J. Mechanical Behavior Analysis of Fully Grouted Bolts under Shear throughout the Entire Elastic–Plastic Process. *Eng. Fail. Anal.* **2025**, *171*, 109340. [CrossRef]
16. Morido-García, G.; De Santos-Berbel, C. Structural Performance of Bolted Lateral Connections in Steel Beams under Bending Using the Component-Based Finite Element Method. *Appl. Sci.* **2024**, *14*, 3900. [CrossRef]
17. Wu, K.; Shao, Z.; Qin, S.; Wei, W.; Chu, Z. A Critical Review on the Performance of Yielding Supports in Squeezing Tunnels. *Tunn. Undergr. Space Technol.* **2021**, *115*, 103815. [CrossRef]
18. Wang, T.; Liu, H.; Kang, M.; Zhao, B.; Shen, J.; Li, Y.; Yang, Y. Study on the Synergistic Effect of Primary Support and Surrounding Rock of Large Buried Depth Tunnel in Soft and Fractured Strata. *Appl. Sci.* **2024**, *14*, 2028. [CrossRef]
19. Sun, Y.; Xu, N.; Xiao, P.; Sun, Z.; Li, H.; Liu, J.; Li, B. Characterizing Large Deformation of Soft Rock Tunnel Using Microseismic Monitoring and Numerical Simulation. *J. Rock Mech. Geotech. Eng.* **2025**, *17*, 309–322. [CrossRef]
20. Deng, P.; Liu, Q.; Liu, B.; Lu, H. Failure Mechanism and Deformation Prediction of Soft Rock Tunnels Based on a Combined Finite–Discrete Element Numerical Method. *Comput. Geotech.* **2023**, *161*, 105622. [CrossRef]
21. Xu, J.; Wen, H.; Sun, C.; Yang, C.; Rui, G. Numerical Simulation of Non-Stationary Parameter Creep Large Deformation Mechanism of Deep Soft Rock Tunnel. *Appl. Sci.* **2022**, *12*, 5311. [CrossRef]
22. Peng, W. *FLAC3D Practical Tutorial*; China Machine Press: Beijing, China, 2014. (In Chinese)
23. Li, W.; Wang, L.; Zhang, C.; Yang, X.; Mei, Y.; Shao, X.; Sun, B. Numerical Investigation Study on Tensile-Shear Failure Behavior of Rock Bolts in Inclined Strata Mining Tunnels. *Eng. Fail. Anal.* **2024**, *162*, 108393. [CrossRef]
24. Yan, X.; Liu, Z.; Zhao, Y.; Li, Y.; Chen, W.; Li, M. Experimental and Numerical Study on the Effect of Load Direction on the Bolt Loosening Failure. *Eng. Fail. Anal.* **2024**, *163*, 108574. [CrossRef]
25. Jiang, W.; Xu, G.; Ma, C. Numerical Simulation on Pull-Tests of a Cable by FLAC3D. *J. Harbin Inst. Technol.* **2009**, *41*, 129–133. (In Chinese)
26. Bahrani, N.; Hadjigeorgiou, J. Explicit Reinforcement Models for Fully-Grouted Rebar Rock Bolts. *J. Rock Mech. Geotech. Eng.* **2017**, *9*, 267–280. [CrossRef]
27. Saadat, M.; Taheri, A. A Numerical Study to Investigate the Influence of Surface Roughness and Boundary Condition on the Shear Behaviour of Rock Joints. *Bull. Eng. Geol. Environ.* **2020**, *79*, 2483–2498. [CrossRef]
28. Zhang, S.F.; Feng, J.X.; Han, R. FLAC3D Numerical Simulation Study on the Failure Developing Law of Interface for Geotechnical Prestressed Anchorage Bolt. *Adv. Mater. Res.* **2013**, *671*, 50–53. [CrossRef]
29. Zhang, S.; Jiang, Y.; Zheng, H.; Wu, X.; Chen, H. Shear Characteristics and Damage Mechanisms of the Bolt–Grout Interface under Cyclic Shear Loading. *Comput. Part. Mech.* 2025, *Early Access*. [CrossRef]
30. Li, D.; Ma, S.; Lane, M.; Chang, P.; Crompton, B.; Hagen, S.A. Laboratory Investigations into the Failure Mechanisms of New Yielding and Inflatable Rockbolts Under Axial and Shearing Loading Conditions. *Rock Mech. Rock Eng.* **2023**, *56*, 565–587. [CrossRef]
31. Ma, L.; Xing, L.; Liu, C.; Cui, T.; Qiao, X.; Miao, W.; Kong, P. Research on the Instability Mechanism and Control Technology of Gob-Side Entry in Deep Mines with Soft Rock. *Buildings* **2025**, *15*, 19. [CrossRef]
32. Warren, M.; Antoniou, A.; Stewart, L. A Review of Experimentation and Computational Modeling of Dynamic Bolt Fracture. *J. Constr. Steel Res.* **2022**, *194*, 107293. [CrossRef]
33. Chen, Y.; Wen, G.; Hu, J. Analysis of Deformation Characteristics of Fully Grouted Rock Bolts Under Pull-and-Shear Loading. *Rock Mech. Rock Eng.* **2020**, *53*, 2981–2993. [CrossRef]
34. Wu, X.; Jiang, Y.; Wang, G.; Gong, B.; Guan, Z.; Deng, T. Performance of a New Yielding Rock Bolt Under Pull and Shear Loading Conditions. *Rock Mech. Rock Eng.* **2019**, *52*, 3401–3412. [CrossRef]
35. Shan, Z.; Guo, Q.; Liang, K.; Ji, X.; Chen, L.; Li, L.; Xu, Z. Design Approach and Its Optimization for Bolted Steel Plate Strengthened RC Beams Considering the Effect of Connection Bolt Slip. *Structures* **2023**, *57*, 105045. [CrossRef]
36. Wang, C.; Zheng, X.; Xin, W.; Wang, J.; Liu, L. Investigation of Bolt Support Mechanisms and Parameter Optimization for Hard Roof Control in Underground Mining. *Processes* **2025**, *13*, 94. [CrossRef]

37. Croccolo, D.; De Agostinis, M.; Fini, S.; Khan, M.Y.; Mele, M.; Olmi, G. Optimization of Bolted Joints: A Literature Review. *Metals* **2023**, *13*, 1708. [CrossRef]
38. Ministry of Housing and Urban-Rural Development of the People's Republic of China. General Administration of Quality Supervision, Inspection and Quarantine of the People's Republic of China. In *Standard for Engineering Classification of Rock Masses*; Ministry of Housing and Urban-Rural Development of the People's Republic of China: Beijing, China, 2014. (In Chinese)
39. Aygar, E.B. Evaluation of New Austrian Tunnelling Method Applied to Bolu Tunnel's Weak Rocks. *J. Rock Mech. Geotech. Eng.* **2020**, *12*, 541–556. [CrossRef]
40. Sugimoto, M.; Chen, J.; Sramoon, A. Frame Structure Analysis Model of Tunnel Lining Using Nonlinear Ground Reaction Curve. *Tunn. Undergr. Space Technol.* **2019**, *94*, 103135. [CrossRef]

applied sciences

MDPI

Article

Simulation Tests on Granite Pillar Rockburst

Xinmu Xu [1,2,3,*], Peng Zeng [1,4], Kui Zhao [1,4,*], Daxing Lei [2,3], Liangfeng Xiong [1,4], Cong Gong [1,4] and Yifan Chen [5]

1 School of Resources and Environment Engineering, Jiangxi University of Science and Technology, Ganzhou 341000, China; zengpeng@jxust.edu.cn (P.Z.); xsdream@hotmail.com (L.X.); gongcong041@163.com (C.G.)
2 School of Resources and Civil Engineering, Gannan University of Science and Technology, Ganzhou 341000, China; daxinglei17@163.com
3 Ganzhou Key Laboratory of Mine Geological Disaster Prevention and Control and Ecological Restoration, Ganzhou 341000, China
4 Jiangxi Provincial Key Laboratory of Mining Engineering, Ganzhou 341000, China
5 School of Resources and Safety Engineering, Central South University, Changsha 410000, China; 223123@csu.edu.cn
* Correspondence: xuxinmu90@163.com (X.X.); kuizhao@jxust.edu.cn (K.Z.)

Abstract: Parallelepipeds specimens were made to further investigate the rockburst occurrence mechanism of ore pillars in underground mining units. The investigation was carried out with uniaxial compression systems and real-time testing systems, such as stress, video, and acoustic emission, combined with digital image correlation (DIC) and SEM electron microscope scanning technology, to systematically analyze the evolution of rockburst of ore pillars, strain field characteristics, acoustic emission characteristics, mesoscopic characteristics of the rockburst fracture, morphology of the bursting crater, and debris characteristics. The findings demonstrate that the pillar's rockburst process went through four stages, including the calm period, the particle ejection period, the block spalling period, and the full collapse period. According to DIC digital image correlation technology, the development of cracks in the rock is not obvious during the calm period, but during the small particle ejection and block spalling periods, the microcracks started to form and expand more quickly and eventually reached the critical surface of the rock, resulting in the formation of a complete macro-rockburst rupture zone. During stage I of the test, the rate of acoustic emission events and energy was relatively low; from stages II to IV, the rate gradually increased; and in stage V, the rate of acoustic emission events and energy reached its maximum value at the precise moment the rock exploded, releasing all of its stored energy. The specimen pit section primarily exhibits shear damage and the fracture exhibits shear fracture morphology, while the ejecta body primarily exhibits tensile damage and the fracture exhibits tensile fracture morphology. The location of the explosion pit is distributed on the left and right sides of the middle pillar of the specimen, and the shape is a deep "V". The majority of the rockburst debris is greater than 5 mm, and it mostly takes the shape of thin plates, which is comparable to the field rockburst debris's shape features.

Keywords: pillar rockburst; rectangular chamber; DIC technique; acoustic emission; rockburst crater; rockburst debris

1. Introduction

Along with the advancement of deep underground mining project construction, pillars and point pillars left by mining methods such as room-and-pillar mining, slicing, and subsequent filling mining suffer stress redistribution and stress concentration damage, which cause a high-stress environment and the rock explosion phenomenon [1–3]. A schematic representation of the high-stress occurrence on the ore pillars in a deep hard rock mining system is shown in Figure 1. The mechanism causing rockbursts is incredibly complicated, and there is not a single, well-defined explanation to account for the phenomenon caused by underground ore body mining. An essential component of the deep underground ore body excavation to pay attention to is the study of various underground rock body properties, and the design of mining structure parameters prompted by the pillar-type rockburst problem.

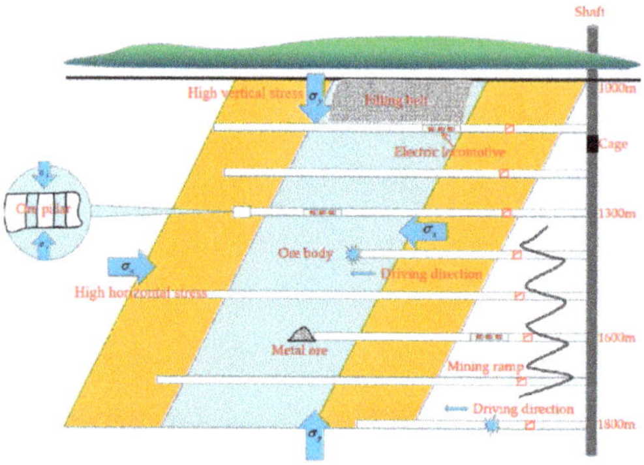

Figure 1. High-stress phenomenon of pillars in deep hard rock mining system.

Since Cook, Singh, Tang and Wang, and Wang et al. used uniaxial compression experiments to investigate rockbursts in the 1960s, many researchers have used various types of mechanical experiments to study the rockburst characteristics of rocks with different mechanical properties, fracture indexes, and structural characteristics in order to study the occurrence mechanism of rockbursts [4–9]. Pei et al. used a GAW-2000 uniaxial electro-hydraulic rigidity testing machine to carry out conventional uniaxial loading and cyclic loading and unloading tests on gabbro and granite, evaluate rockburst tendency, and reveal the energy evolution and sound emission characteristics of rocks with different rockburst tendencies during loading [10]. Huang et al. studied the rockburst process and deformation evolution law of specimens under different crack parameters in detail and discussed the comprehensive discrimination method of the rockburst tendency of granite [11]. Yang et al. conducted rockburst inclination tests on andesite and granite samples with different horizontal stresses using a digital drilling instrument and calculated the fractal dimension of rockburst fragments using fractal theory [12]. Wang et al. created three different types of rock materials for the pillar test model, used uniaxial compression testing and high-speed video cameras to obtain the specimen to produce the rockburst process and the velocity of debris ejection, and used finite element software to calculate the energy for the rupture of the pillars and rockburst mechanism research [13]. In recent years, with the continuous improvement of rock test equipment, the true triaxial rockburst experiment that can more accurately describe the displacement and load changes in different directions of underground engineering has attracted the attention of scholars. Using the self-developed true triaxial rockburst experimental system, Li et al. conducted a series of hysteretic rockburst experiments with different numbers of true triaxial fast unloading surfaces, and studied the rockburst energy storage characteristics, excess energy, excess energy release velocity and AE crack evolution characteristics under different unloading surfaces [14]. He et al., Miao et al., He et al., and He et al. took the initiative to carry out the true triaxial conditions of granite explosion simulation containing a circular cavern structure in order to achieve the impact rock explosion phenomenon of the cavern structure [15–18]. Su et al. used the experimental method of "single face space five face loading" to study the acoustic emission evolution of granite strain-type rockbursts and the ejection process of rockbursts [19–21]. Wang et al. investigated the acoustic emission spatial and temporal entropy of the explosion precursor information in the granite explosion experiment of "true triaxial loading—single surface unloading" [22]. Gong et al. and Si et al. conducted numerous simulation experiments of real triaxial rockbursts in rocks with cavernous structures, achieving the slab cracking flexural rockburst phenomenon in circular caverns, straight-walled arched caverns, rectangular caverns, and other caverns, obtaining results that were consistent with the actual tunnel damage in the field [23–25]. Wang et al. studied the strength failure mechanism of granite and its relationship with high temperature and high difference stress through a true triaxial compression test and evaluated its rockburst tendency according to the ultimate energy storage characteristics of deep-buried high geothermal tunnel granite [26]. Chen et al. conducted true triaxial rockburst tests on rectangular prismatic coarse-grained granite and quantitatively analyzed the relationship between the loading rate, fractal dimension, and total energy dissipation [27]. Si et al. used a real triaxial test system to investigate the impact of the loading rate on rockburst in the surrounding rock of circular tunnels [28]. Feng et al. established an orthogonal anisotropic thin plate mechanical model to derive the critical load value of plate fracture buckling rockburst under bidirectional stress circumstances by conducting acoustic emission monitoring tests [29]. Liang et al. carried out tunnel rock explosion acoustic emission monitoring experiments and analyzed the tunnel rock explosion process macro-damage characteristics as well as rock explosion energy accumulation, release patterns, and the time effect of energy [30]. Li et al. conducted true triaxial rockburst tests on granite samples undergoing different initial damage temperatures, quantitatively analyzed the effects of initial temperature damage on the scale parameters, surface morphology, and microscopic cracks of rockburst fragments, and established the relationship between the peak rockburst strength and fractal dimension of fragments [31]. Sun et al. conducted an experimental investigation of the rockburst occurrence mechanism using infrared imaging and acoustic emission technology [32]. Gong et al. conducted theoretical and experimental investigations on the ejection velocity of shattered rock masses during rockbursts by employing dynamic Moore's circle analysis and true triaxial testing in order to anticipate the destructiveness of rockburst disasters [33]. Lin et al. performed triaxial compression rockburst experiments with various stress gradients, and for precise morphological characterization, they used a scanning electron microscope to describe the damaged surface of the specimen [34]. Wang et al. applied the RF-PA2D program to simulate the spatial distribution characteristics and event sequence characteristics of the microfracture-induced macrofracture process and microfracture-related acoustic emission event sources in rockbursts of ore pillars [35]. Wang et al. used granite to carry out rockburst laboratory tests under true triaxial conditions, combined with the deformation characteristics of granite and acoustic emission technology, and revealed the evolution law of microcracks in the process of rockburst of granite [36]. Using the triaxial compression test of a 1000 m deep borehole in a mountain tunnel in Tibet, Liu et al. conducted an experimental study on rockburst potential evaluation of the strength, deformation, failure, and energy evolution characteristics of granite under the coupling effect of high confining pressure and ground temperature and introduced the energy storage index to discuss the rockburst potential of granite [37]. It can be found that the above experimental studies are mostly focused on single-sided unloading strain rock explosion experiments and rock explosion experiments in the cavern and are less involved in the pillar-type rock explosion experiments. After the excavation of the underground ore body, such as the room-and-pillar mining method, the pillars are often on the four sides of the unloading and axial stress state. At the same time, the post-mining residue is often the co-existence of the room and the pillar. Exploring the whole mining structural unit in the pillar rockburst occurrence mechanism can further reflect the actual stress environment of the pillar rockburst phenomenon.

This paper analyzes the behavior of granite specimens containing double rectangular holes, simulating the excavation of the ore body, the formation of two mine rooms, and a pillar of the quarry unit. The tests were carried out with uniaxial compression equipment on the specimen for vertical loading, simulating the four-sided unloading environment of the pillar and observing the rockburst phenomenon of the inter-pillar. Combined with an acoustic emission system, digital image correlation technology (DIC) and SEM electron microscope scanning technology to record the entire experimental process and collect rockburst debris and burst pit data after the test are used to analyze the results of the rockburst experiments for the prevention and control of mine pillar-type rockbursts and provide an engineering reference.

2. Materials and Methods

2.1. Material Characterization

The granite utilized in this investigation was collected from a quarry in Luotian County, Hubei Province, China. Through the testing of the physical and mechanical properties of the granite, it is known that the average longitudinal wave velocity of the batch of specimens at room temperature is 3548 m/s, the average uniaxial strength is 54.39 Mpa, the average modulus of elasticity is 20.95 Gpa, and the average Poisson's ratio is 0.21. The primary elements of this granite, as determined by EDS point scanning, are Si, O, Al, K, Na, etc., as shown in Figure 2a. The principal mineral compositions of the specimen, as seen by polarized light microscopy on its surface, are quartz, plagioclase, and biotite. As shown in Figure 2b, the specimen is homogeneous and has a decent structure.

(**a**) Granite EDS point sweep (**b**) Granite microstructure

Figure 2. Elemental and mineral composition of granite specimens.

2.2. Specimen Preparation and Testing Methods

Granite blocks with good geometric integrity and rock homogeneity were sliced, and the actual engineering structure was machined into a model specimen with double rectangular holes at a scale of 1/1000. Two holes were opened in the middle of the model to simulate the stope, and a pillar was formed between the holes. The specimen holes are 40 mm in length, 60 mm in width, and 50 mm in height; the pillar between the holes is 30 mm in length, 60 mm in width, and 50 mm in height; and the specimen's total dimensions are 200 mm in length, 200 mm in width, and 50 mm in height. Three groups of specimens were carefully cut, ground, and polished in accordance with the test requirements of international standards to make sure that each specimen's non-perpendicularity and non-parallelism were less than 0.02 mm and that its surface was flawless.

The experiments used the RMT-150C rock mechanics testing system for uniaxial compression to simulate the surrounding rock stresses in a deep hard rock high-stress environment. The RMT-150C rock mechanics test system has a maximum axial pressure of 1000 kN, and the deformation rate ranges from 0.0001 to 1.0 mm/s. Combined with the sensor that comes with the RMT-150C rock mechanics test system, the axial deformation and uniaxial compressive strength are recorded in real time. The tests are equipped with four nano30-type acoustic emission sensors and a PCI-II acoustic emission system with sampling rates of 1 MSPS, 40 dB preamplifier gain, and a 40 dB threshold. Utilizing DIC technology along with camera video, the damage and fracture progression of the specimen surface were captured and examined.

In order to ensure that the granite specimen is in flat contact with the end face of the testing machine and that the force is uniform, rigid pads larger than the size of the specimen are installed between the indenter and the support of the testing machine, and a lubricant is applied to the contact surfaces of the pads in order to minimize the end effect and end noise. The loading mode of the test is displacement control, and the loading rate is 0.3 mm/min. A schematic representation of the specimen loading path and uniaxial compression loading facilities is shown in Figure 3.

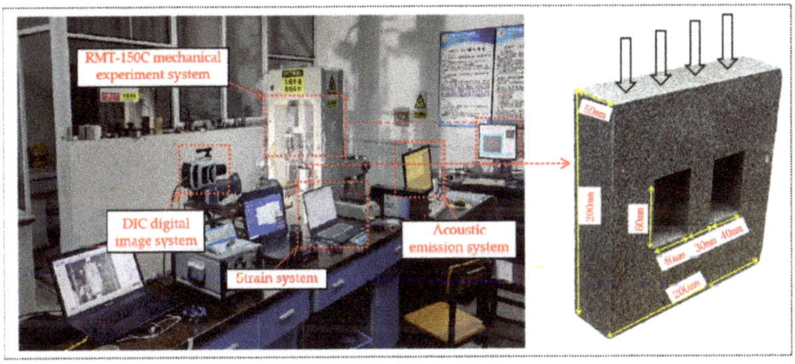

Figure 3. Uniaxial compression loading facility and loading path.

3. Test Results and Analyses

3.1. Evolution of Rockbursts in a Double Rectangular Chamber's Center Pillar

The rockburst process of the middle pillar holding the double rectangular hole specimen went through four phases, as seen in the video of the rockburst test: the calm period, the particle ejection period, the block spalling period, and the full collapse period [23–25]. A thorough explanation of the whole pillar rockburst procedure follows. Until the moment of 242.15 s, the specimen was in a calm period, and the pillar was structurally intact without any damage, as shown in Figure 4a. The phenomenon of the ejection of small particles is shown in Figure 4b,c. When the axial stress continued to be applied at 51.31 MPa, the phenomenon of the ejection of very small particles was seen on the left wall of the pillar. When the axial stress continued to be applied at 51.53 MPa, the phenomenon of the ejection of very small particles was seen on the right wall of the pillar. According to the situation in the calm phase, the pillar structure is still intact, and there have not been any evident damage phenomena as of yet.

According to Figure 4d, the experiment's first instance of block spalling occurred at the pillar's left shoulder at 52.13 MPa; the experimental block spalling event was seen in the right shoulder corner of the pillar at 52.94 MPa, as depicted in Figure 4e. After block spalling occurred in the pillar, the test specimen returned to a stable state after a calm period of about 20 s. The second small particle ejection phenomenon was observed on the left wall of the pillar when the test was loaded to 57.07 MPa, and the small particle ejection phenomenon was observed on the right wall of the pillar when the test was loaded to 57.55 MPa. The second block spalling phenomenon of the experiment was observed in the middle pillar at 57.83 MPa. At 58.50 MPa, the specimen's two holes almost simultaneously witnessed a large number of loose particle fall phenomena. The loose particle fall locations were in the left hole's left and right hole's right shoulder corners, the occurrence of which was accompanied by the production of a large amount of rock debris, which covered the entire hole. The specimen briefly entered a steady state again at this moment. After about 17 S, at 64.21 MPa, the block spalling phenomenon was seen again in the middle pillar of the specimen. At 68.12 MPa, a large area of collapse occurred in the middle column, and the collapsed rock mass covered the whole hole, as shown in Figure 4m. At 70.83 MPa, the rockburst phenomenon occurs in the test, accompanied by a sound of rock cracking, and the phenomena of continuous shot ejection, block collapse, and rockburst appear in the middle pillar in a short time.

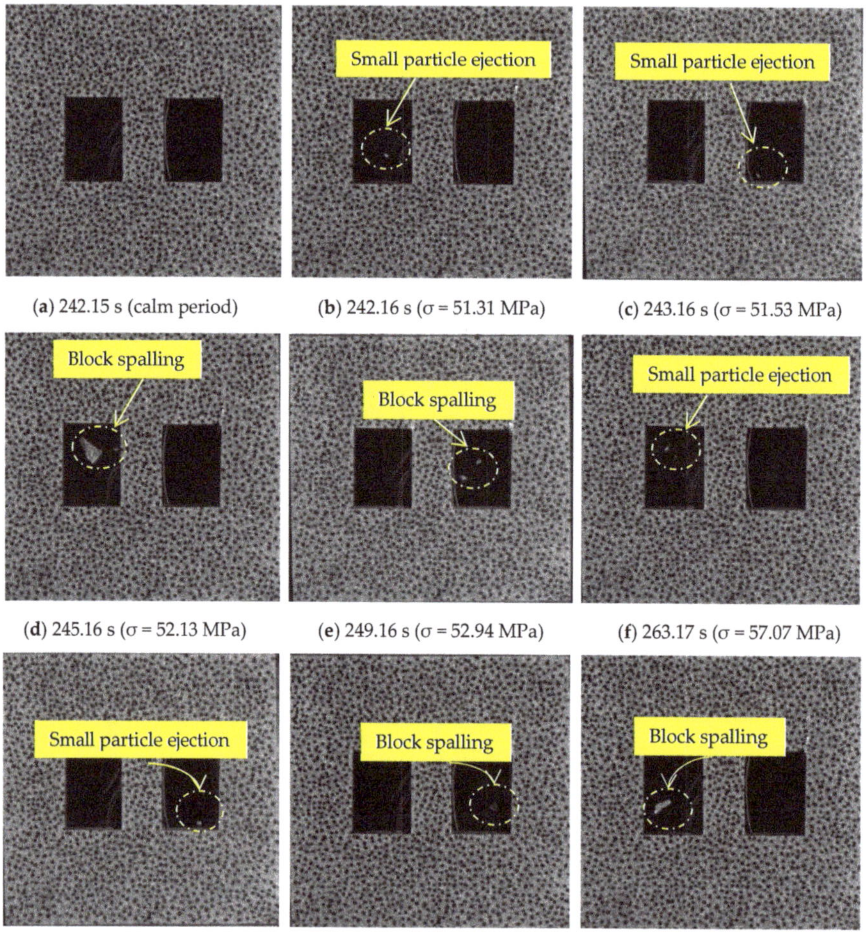

(**a**) 242.15 s (calm period) (**b**) 242.16 s (σ = 51.31 MPa) (**c**) 243.16 s (σ = 51.53 MPa)

(**d**) 245.16 s (σ = 52.13 MPa) (**e**) 249.16 s (σ = 52.94 MPa) (**f**) 263.17 s (σ = 57.07 MPa)

(**g**) 264.17 s (σ = 57.55 MPa) (**h**) 265.17 s (σ = 57.83 MPa) (**i**) 266.17 s (σ = 58.30 MPa)

Figure 4. *Cont.*

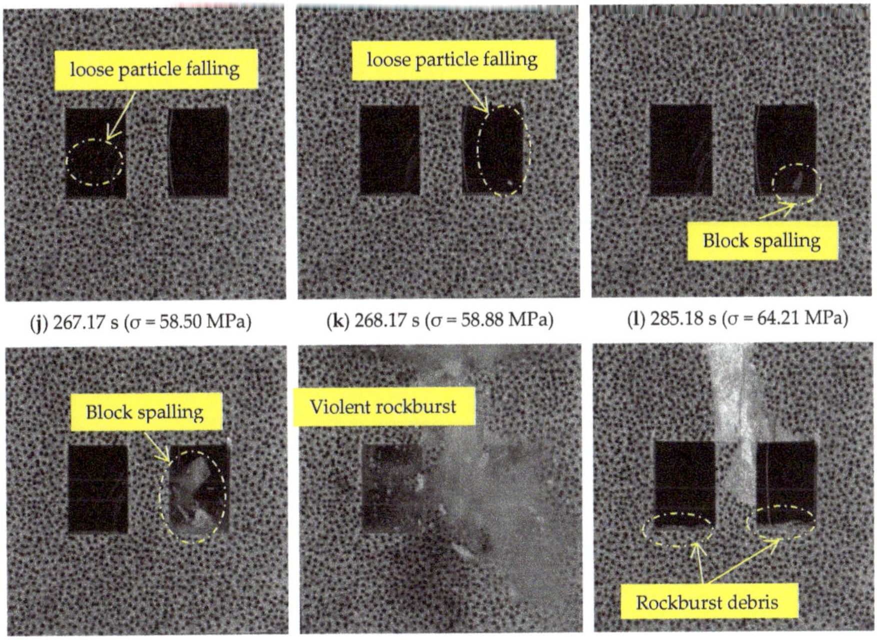

(**j**) 267.17 s (σ = 58.50 MPa) (**k**) 268.17 s (σ = 58.88 MPa) (**l**) 285.18 s (σ = 64.21 MPa)

(**m**) 297.19 s (σ = 68.12 MPa) (**n**) 307.19 s (σ = 70.83 MPa) (**o**) 309.30 s (post-rockburst period)

Figure 4. Rockburst damage process of the pillar between double rectangular holes.

3.2. Characterization of the Strain Field

The DIC technology analysis system monitored the changes in characteristics of the strain field on the specimen surface in real time, and XT-DIC software was used to calculate the spatial distribution of strain and the values of all the strain points of the rock samples throughout the loading process. As shown in Figure 5, the maximum principal strain field cloud diagrams of the five phases of the calm period, particle ejection period, block spalling period, full collapse period, and post-rockburst period produced by the rockburst of the ore pillar are chosen for representation in this paper to aid in analysis.

The contour of the maximum main strain at the surface can effectively characterize the shape and extension behavior of the crack, as shown by the chromatographic bars in the image. The following are the outcomes of strain field calculations used to analyze the specimen during rockburst generation. In Figure 5a, it can be seen that the rock samples' surface deformation is equally distributed throughout the calm period; none of the rock's maximum strains surpass 0.2, and the region of maximum major strain is primarily located in the pillar and at the shoulder corners of the holes. During the particle ejection period, when primary defects and fissures within the rock have largely closed as the stress increases, the pillar starts storing elastic strain energy at a faster pace, and strain aggregation happens. The region of maximum principal strain is at the shoulder corner of the pillar, which is symmetrically distributed, and the specimen begins to show microcracks, which are winged and expand in the direction of maximum principal stress, as shown in Figure 5b. As the granite enters the block spalling phase, the rapid development of localized microfractures created at the shoulder corners makes the macroscopic growth of microcracks more visible. The region of the greatest strain field is banded and corresponds with the site of crack formation in the rock, as shown in Figure 5c. As shown in Figure 5d, the fracture zone has extended through to the critical surface of the rock during the full-scale collapse phase with the unstable and accelerated expansion of microfractures, and a complete macroscopic rockburst rupture zone is formed in and above the pillar with complete failure of the rock sample.

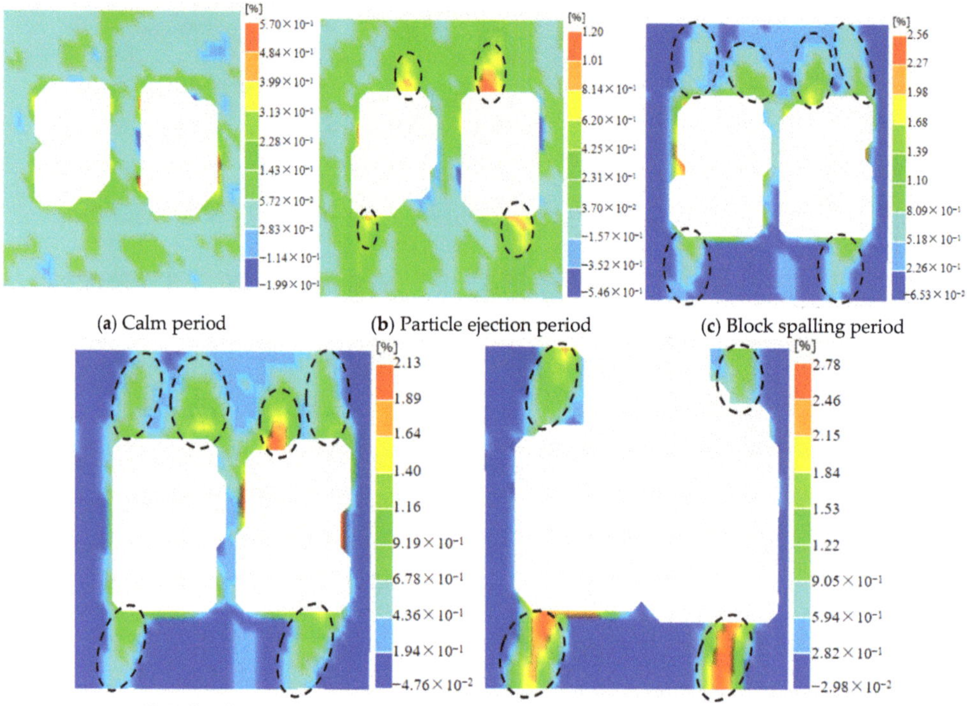

(a) Calm period (b) Particle ejection period (c) Block spalling period

(d) Full collapse period (e) Post-rockburst period

Figure 5. Cloud image of a rockburst strain field (Black circles indicate areas of maximum strain).

3.3. Analysis of Acoustic Emission

Even though the rockburst video and the stress–strain measurement system of the testing apparatus can observe the macroscopic damage relationship of the test from the calm period to the occurrence of the rockburst, the damage process inside the specimen is difficult to reflect. However, with the aid of an acoustic emission system, the internal damage law of the rock can be accurately represented [38,39]. In this section, we investigate the pillar's rockburst process by assessing the number of events and the energy evolution law of acoustic emission, analyzing the rockburst stage information and establishing whether or not the test rock is in a rupture state. As shown in Figures 6 and 7, the event rate and energy rate of the acoustic emission test periodically increase and subsequently decline, revealing a multi-peak condition. The total number of acoustic emission events and cumulative energy grow gradually, reflecting the damage form, damage state, and damage degree at various stages of the rockburst process.

When exposed to uniaxial compression, rocks can be split into five stages, according to Bieniawski's definition of stress–strain curves of rocks under compression: crack closure stage (I), linear elastic deformation stage (II), crack initiation and stable extension stage (III), unstable extension of the crack stage (IV), and destruction and post-peak stage (V) [40,41]. Each of these two stages is separated by a distinct stress value associated with crack evolution, with stress levels corresponding to crack closure stress σ_{cc}, initiation stress σ_{ci}, damage stress σ_{cd}, and peak stress σ_c. In this study, we utilize the volumetric strain method to calculate the typical stresses for each of the two phases, which are 24.3 MPa, 39.5 MPa, 53.1 MPa, and 70.83 MPa [42].

Using the example of Figures 6 and 7, stage I denotes the gradual closure of the specimen's original cracks, porosity, and other flaws under external strain. At this point, only the specimen's surface is distorted to generate microcracks, resulting in a limited number of acoustic emission events and a gradual rise in the total energy of the emissions. New cracks begin to form and spread throughout the specimen as the loading increases in stages II and III. This causes a gradual increase in the number of acoustic emission events and multiple peaks of acoustic emission events, as well as a steady and continuous rise in the total energy of acoustic emission events. In stage IV, stress loading causes the specimen's crack and fracture structure surfaces to rapidly expand. As a result of the quick crack formation, the number of acoustic emission events quickly approaches the maximum value, and the energy rate also sharply approaches the maximum value. The specimen's number of acoustic emission events and energy rate have reached their maximum values in stage V, and the specimen's cumulative number of events and cumulative energy have also reached their maximum values. At this point, the specimen's rockburst phenomenon occurs, and the original stored energy is fully released.

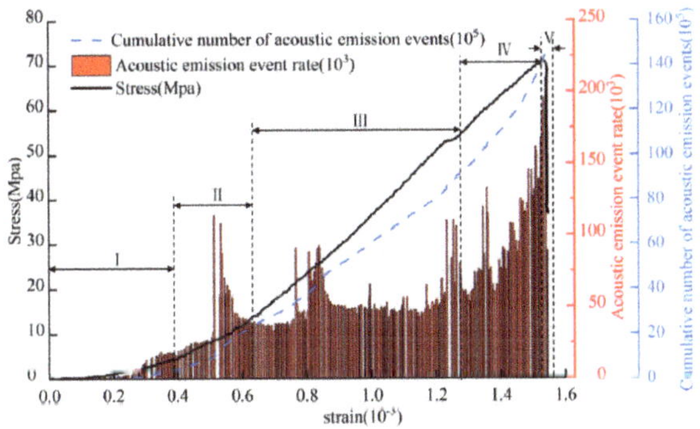

Figure 6. Load path and acoustic emission count.

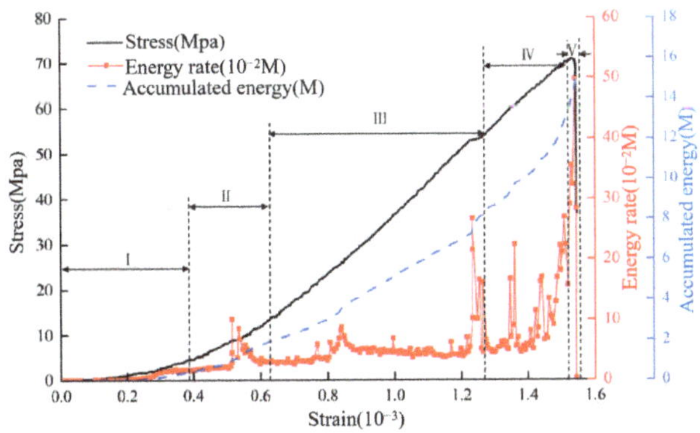

Figure 7. Load path and count of acoustic emission energy.

3.4. *Analysis of Microscopic Characteristics of Pillar Failure*

In order to investigate the mesoscopic fracture mechanism in pillar rockbursts, the kind of grain rupture of rock samples is screened in this section using SEM electron microscope scans [43]. The SEM grayscale maps of the rockburst damage fracture of the pillar and the section of the ejected debris revealed that crystalline powder debris was distributed throughout the majority of the fracture of the specimen burst crater, especially at low-lying scratches or staggered gaps. It mostly occurs as shear damage and exhibits significant shear fracture morphological characteristics, as seen in Figure 8a. The fracture on the rockburst-expelled debris is a tensile fracture, which primarily appears in tensile damage because it is tidy and rarely littered with crystalline powder debris, as seen in Figure 8b. The following SEM grayscale maps show the fine-scale breakage of each mineral. Quartz mainly undergoes intergranular and transgranular fracture, with the fracture morphology mainly showing step and honeycomb shapes, as shown in Figure 9a. Potassium feldspar mainly undergoes intergranular and transgranular fracture damage, with the fracture morphology mainly showing honeycomb and flat shapes, as shown in Figure 9b. Plagioclase mainly undergoes transgranular fracture, and the fracture morphology is mainly ridged, as shown in Figure 9c. Mica mainly undergoes transgranular fracture, and its fracture morphology is mainly stepped and lamellar, as shown in Figure 9d.

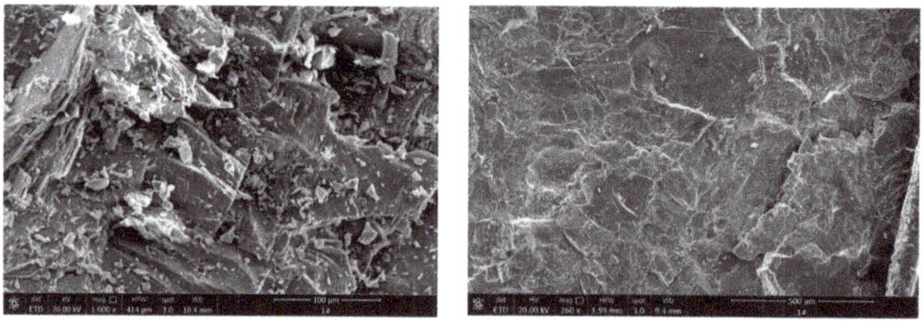

(**a**) Crater fracture (**b**) Ejection debris fracture

Figure 8. SEM typical patterns of a fracture in the rockburst zone.

(**a**) Quartz section morphology (**b**) Potassium feldspar section morphology

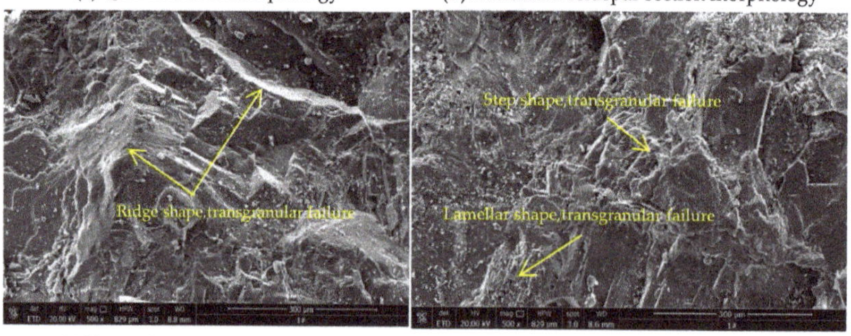

(**c**) Plagioclase section morphology (**d**) Mica section morphology

Figure 9. Typical mineral microfracture mode patterns in rockburst samples as observed by SEM.

3.5. Characterization of Burst Crater Morphology

Figure 10 shows the whole specimen as well as the damage to the pillar after the rockburst test. Based on the figure, it can be seen that the specimen mostly damages the middle pillar and the rock body above the pillar, which is quite similar to a phenomenon that occurs in the field of engineering. It is quite simple to destroy the inter-pillar when the specimen is subjected to axial vertical stress and the intermediate pillar of the mining unit structure is left too small. The next destructive locations are the sidewalls and corners of the holes, both of which are highly susceptible to stress concentrations. The pillar still included some undropped rock at the conclusion of the rockburst test, and a significant amount of rock and debris had gathered at the cave chamber's bottom. The main damage location of the test is in the form of an obvious deep "V", which is comparable to the breakage pattern of the actual ore pillars in the field, as shown in Figure 11. The rockburst test results show that the damage process for the rockburst occurs from the surface to the interior of the specimen, as evidenced by the chamber from the hole corner sprouting cracks, which eventually extended through to the specimen's boundary and split the entire specimen into six sizable blocks. The internal wall of the hole undergoes a process of destruction of the surrounding rock when the loading level is low, from the calm period to the spalling of the rock mass, in which small particles and blocks are displaced from the parent body. At the same time, the internal part of the test specimen also experiences a process of fissure budding, fissure expansion and extension, and fissure infiltration inward. The cracks formed in the pre-test stage expand immediately under higher loading levels, and a large range of shear failure is formed in the middle column, resulting in a rockburst phenomenon. The damage degree in the rockburst area is great, and the rockburst occurs for a short time. As can be seen, the specimen produced a fissure in the rock body that provided the necessary conditions for a rock explosion to occur. Additionally, the fissure's spatial location in relation to the explosion crater's location also influences the crater's location and size.

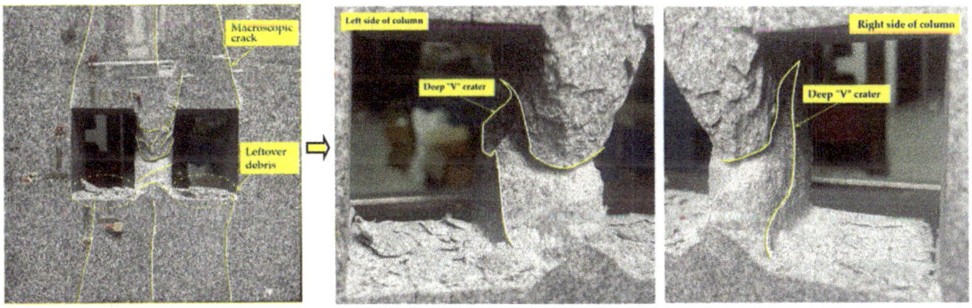

Figure 10. Illustration of granite rockburst damage.

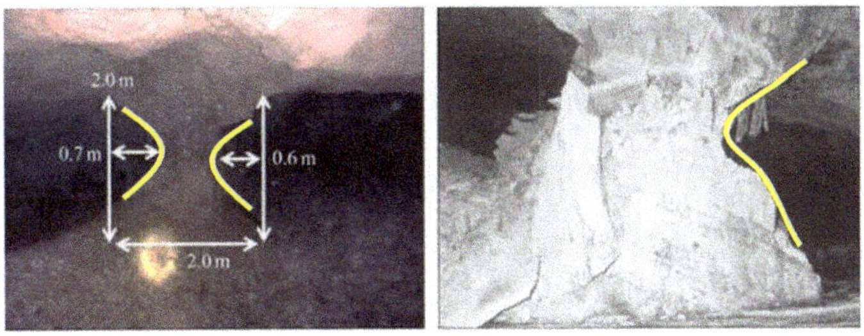

Figure 11. Illustrations of the location's damaged rock pillars (Yellow lines indicate breakage patterns).

3.6. Analysis of Debris Characteristics

To some extent, the rockburst intensity can be determined from the bulk and grain size distribution of the debris [44,45]. In order to conduct this test, debris from the holes left behind after the rockburst of the specimen and from additional injection sites was collected. The debris was then processed using a sifter and was sieved and weighed for each grain size range. The test result debris was divided into seven groups and sieved into sizes of 10, 5, 2, 1, 0.5, 0.25, and 0.075 mm, as illustrated in Figure 12. As shown in Table 1, the debris was also weighed for each grain size range to determine the quality of rockburst debris and the proportion of the microfine quality to the total quality. The table indicates that the majority of the debris is made up of huge fragments larger than 5 mm, which make up more than 81% of the average quality. It suggests that during the destructive process of rock bursting, huge pieces of debris predominate the debris field, and the shape of the debris is primarily that of thin plates, which is comparable to instances in the engineering industry, as seen in Figure 13.

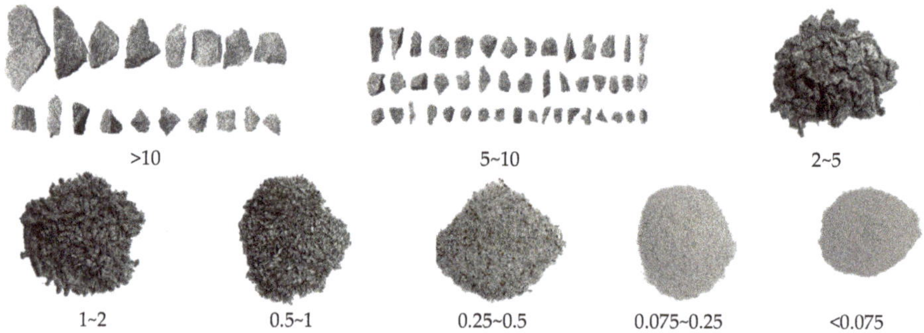

Figure 12. Rockburst debris.

Table 1. Quality distribution of rockburst debris.

Specimen Number	Quality Ratio/%							
	<0.075	0.075~0.25	0.25~0.5	0.5~1.0	1~2	2~5	5~10	>10
S-1	0.015	0.032	0.022	0.0347	0.0345	0.0507	0.1339	0.6772
S-2	0.012	0.015	0.032	0.0447	0.0045	0.0607	0.1239	0.7072
S-3	0.007	0.01	0.062	0.0427	0.0245	0.0407	0.1439	0.6692
Average value	0.0113	0.019	0.0386	0.0407	0.0211	0.0507	0.1339	0.6845

Figure 13. Comparison between test rockburst debris and engineering debris.

The particle size distribution of the debris can be fitted and statistically analyzed with the Weibull distribution function, which is expressed as follows [46]:

$$G_{weibull}(x, \lambda, k) = 1 - \exp\left[-\left(\frac{x}{\lambda}\right)^k\right] \quad (1)$$

where λ is the shape parameter, k is the size parameter, and x is the maximum length value of the debris particle group. The more fragmented the rockburst debris, the smaller the value of the size parameter k, which indicates the average value of the size of the debris. Figure 14 displays the fitted curve of the Weibull distribution of rockburst debris, and Table 2 displays the size parameter k derived from the fitted curve of the Weibull distribution.

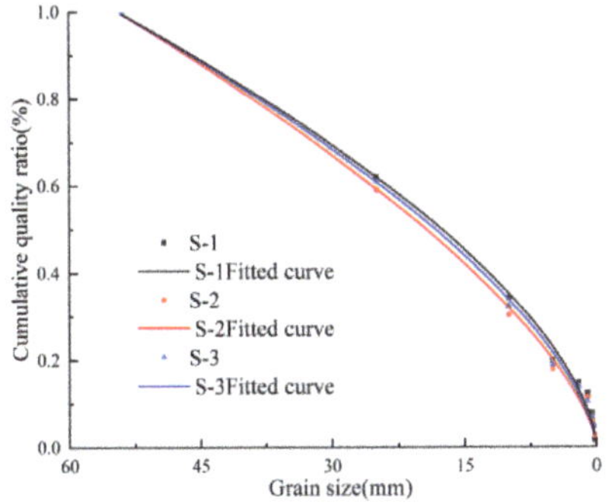

Figure 14. Weibull distribution curve of rockburst debris.

Table 2. Curve parameters fitted by the Weibull function.

Specimen Number	Parameter Combination	K Value
S-1	Granularity—quantity	0.61
S-2	Granularity—quantity	0.64
S-3	Granularity—quantity	0.67
Average value	Granularity—quantity	0.64

4. Discussion

An essential condition for ensuring safe mining in underground mines is the design of suitable structural parameters for the stope. We analyzed the rockburst occurrence mechanism and parts of ore pillars in different span–height ratio stopes, which can provide a strong basis for the prevention and control of rockbursts in the stope.

The study of the rectangular aperture was carried out using the mapping conditions in Figure 15, and the span–height ratios a/b were selected to be 1, 1.5, 2, 3, and 5. The left hole is used as an illustration to investigate the impact of the span–height ratio on the stress distribution of rectangular chamber mining, and the position of the center pillar of the stope is shown by the AB connecting line. Take the midpoint E (where $x = 0$) of the side as the origin; the θ variable is the horizontal axis, and the stress concentration factor σ is the vertical axis. The calculations in the literature are analyzed to obtain the stress concentration factors at θ = 0°, 15°, 30°, 45°, 60°, 75°, and 90° [47], and the fitted curves are plotted in Figure 15. It is feasible to analyze the stresses in the intermediate pillar and the rest of the chamber by choosing the pore-side stress distribution in the first quadrant because the loading conditions, section geometry, and elastic analytical solutions for the pore-side stresses are all symmetric about the x and y axes.

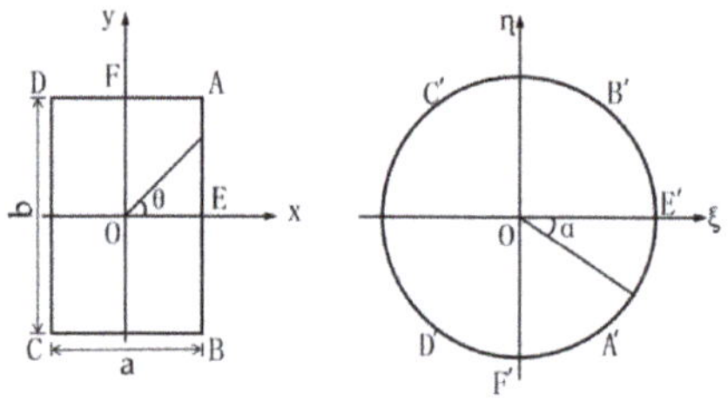

Figure 15. Mapping the relationship between the rectangular stope and the unit circle (The letters A, B, C, D, E and F represent six points in different directions of the rectangle, and A′, B′, C′, D′, E′ and F′ represent A, B, C, D, E and F mapping points in the circle).

It is necessary to make a clarification here. In order to avoid creating an infinite situation of stress calculations at the corner points, the corner points have been processed with rounded transitions. The absolute dimensions of the rectangular sections of the five selected span–height ratios are different, but the mapping angle of each analyzed point ranges from 0 to 90°, and the values of the rectangular corner points are different.

According to Figure 16, the stresses on the pillars in the case of uniaxial compression rise with the span–height ratio and are mainly manifested as compressive stresses. Contrarily, the stress in the top plate changes from compressive to tensile stress as θ increases, which shows that the span–height ratio of the cavern has a great influence on the stress distribution in all directions of the cavern. With an increase in the span–height ratio, the stress concentration factor at the midpoint of the intermediate pillar (when the mapping angle θ = 0°) rises, and the stress concentration coefficient at the midpoint of the roof plate of the rectangular chamber (when the mapping angle θ = 90°) slightly falls, but the absolute value of the stress is higher, which is basically in agreement with the laws reflected in the literature [48].

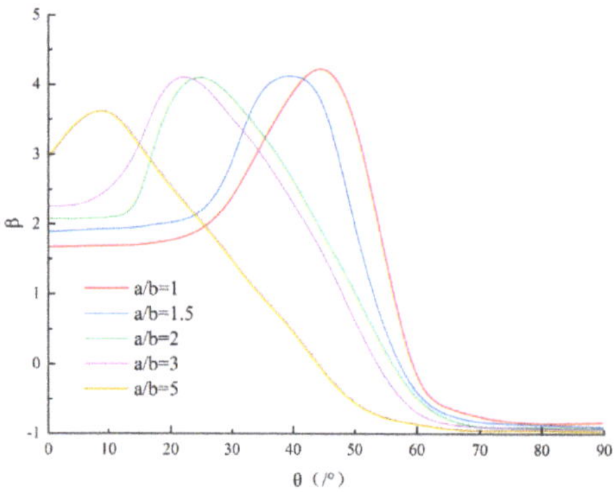

Figure 16. Stress concentration factor around the rectangular cavern with the different span–height ratio.

5. Conclusions

In this study, uniaxial compression rockburst tests were carried out on rock specimens containing a double rectangular cavern structure. By analyzing and discussing the real-time video, acoustic emission, bursting crater, debris, and other information obtained during the test, the main conclusions are as follows:

(1) The test simulated the underground mine pillars rockburst phenomenon. The real engineering structure was scaled down to 1/1000 and processed into a model specimen with a double rectangular chamber. The simulation test results and the real rockburst

phenomenon were in agreement, demonstrating the validity of the experimental model and experimental methodology.

(2) The rockburst of the ore pillars is divided into four stages, the calm period, the small particle ejection period, the block spalling period, and the full collapse period, and the displacement field monitored by DIC corresponds to the above four stages. Starting from the small particle ejection phase, the rock fissures gradually developed into macrorupture zones through the processes of sprouting, extending, and penetrating inward, and finally formed the rockburst crater contour, which demonstrated the reliability of the simulation experiments.

(3) The event rate and energy rate of the acoustic emission test repeatedly increased and then decreased, exhibiting a multi-peak state. The cumulative number of acoustic emission events and the cumulative energy displayed stepwise growth. The test's acoustic emission event rate and energy rate are relatively low in stage I, gradually increase from stage II to stage IV, and reach their maximum values in stage V.

(4) Large debris fragments greater than 5 mm make up the majority of rockburst debris, and the debris's morphology primarily resembles a thin plate, which is comparable to the shape features of engineering rockburst debris. The gradation of the rockburst debris is uniform, and the average value of the fractal dimension is 2.02. Burst crater damage is mainly in the middle pillar and the pillar above the part, the middle pillar after the destruction of the obvious deep "V" shape.

Author Contributions: X.X.: conceptualization, methodology, writing—original draft, investigation, and data Curation; K.Z.: supervision, conceptualization, and writing—review and editing; P.Z.: resources and data curation; L.X.: conceptualization and methodology; D.L.: investigation and formal analysis; C.G.: resources, data curation, and writing—review and editing; Y.C.: software and visualization. All authors have read and agreed to the published version of the manuscript.

Funding: The authors gratefully acknowledge financial support from the National Natural Science Foundation of China (52164004, 52104086), the key research and development program of Jiangxi Province, China (20212BBG71009), the Science and technology project of Jiangxi Provincial Department of Education, China (GJJ2203604, GJJ2200824), and the post-doctoral research program of Jiangxi Province, China (2020KY39). We thank the anonymous reviewers for their comments on our paper.

Institutional Review Board Statement: Not applicable.

Informed Consent Statement: Not applicable.

Data Availability Statement: Some or all data or models that support the findings of this study are available from the corresponding author upon reasonable request.

Conflicts of Interest: The authors declare no conflicts of interest.

References

1. He, M.C.; Xie, H.P.; Peng, S.P. Study on rock mechanics in deep mining engineering. *Chin. J. Rock Mech. Eng.* **2005**, *24*, 2803–2813.
2. Xie, H.P.; Gao, F.; Ju, Y. Research and development of rock mechanics in deep ground engineering. *Chin. J. Rock Mech. Eng.* **2015**, *34*, 2161–2178.
3. Li, X.B.; Zhou, J.; Wang, S.F. Review and practice of deep mining for solid mineral resources. *Chin. J. Nonferrous Met.* **2017**, *27*, 1236–1262.
4. Cook, N.G.W. The basic mechanics of rockbursts. *J. S. Afr. Inst. Min. Metall.* **1963**, *63*, 71–81.
5. Singh, S.P. The infuence of rock properties on the occurrence and control of rockbursts. *Min. Sci. Tech.* **1987**, *5*, 11–18. [CrossRef]
6. Singh, S.P. Burst energy release index. *Rock. Mech. Rock Eng.* **1988**, *21*, 149–155. [CrossRef]
7. Singh, S.P. Classifcation of mine workings according to their rockburst proneness. *Min. Sci. Tech.* **1989**, *8*, 253–262. [CrossRef]
8. Tang, L.Z.; Wang, W.X. New rock burst proneness index. *Chin. J. Rock. Mech. Eng.* **2002**, *21*, 874–878.
9. Wang, J.A.; Park, H.D. Comprehensive prediction of rockburst based on analysis of strain energy in rocks. *Tunn. Undergr. Space Technol.* **2001**, *16*, 49–57. [CrossRef]

10. Pei, F.; Ji, H.; Zhao, J.; Geng, J. Energy evolution and AE failure precursory characteristics of rocks with different rockburst proneness. *Adv. Civ. Eng.* **2020**, *2020*, 8877901. [CrossRef]
11. Huang, D.; Li, Y. Experimental investigation and failure characteristics of slit-cut method in rockburst prevention. *Eng. Fail. Anal.* **2023**, *150*, 107344. [CrossRef]
12. Yang, B.; He, M.; Xiao, Z.; Zhao, J.; Zhang, Y. Effect of horizontal stress on fractal characteristics of rockburst fragments in coal mining. *Energy* **2023**, *281*, 128181. [CrossRef]
13. Wang, Z.D.; Li, L.Y.; Chen, T. Study of energy release in model tests on pillar rockburst. *Rock Soil. Mech.* **2018**, *39*, 177–185.
14. Li, J.Y.; Liu, D.Q.; He, M.C.; Yang, J.S. Experimental study on excess energy characteristics of true triaxial multi-surface fast unloading rockburst. *J. Cent. South. Univ.* **2024**, *31*, 1671–1686. [CrossRef]
15. He, M.C.; Miao, J.L.; Li, D.J. Experimental study on rockburst processes of granite specimen at great depth. *Chin. J. Rock Mech. Eng.* **2007**, *26*, 865–876.
16. Miao, J.L.; Li, D.J. Acoustic emission characteristics of granite under strain rockburst test and its micro-fracture mechanism. *Chin. J. Rock Mech. Eng.* **2009**, *28*, 1593–1603.
17. He, M.C.; Liu, D.Q.; Gong, W.L. Development of a testing system for iMPact rockbursts. *Chin. J. Rock Mech. Eng.* **2014**, *33*, 1729–1739.
18. He, M.C.; Li, J.Y.; Ren, F.Q. Experimental investigation on rockburst ejection velocity of unidirectional double-face unloading of sandstone with different bedding angles. *Chin. J. Rock Mech. Eng.* **2021**, *40*, 433–447.
19. Su, G.S.; Jiang, J.Q.; Feng, X.T. Experimental study of ejection process in rockburst. *Chin. J. Rock Mech. Eng.* **2016**, *35*, 1990–1999.
20. Su, G.S.; Hu, L.H.; Feng, X.T. True triaxial experimental study of rockburst process under low frequency cyclic disturbance load combined with static load. *Chin. J. Rock Mech. Eng.* **2016**, *35*, 1309–1322.
21. Su, G.S.; Yan, S.Z.; Yan, Z.F. Evolution characteristics of acoustic emission in rockburst process under true-triaxial loading conditions. *Rock Soil. Mech.* **2019**, *40*, 1673–1682.
22. Wang, C.L.; Liao, Z.F.; Li, C.F. Experimental investigation of dynamic characteristics of AE spatio-temporal entropy for granitic rockburst. *J. Min. Saf. Eng.* **2019**, *36*, 626–633.
23. Gong, F.Q.; Luo, Y.; Si, X.F. Experimental modelling on rockburst in deep hard rock circular tunnels. *Chin. J. Rock Mech. Eng.* **2017**, *36*, 1634–1648.
24. Gong, F.Q.; Luo, Y.; Liu, D.Q. Simulation tests on spalling failure in deep straight-wall-top-arch tunnels. *Chin. J. Rock Mech. Eng.* **2019**, *41*, 1091–1100.
25. Gong, F.Q.; Wu, W.X.; Li, T.B. Simulation experimental study of spalling failure of surrounding rock of rectangular tunnel of deep hard rock. *Rock Soil. Mech.* **2019**, *40*, 2085–2098.
26. Wang, H.; Liu, Z.; Lin, C.; Yang, Q.; Hamdi, E. Non-monotonic effect of differential stress and temperature on mechanical property and rockburst proneness of granite under high-temperature true triaxial compression. *Geomech. Geophys. Geo-Energy Geo-Resour.* **2024**, *10*, 162. [CrossRef]
27. Chen, Z.Y.; Su, G.S.; Jiang, J.Q. Experimental study on energy dissipation of fragments during rockburst. *Bull. Eng. Geol. Environ.* **2019**, *78*, 5369–5386. [CrossRef]
28. Si, X.F.; Huang, L.Q.; Gong, F.Q.; Liu, X.L.; Li, X.B. Experimental investigation on infuence of loading rate on rockburst in deep circular tunnel under true-triaxial stress condition. *J. Cent. South. Univ.* **2020**, *27*, 2914–2929. [CrossRef]
29. Feng, F.; Li, X.B.; Li, D.Y. Mechanism and control strategy of buckling rockbursts of orthotropic slab. *Chin. J. Geotech. Eng.* **2017**, *39*, 1302–1311.
30. Liang, P.; Zhang, Y.B.; Tian, B.Z. Experimental study on energy evolution characteristics of rockburst process in roadway l of Rock. *Chin. J. Rock Mech. Eng.* **2019**, *38*, 736–746.
31. Li, C.; Li, D.; Liu, D.; Hu, B.; Du, R. Experimental investigation on fractal characteristics of rockburst fragments for granite experiencing initial temperature damage. *Eng. Fail. Anal.* **2024**, *166*, 108842. [CrossRef]
32. Sun, X.M.; Xu, H.C.; He, M.C. Experimental investigation of the occurrence of rockburst in a rock specimen through infrared thermography and acoustic emission. *Int. J. Rock Mech. Min. Sci.* **2017**, *93*, 250–259. [CrossRef]
33. Gong, W.L.; Wang, H.; He, M.C. Theoretical and experimental study on rock block ejection velocityfor rock burst found in deep mining. *J. China Coal Soc.* **2015**, *40*, 2269–2278.
34. Lin, M.Q.; Zhang, L.; Liu, X.Q.; Xia, Y.Y.; Zhang, D.J.; Peng, Y.L. Microscopic analysis of rockburst failure on specimens under gradient stress. *Rock Soil. Mech.* **2020**, *41*, 2984–2992.
35. Wang, S.Y.; Tang, C.A.; Xu, T. Numerical simulation on acoustic emission during pillar rock burst. *Chin. J. Nonferrous Met.* **2003**, *13*, 754–759.
36. Wang, C.L.; Cao, C.; Li, C.F.; Chuai, X.S.; Zhao, G.M.; Lu, H. Experimental investigation on synergetic prediction of granite rockburst using rock failure time and acoustic emission energy. *J. Cent. South. Univ.* **2022**, *29*, 1262–1273. [CrossRef]
37. Liu, Z.; Wang, H.; Li, Y.; Wang, X.; Selvadurai, A.P.S. Triaxial compressive strength, failure, and rockburst potential of granite under high-stress and ground-temperature coupled conditions. *Rock Mech. Rock. Eng.* **2023**, *56*, 911–932. [CrossRef]

38. Zhao, K.; Xiang, W.B.; Zeng, P.; Yang, D.X.; Wu, W.K.; Gong, C.; Yang, X.D. The status and prospects of rock acoustic emission Kaiser effect. *J. Met. Mine* **2021**, *2021*, 94–105.
39. Zeng, P.; Cao, W.; Zhao, K.; Gong, C.; Song, L.; Liu, G.Z.; Li, Y.D. Strength evolution and acoustic emission characteristics of tailings consolidated backfill mixed with mineral processing wastewater. *J. China Coal Soc.* **2024**, *49* (Suppl. S1), 167–181.
40. Bieniawski, Z.T. Mechanism of brittle fracture of rock: Part I—Theory of the fracture process. *Int. J. Rock Mech. Min. Sci. Geomech. Abstr.* **1967**, *4*, 395–406. [CrossRef]
41. Bieniawski, Z.T. Mechanism of brittle rock fracture. *part II. experimental studies. Int. J. Rock Mech. Min. Sci.* **1967**, *4*, 407–423. [CrossRef]
42. Martin, C.D. The progressive fracture of Lac du Bonnet granite. *Int. J. Rock Mech. Min. Sci. Geomech. Abstr.* **1994**, *31*, 643–659. [CrossRef]
43. Liu, Y.P.; Deng, H.; Huang, R.Q. Rock mechanics test of plate fracture structure and characterization of fine morphology of rupture fracture. *Chin. J. Rock Mech. Eng.* **2015**, *34*, 3852–3861.
44. Xia, Y.Y.; Lin, M.Q.; Liao, L.L. Fractal characteristic analysis of fragments from rockburst tests of large-diameter specimens. *Chin. J. Rock Mech. Eng.* **2014**, *33*, 1358–1365.
45. He, M.C.; Wang, Y.; Su, J.S. Analysis of fractal characteristics of fragment of sandstone iMPact rock burst under static and dynamic coupled loads. *J. China Univ. Min. Technol.* **2018**, *47*, 699–705.
46. Li, D.J.; Jia, X.N.; Miao, J.L. Analysis of fractal characteristics of fragment from rockburst test of granite. *Chin. J. Rock Mech. Eng.* **2010**, *29* (Supp. S1), 3280–3289.
47. Shi, G.P.; Zhu, J.H.; Li, B.H. Elastic analysis of stresses at hole edges of rectangular roadways. *Geotechnics* **2014**, *35*, 2587–2601.
48. Li, L.F.; Yu, X.B.; Hou, K.P. Concentration factor of stress around noncircular underground roadways. *Met. Mine* **2008**, *3*, 80–82.

applied sciences

MDPI

Article

Damage Evolution and Failure Precursor of Rock-like Material Under Uniaxial Compression Based on Strain Rate Field Statistics

Jin Jin [1,2], Ping Cao [3], Jun Zhang [2,*], Yanchao Wang [2], Chenxi Miao [1,2], Jie Li [4] and Xiaohong Bai [1,*]

[1] College of Civil Engineering, Taiyuan University of Technology, Taiyuan 030024, China; jinjin@tyut.edu.cn (J.J.); miaochenxi@tyut.edu.cn (C.M.)
[2] Transportation Industry Key Laboratory of Highway Construction and Maintenance Technology in Loess Area, Shanxi Province Transportation Technology Research and Development Co., Ltd., Taiyuan 030032, China; wangyanchao0602@hotmail.com
[3] School of Resources and Safety Engineering, Central South University, Changsha 410083, China; pcao_csu@sina.com
[4] College of Vehicle and Transportation Engineering, Taiyuan University of Science and Technology, Taiyuan 030024, China; tycomelj@163.com
* Correspondence: zj_sxjt@hotmail.com (J.Z.); bxhong@tyut.edu.cn (X.B.)

Abstract: In rock engineering, it is crucial to collect and analyze precursor information of rock failure. This paper has attempted to study the strain rate field of rock-like material to obtain the precursor information of its failure. Based on the available laboratory experiments, the intact BPM (bonded-particle model) and other BPMs with a single open prefabricated flaw were simulated by PFC (Particle Flow Code). The volume strain rate field data before the peak stress have been obtained from two hundred measurement circles across each model. The strain rate field data have been firstly statistically analyzed to explore the failure precursor based on the intact model and 45° flaw model and then compared to find the influence of the pre-existing flaw on the damage evolution and precursor signal. The results indicate that (1) all types of statistical data are positively correlated with the increment of microcracks; (2) corresponding to the fluctuation patterns of statistical data, the damage evolution of BPMs in the pre-peak stage can be divided into three parts; (3) the pre-existing flaw would accelerate the damage evolution; (4) the location and evolution rate of damage could be determined by comprehensively analyzing the average deviation curve, the coefficient of variation, and the contour maps of the strain rate field. These analyses of the particle displacement field can be used to distinguish the impacts of the flaw angle and provide some assistance for the failure forecast.

Keywords: rock-like material; strain rate field; damage evolution; failure precursor; particle flow code

1. Introduction

The damage evolution and failure precursor of rock-like material are two significant topics in rock mechanics. Under quasi-static compression, the strength, elastic modulus, rock type, defect, and anisotropy of rock and rock-like material can all affect the damage evolution and failure process. Many studies show that the load capacity loss for rocks is a gradual and consequent process under quasi-static compression, and the damage development has three stages (distributed damage, localized damage, and catastrophic failure) [1–6]. Micro-cracks, flaws, and pores, existing as defects in rocks and rock-like

 https://doi.org/10.3390/app15020686

materials, considerably affect the behaviors of material properties (physical properties, mechanical properties, etc.) and evolve to macro-failure under external loads.

The damage evolution during the failure process of rock-like material is always accompanied by changes in physical properties (electric pulse, electromagnetic radiation, microwave radiation, infrared radiation, acoustic emission, etc.) and surface deformation. Various physical signals, delivered by sophisticated geophysical methods, including magnetometry, gravimetry, geoelectric, electromagnetic, seismic, geothermal, and radiometric, can indirectly characterize the physical and mechanical properties and reflect the damage location [7]. Different physical signals were used to describe the damage accumulation process of geomaterials among multi-scale analyses and found corresponding information of failure precursors [8–14]. Additionally, the study has shown that the damage localization has been viewed as a precursor to geomaterial failure [15].

In related studies, different types of physical information on damage localization have been investigated by laboratory tests, numerical simulation, and theoretical models on recognizing methods, field characteristics, geometric features, development pattern, and other different features in localized damage regions. In studies about the strain field of rock samples, different manifestations of the surface strain field, including linear strain field, shear strain field, volumetric strain field, and apparent equivalent strain, have been adopted to study the strain localization and damage process. Hao et al. [16] used the digital speckle correlation method (DSCM) to obtain the spatio-temporal pattern of the first principal strain and the width of the damage localization zone and develop the local-mean-field model to calculate the rupture strain. Cheng et al. [17] used the axial and lateral strain field, from digital image correlation (DIC), to describe the rock formation deformation in bedded composite rocks, revealing the failure mechanics of bedded composite rocks with different bedding dip angles. Based on the shear strain field, Ma et al. [18] have divided the failure phase into five stages (with the DSCM), and Cheng et al. [19] have provided the spatial index and the degree index of the strain field to describe the failure process of pre-existing crack sandstone. Song et al. [20] proposed a damage factor, based on the statistic parameter of the apparent equivalent strain, to describe the level of deformation localization. Munoz et al. [21] and Sun et al. [22] investigated the deformation field evolution characteristics of heterogeneous materials both in pre-peak and post-peak regimes under uniaxial compression.

The existence of discontinuous surfaces in rock material could lead to the deterioration of mechanical properties and accelerate its failure process. Thus, it is significant to study the influence of flaws on various types of physical information during the damage accumulation and at the critical rupture. To investigate the fracture dynamics of SiO_2-based materials, Salje et al. [23], Nataf et al. [24], and Zhao et al. [25] have collected and analyzed acoustic emission (AE) signals (the AE activity, amplitude, duration, and energy) in a compression experiment and found the crackling noise distribution of the avalanche criticality. Ma et al. [26] and Gao et al. [27] have investigated the crack development patterns of rocks under compression stress, including the fracture process zone (FPZ) and the relevant fracture parameters, by measuring the surface strain fields of rock materials. Xianyu et al. [28] have investigated the strain feature of single-flaw rock samples with the particle imaging velocimetry (PIV) technique and found that the displacement direction lines significantly changed in the region with flaws. Zhao et al. [29] have explained the initiation mechanism of secondary cracks by analyzing the surface strain field, AE event, and distribution during the uniaxial compression tests of rock-like specimens with an internal open-type flaw. Ding et al. [30] have observed the AE and charge signal of coal compression tests and found that the richness of the signal increases with the deterioration of coal. Li et al. [31] described the strain dispersion of complex fractured rock masses by the standard deviations of the strain field and quantitatively identified the crack initiation stress (CIS). Misra et al. [32] have built an elastic–plastic rheological model to calculate the specific crack orientation resulting in different types of crack propagation based on the variability of deformation localization near pre-existing shear cracks. Zhou et al. [33] have built a micromechanics-based model, taking the interaction among sliding cracks into account, for analyzing the damage and deformation localization of brittle rock.

Furthermore, the strain field of rock-like material has been statistically analyzed to describe the damage accumulation and some precursor signals of the final collapse. Wang et al. [34] explored that the variation coefficients of the vertical linear strain and the maximum shear strain might be the failure precursors of the coal specimens under uniaxial compression. Zhang et al. [35] have found that the differential rate of the strain field–axial strain could be used to recognize the precursor before the failure of the fractured rock mass. Through analyzing the absolute standard deviation (ASD) and absolute variation coefficient (AVC) of the strain field obtained from laboratory tests and numerical simulation, Wang et al. [36] found that the variation laws of the ASD and AVC of the shear strain field were highly correlated with the rock fracture evolution and obtained the precursory indices of full-field strain information. Some researchers developed a damage variable based on the standard deviation of the apparent equivalent strain fields to evaluate the damage degree of rock samples [20,26,37]. Zhang et al. [38,39] have scanned the full-field view of axial and circumferential strains on the sandstone specimen surfaces by the distributed fiber optic strain sensing (DFOSS) technology and detected the initialization of microcrack nucleation.

This paper has used two-dimensional (2D) discrete element models to simulate the whole damage process of the intact and fractured rock based on the compression tests of the rock-like material samples. The volumetric strain rate field was obtained by 200 measurement circles covered on each BPM. The statistical data of the strain rate fields, including the mean, variance, simple mean deviation, and correlation coefficient, were analyzed in relation to the characteristics of the damage evolution and failure precursor among the pre-peak stage of compression tests. Moreover, the results of the intact BPM were compared with the results of the BPM with 45° flaw to find the influence of pre-existing flaws on those characteristics. Based on the results of the 2D models, the evolution characteristics of localized deformation have been quantitatively analyzed. Meanwhile, combined with strain rate contour map, the precursory characteristics of the instability of intact rock mass and fractured rock mass are analyzed comparatively.

2. Numerical Simulations

According to the laboratory test results in the paper by Jin et al. [40], a set of numerical models were generated by Particle Flow Code (PFC). As a discrete element method, the particles and bonds determine the behavior of the macro material, and it can indicate a similar simulation result in the macro-phenomenon of specimen's failure [41]. The particle discrete element model provides support to comprehend the failure process of rock-like material from macroscopic-scale phenomena and statistics. In this paper, the parallel bond was used to generate the bonded-particle models (BPMs).

In the laboratory test, cement, sand, and water were used to make samples with a size of 50 mm (length) × 100 mm (height) × 30 mm (width) [40]. The ratio of the sample length to height complies with the ISRM standards. Following the block size in the laboratory test, the simulated specimens in the PFC were 100 mm high and 50 mm long. We adopted the two-dimensional models to investigate the surface deformation characteristics. Therefore, the results and conclusions only involve two-dimensional situations and might be slightly different from actual situations or three-dimensional models.

The intact BPM has a uniform particle-size distribution bounded by R_{min} (0.25 mm) and R_{max} (0.415 mm). The micro-parameters of the parallel bonds were determined by the calibration process of the simulated uniaxial compression test of the intact model. The parameters of the parallel bonds were adopted when the simulated results reproduced the laboratory test values, including ball–ball contact modulus (E_c) of 2.5 GPa, normal strength (σ_n) of 19 MPa, shear strength (σ_s) of 27.1 MPa, and ball stiffness ratio (k_n/k_s) of 2.5. The macro properties of the intact specimens in the physical experiment and numerical simulation are listed in Table 1, which proves the rationality of the numerical model. The uniaxial compressive strength of the cement specimens corresponds to some kinds of homogeneous and isotropic sandstone and argillite shale. After the reproduction of the intact sample, the specimen with 45° flaw was created by deleting some particles in a specific area. The defined flaw in the model had the same geometry as that in the experimental samples.

Table 1. Material properties of physical experiment and numerical study.

	Intact Specimen		**45° Flaw Specimen**	
	BPM	**Physical Experiment**	**BPM**	**Physical Experiment**
Uniaxial compressive strength (UCS)/MPa	27.58	26.96	18.9	19.12
Elastic modulus/GPa	3.2	3.0	2.9	3.0
Poisson's ratio	0.24	0.17	0.28	--

Figure 1 shows the stress–strain curves of intact and flawed samples in the laboratory test and numerical simulation. It is apparent from Figure 1 that the stress–strain curve of the BPM has an obvious similarity with those of the physical test, both in the intact sample and the flawed sample, indicating the models are feasible.

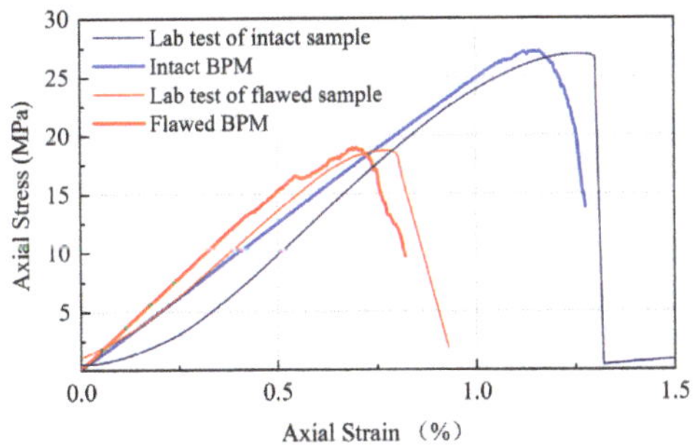

Figure 1. Stress–strain curves of PFC models and tests.

The numerical models and their failure modes are shown in Figure 2. The regularly arranged small circles on the model surface are measurement circles (MCs), which are used to obtain a further investigation about the process of the deformation localization of the specimen. The final collapses of the intact specimens in Figure 2a show a shear failure with shear cracks (blue dot) and a few tensile cracks (red dot). At 45° tips, small slippages emerged, and tensile cracks interpenetrated each other, while shear cracks quickly developed to lead to failure, as shown in Figure 2b.

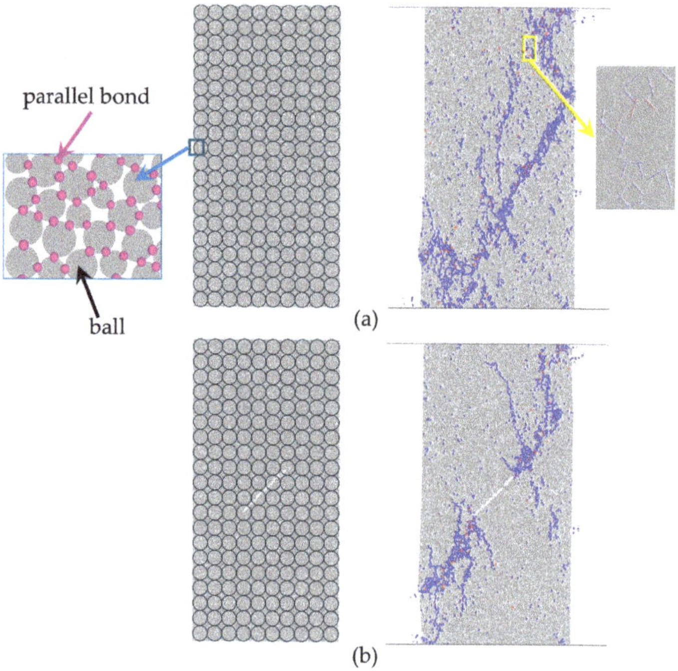

Figure 2. Measurement circle arrangement (**left**) and failure modes of numerical simulations (**right**). (**a**) Intact BPM; (**b**) 45° flaw BPM.

3. Results and Discussion

3.1. Case 1: Intact Specimen

During the process of compression, the volume strain rate of each measurement circle has been automatically recorded every 200 steps. Statistical analysis of the volume strain rates were made to get the average, the standard deviation (SD), the range, and the average deviation (AD) at each recorded time. From the statistics, it has been found that the occurrence of floater particles can cause extremely high values of the MC strain rate and significantly impact the statistics at corresponding moments. In subsequent analysis, the measurement circle with floater particles was removed from the data array.

The extremum MC strain rates before the peak stress are plotted in Figure 3. The black curve and blue curve indicate the maximum and minimum value of the strain rate, respectively; the red lines represent the crack increment; and the purple curve represents the average deviation of the strain rate. The bottom axis reflects the normalized strain (from 0% to 100%); and the 100% normalized strain matches the total strain at the peak strength of the intact numerical model.

In the magnified part of Figure 3, the corresponding gaps between the strain rate extremum curves, i.e., the strain rate ranges, are obviously larger when the micro-cracks increase. This indicates that the variation patterns of the strain rate largely reflect the micro-crack generation and damage evolution of the BPM. Thus, the statistical analysis of the MC strain rates might be used to identify the failure process of specimens and to explore methods for failure predicting.

In Figure 3, before the 30% peak strain, the micro-crack increment emerges occasionally; between 30 and ~60% strain, its value and frequency both change a little; in the last section, its curve has sharp rises. The AD curve is similar to the micro-crack increment. All curves in Figure 3 indicate that the damage accumulation of the intact specimen accelerates after 60% strain and becomes much larger after 80% strain.

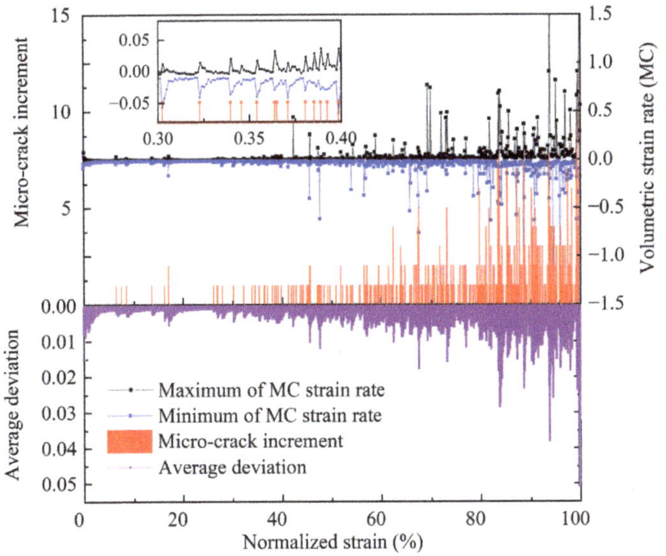

Figure 3. Extremum values and average deviation of MC strain rate and micro-crack increment (case 1).

Figure 4 shows the average and standard deviation of the MC strain rate at the pre-peak part of the intact specimen simulation. The SD curve (the green one) has larger fluctuations, compared with the average curve. Based on the tendency of the green curve, the SD variation can be split into four phases: phase 1, at 0~40% strain, the values of the SD are close to zero; at 40~60% strain, the green curve has some peaks; at 60~80% strain, fluctuations of the SD curve become denser and larger than before; at 80~100% strain, the larger fluctuations emerge intensively.

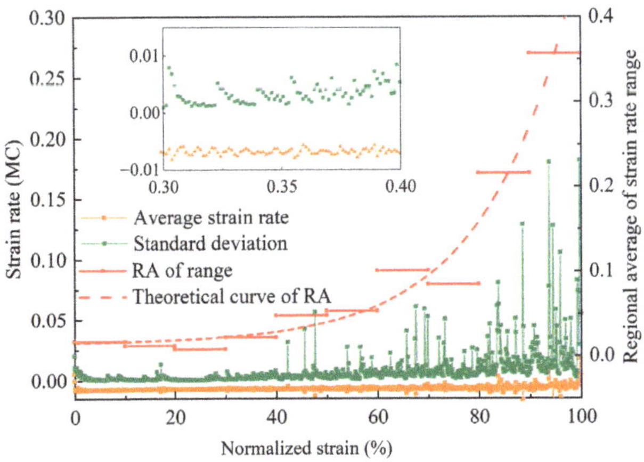

Figure 4. The standard deviation, average, and range regional averages of MC volume strain rate (case 1).

For a further investigation of the damage stage of the intact BPM, the regional averages for the MC strain rate ranges (RA) have been calculated at each 10% strain, which appear as red segments in Figure 4. Those segments can be characterized as four groups: 0~40% strain is group 1, in which part of the regional average has slightly changed; 40~60% strain is group 2, in which part of the red segments gently increase; 60~80% strain is group 3, in which part of the RA has an obvious integral uplift; and 80~100% strain is group 4, in which part of the RA has a significant growth. By comprehensively analyzing the trends of the RA and SD curves, it can be concluded that the damage evolution of the intact BPM can be seen as three sections: the first section is between 0 and ~40% strain; the second section is between 40 and ~80%; the section is between 80 and ~100% strain.

Through data fitting, it has been found that the relationship of the RA and loading strain fit into the exponential function. This can be described by the following exponential function:

$$f(x) = 0.00139 * e^{x/0.17259} + 0.01527 \tag{1}$$

It is a one-phase exponential decay function with a time constant parameter. The theoretical curve is shown as the red dashed line. The fluctuations of the SD curve in Figure 4 indicate that the MC strain rates have a larger dispersion degree as the total strain reaches closer to the peak stress. Therefore, there is a significant correlation between the dispersion of the MC strain rate and the degree of specimen failure.

The coefficient of variation (CV) for each strain rate field was calculated and plotted as a red curve in Figure 5. Then, the averages of 20 contiguous CVs were calculated and plotted as a blue curve in Figure 5. These two curves could be used to analyze the relationship between the dispersion of the strain rate data and the damage evolution of the intact specimen.In this figure, the blue curve has more obvious characteristics that can distinguish the damage stages of pre-peak loading. From 30% to 70% strain, the curve of the strain-contrast RA of the CVs ascends slowly with some slight waves; from 70% to 80% strain, the waves become larger; between 0.8 and ~1.0 strain, the extremely volatile growth of the blue curve correspond with the rapid damage accumulation in the BPM.

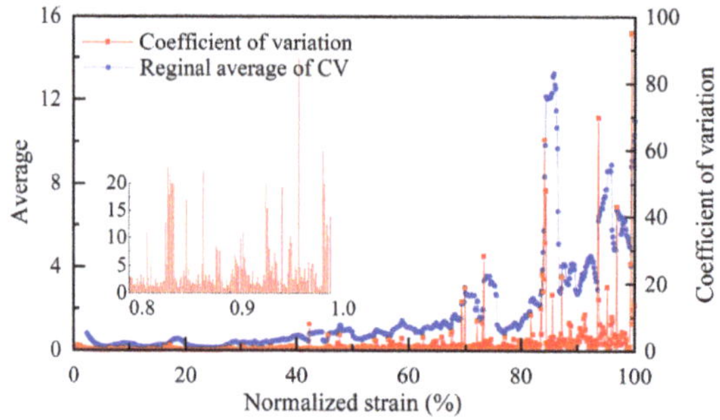

Figure 5. Coefficient of variation of strain rate field and its regional average (case 1).

To further analyze the evolution of the MC strain rate distribution during the loading process, eight contour maps of the strain rate field at different moments were investigated, as shown in Figure 6. The Z-axes values illustrate the volume strain rate of each measurement circle. A positive strain rate indicates an increase in volume strain, and vice versa. Under uniaxial compression, negative strain rates are the main component of the strain rate field; however, a positive strain rate might occur in situations such as particle ejection at the edge of the specimen, shear dilation, and tensile failure of bonds.

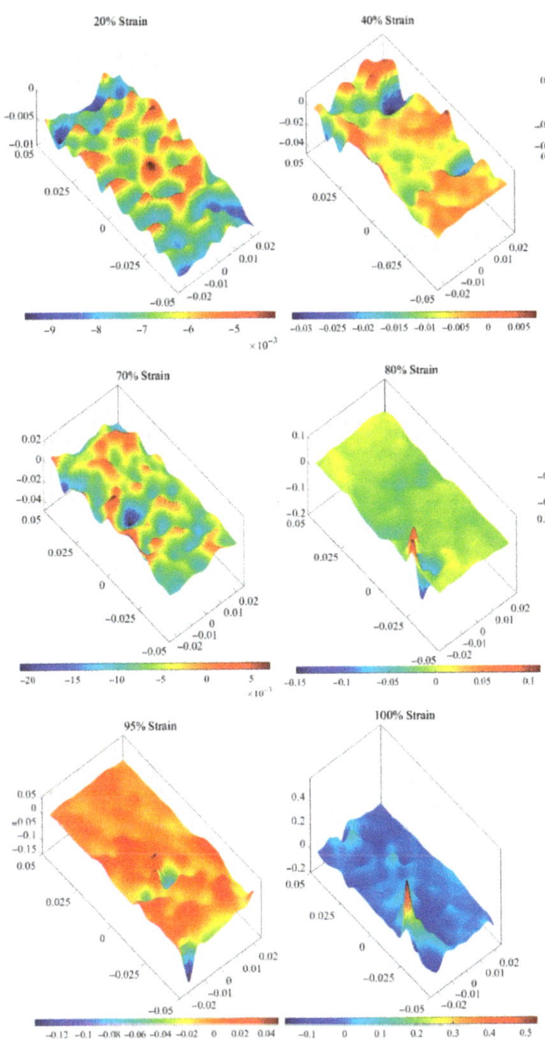

Figure 6. Strain rate field at different strains (case 1).

In Figure 6, at 20% peak strain, the strain rate field has slight waves uniformly distributed over the whole surface with negative values. At this moment, the intact specimen is in the stage of crack closure. At 40%, 60%, and 70% strain, a few high ratios of MC strain rates randomly emerge in the strain rate field, corresponding to the growth pattern of microcracks in Figure 3. From 40% to 70% strain, the specimen is in the stage of stable crack growth. From 80% to 100% strain, the extreme values of the MC strain rate are larger than ever before and concentrate in the left-bottom side of the sample. This indicates that the damages concentrate in that particular area, and the specimen moves into the stage of unstable crack growth.

Figure 6 represents the failure process of this intact BPM. From scattered to centralized, the damage distribution changed by the load increment and finally collapsed because of damage localization at the left-bottom side of the BPM. The variation of these contour maps confirms the micro-crack increase tendency and other statistical data curves.

As mentioned above, the pre-peak failure process of the intact sample could be divided into three or four sections: the first section is before 30% peak strain; section 2 can be seen as two parts: one is between 30 and 60% peak strain and the other is between 60 and 80% peak strain; the last section is between 80 and 100% strain.

In section 1, the micro-crack increment occasionally and separately emerged; all three statistics (average, SD, and CV) of the MC strain rates are close to zero and have minor up and down patterns; the RA values of the range are decreasing and close to zero; the strain rate field displays regular undulation with negative values. These performances indicate that section 1 corresponds to the crack closure stage and linear elastic deformation stage in which the damage accumulation is essentially close to zero.

In the first stage of section 2 (section 2-1), approximately from 30% strain to 60% strain, the micro-crack increment has shown relatively stable growth; the SD values have some peaks; the RA values of the range and CV both slightly increase; and the strain rate fields show a few randomly independent peaks at 40% and 60% peak strain.

In the second stage of section 2 (section 2-2), approximately from 60% strain to 80% strain, the micro-crack growth basically keeps a constant speed, which is higher than the previous stage of section 2. Meanwhile, the values of the RA of the range and CV both become larger than part one of this section. These phenomena indicate that section 2 corresponds to the stage of stable crack growth. Additionally, in this section, the damages steadily scattered and increased in the specimen.

In section 3, the micro-crack increment is significantly higher than those of section 1 and 2; all statistics and regional average of the CV of the strain rate field are extremely volatile; and the RA of the range increases rapidly. Meanwhile, from 90% to 100% peak strain, the strain rate extremums concentrate near the left-bottom side of the intact model. These phenomena indicate that section 3 corresponds to the stage of unstable crack growth and failure. In this section, the localization deformation section, the damage rapidly grows in a centralized area till the BPM collapses.

3.2. Case 2: Specimen with an Open Flaw

To study the influence of the pre-existing flaw on damage procession, the corresponding data of specimen with 45° flaw have been calculated and statistically analyzed.

Figure 7 has shown the strain rate extremums of the MC and micro-crack increment in the pre-peak stage of the flawed specimen, with the same legends as Figure 3. The micro-crack increment (red bars) appears intermittently before 50% peak strain; then, the red bars become crowded till 80% strain; at last, the red bars become much longer and more intense. Between 80% strain and 100% strain, there are four peaks on the maximum curve, accompanied by the rapid growth of micro-cracks.

The average, standard deviation (SD), and range RA of the MC strain rate at the pre-peak part of the flawed specimen are shown in Figure 8 (the legends are consistent with Figure 3). The SD of the strain rate field is close to zero before 60% strain and has three peaks between 60 and ~100% strain. The average curve shows the same tendency as the SD curve.

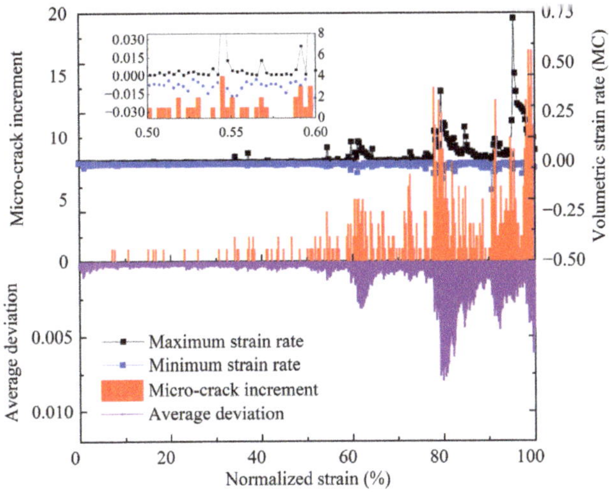

Figure 7. Extremum values and average deviation of MC strain rate and micro-crack increment (case 2).

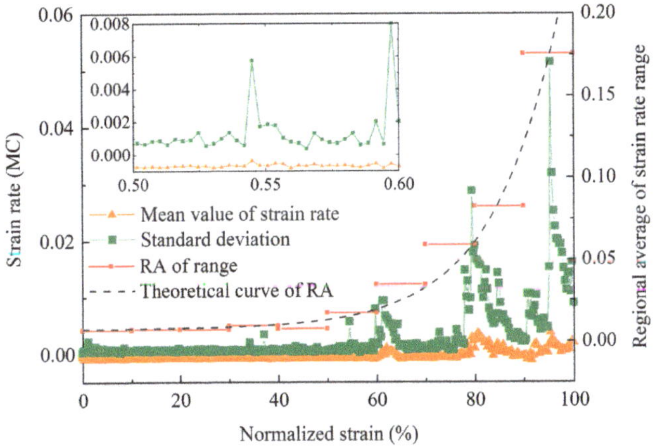

Figure 8. The standard deviation, average, and range regional averages of MC volume strain rate (case 2).

In Figure 8, the rise height of the range RA segments becomes higher as the loading strain increases, and the skip distance between the last two red segments is obviously larger than before. Compared to the intact BPM, the red segments of the flawed BPM become well regulated and have more uplifts after 50% strain. By fitting the range RA data with an exponential function, we obtain the following exponential function:

$$f(x) = 0.0006 * e^{x/0.17259} + 0.00774 \quad (2)$$

The regional average data fits into the exponential function, and the theoretical curve is shown as the black dash curve. Equations (1) and (2) both are a one-phase exponential decay function with a time constant parameter. These two equations have the same time constant 0.17259, but the parameters of the offset and amplitude both reduce to 50% in Equation (2). Compared with the theoretical curve of the single-flaw specimen in Figure 8, the curve of the intact model in Figure 4 is steeper. It indicates that the regional averages of the strain rate range of the intact model are higher than that of the flawed model when the strain proportions were coincided in the two cases.

Figure 9 shows the coefficient of variation in case 2 (the red curve) and the average of 20 contiguous CVs (the blue curve). The blue curve can be loosely divided into two parts: the line part before 60% US strain and the wave part with three waves.

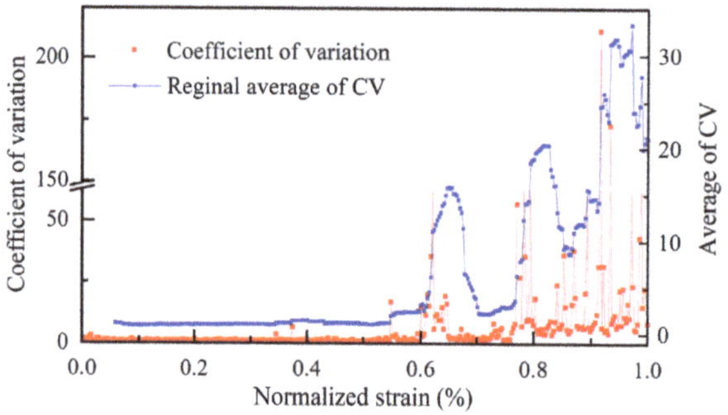

Figure 9. Coefficient of variation of strain rate field and its region average (case 2).

In Figure 9, between 60% and 90% strain, the blue curve has two waves with clear peaks and troughs; meanwhile, the CV curve has fluctuations that appear at intervals. From approximately 87% strain to 100% strain, the last wave on the blue curve has fluctuated upwards higher than ever before. The blue curve shape in case 2 is significantly different from the corresponding curve in case 1. The existence of the flaw has made the variation pattern of the CV and its regional average more regular and more distinguished.

Eight contour maps of MC strain rates at critical moments are shown in Figure 10, providing help to determine the damage localization of the flawed sample. In contour maps of 20% and 40% strain, the high ratio strain rates surround the pre-existing flaw, and there is basically no change in other areas. This shows that the strain localization appears near the flaw during the initial stage of loading. It is different from the first stage of case 1 that the damage concentration begins from the initial loading period, omitting the scattered damage stage.

In the contour map of 60% peak strain, the areas with high ratio strain rates move to the flaw tips. When the strain reaches 70% and 80% peak strain, the extreme values of the strain rate field move with crack propagation. This indicates that the damage accumulates with crack propagation, and some tensile micro-cracks form at the crack tips.

The strain rate field at 90% strain shows a relatively smooth surface during the later stage of the pre-peak loading, with peaks dispersed along the wing crack boundaries. At 90% strain, the micro-crack increment and all preliminary statistics are close to zero.

In 95% and 100% strain, the fluctuations of the strain rate field are distributed along the wing cracks and become more congested. From Figure 10, it can be seen that the deformation localization started around flaw and then moved to the edges of cracks with the loading increasing.

By comprehensively analyzing the above results, we can tell that the pre-existing flaw has made the strain localization form early and the variation patterns of the statistical data more regular and distinguished. The pre-peak failure process of the flawed sample could be divided into three sections: section 1 is before 50% peak strain; section 2 is between 50% and 90% peak strain; section 3 is between 90% and 100% peak strain.

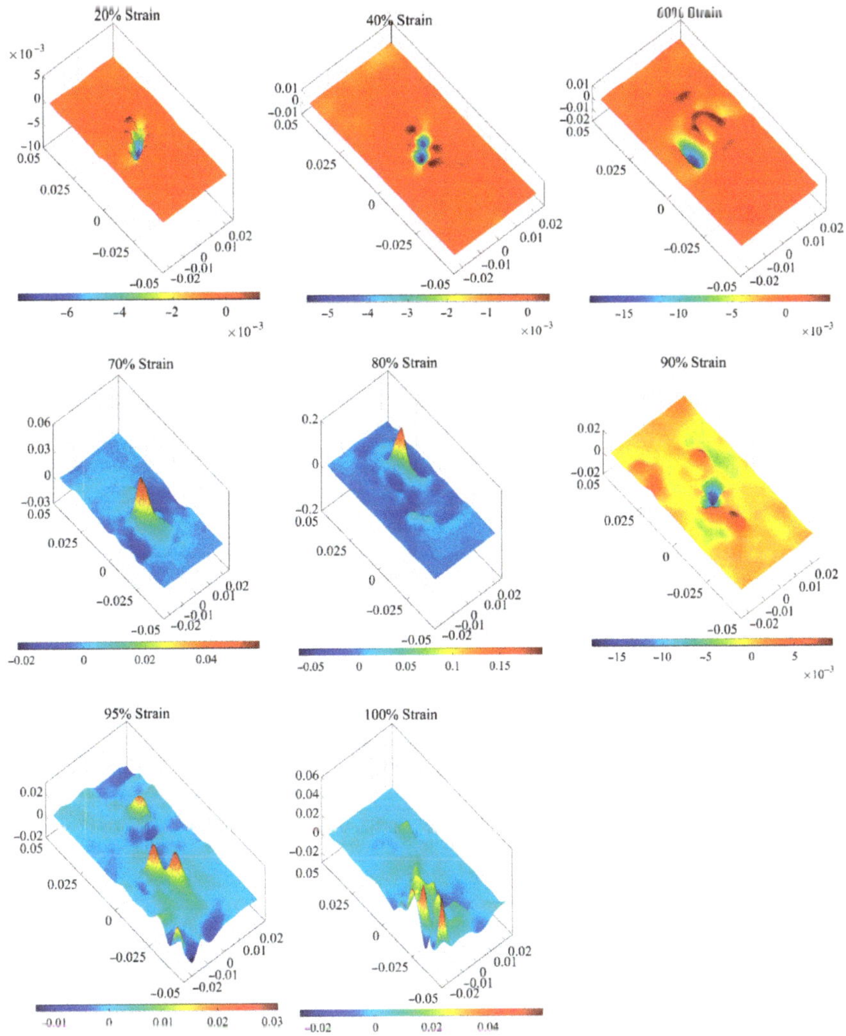

Figure 10. Strain rate of MC at different loading strains (case 2).

In section 1, the micro-crack increment occasionally emerged, as did peaks in the strain rate field around the flaw; meanwhile, all of the statistic curves of the MC strain rates are basically straight lines approaching zero. These phenomena indicate that section 1 corresponds to the stage of the crack closure and the linear elastic deformation stage in which the deformation localization occurs around the pre-existing flaw. The damage evolution has skipped the stage of distributed damage.

In section 2, approximately from 50% strain to 90% strain, the micro-crack growth rate becomes larger; the strain rate peaks follow the tips of the crack propagation; the curves of three statistics and the coefficient of variation all have two waves around 60% and 80% strain, corresponding to the crack initiation and stable propagation; the rise height of the range RA segments gradually become larger as the loading strain increases. These performances indicate that section 2 corresponds to the stage of crack initiation and stable crack growth in which the localized damage accumulation is relatively steady. However, the micro-crack growth presents staged and whitespace between two peaks.

In section 3, approximately from 90% strain to 100% strain, the values of SD and CV both have their largest fluctuations; the RA of the range surges at the last segment; the RA of the CV fluctuates to the highest peak; the strain rate extremums significantly increase and concentrate along the crack propagation path. These phenomena indicate that cracks have unstable growth and coalescence in section 3. They show that section 3 corresponds to the later stage of localized damage and the stage of catastrophic failure in which the damage rapidly grows in a centralized area till the BPM collapses.

3.3. Comparative Analysis

The correlation coefficients between the micro-crack increment and four types of statistics (the range, average, average deviation, and standard deviation) on the MC strain rates have been calculated and plotted in Figure 11. All of the correlation coefficients are positive, which illustrates that the crack increment is positively correlated with the four types of statistics both in case 1 and case 2. Three types of statistics in case 2 are lower than those in case 1, meaning that the existence of a flaw would degrade the correlation between the strain rate field and damage evolution and reduce the dispersion of the strain rate field.

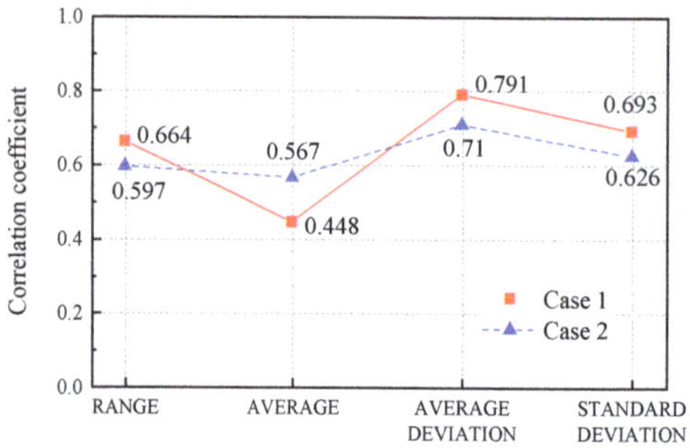

Figure 11. Correlation coefficient between micro-crack increment and statistical data of MC strain rate.

In Figure 11, these two curves have a similar tendency: the minimum of each case is the correlation coefficient between the strain rate average and crack increment; the maximum of each case is the correlation coefficient between the strain rate average deviation and crack increment. This means that the crack growth is highly correlated with the average deviation. Therefore, the average deviation of the MC strain rate field is a better choice for making discriminating judgments and evaluations of specimen damage evolution.

The volume strain rates of MCs and their statistical data could illustrate the damage evolution during the loading process from varying points and degrees. For integrating preceding analysis, the applications of each parameter have been collected and listed in Table 2. In this table, the plus sign (+) means a parameter could facilitate the corresponding function, and the minus sign (−) has the opposite meaning.

Summarizing the above work, Table 2 shows the following results: the MC strain rate contour maps can tell the damage position; the average deviation can judge the crack increment; the average fluctuation and other statistics can indicate the damage stage of samples. Therefore, comprehensive analysis of the MC strain rates and the corresponding statistics can be used for the collapse prediction of rock-like material.

Table 2. Parameter indication on different applications.

	Crack Increment	Damage Discriminating Judgment	Damage Position
Range	+	+	−
Average	+	+	−
Average deviation	+++	+	−
Standard deviation	+	+	−
Coefficient of variation	+	+	−
Range RA	+	+++	−
CV RA	+	+++	−
Contour map of strain rate field	+	+	+++

4. Conclusions

Based on the laboratory uniaxial compression test of the rock-like material, this paper builds the corresponding numerical models for an intact sample and a sample with a 45° flaw. In the numerical simulations, the measurement circles are uniformly distributed on numerical specimens to monitor the variation of the volumetric strain rate field. We analyzed the five types of statistics (average, range, average deviation, standard deviation, and coefficient of variation) and drew the contour maps of the strain rate field. We obtained the following conclusions:

1. Several relationships between the damage evolution of the rock-like material and statistical data of the volumetric strain rate field have been built in this paper. The micro-crack increasement, or the damage accumulation, corresponds to large gaps between the strain rate extremums. In the stage of the pre-peak loading, statistical data always have growth and volatility, and the corresponding regional averages rise higher as the strain increases. The results showed that all types of statistical data positively correlate with the microcrack increment, and the regional average of the strain rate range has an exponential raise as the strain increases.
2. Based on the comprehensive analysis of the variation characteristics of all MC strain rate statistic data, the damage evolution of the pre-peak loading stage could be separated into three sections. In section 1, the strain rate field is relatively flat, and all the statistical data are flat. In section 2, all types of statistic data of case 1 have different variation patterns from those of case 2. However, in section 3, relatively large - scale local fluctuations occur in the strain rate field,, and all the statistical data rapidly increased. Among all types of statistics, the standard deviation, the regional average of range, and the CV are more distinguished than others.
3. Due to the pre-existing flaw, the development of the strain localization of case 1 and case 2, the intact sample, and the flawed sample is different from each other. The pre-existing flaw accelerates the damage evolution. The intact specimen in section 1 has a strain rate field with a regular fluctuation pattern, and, in section 2-1, the strain rate field has peaks randomly distributed. The damage evolution of the intact sample developed to the stage of the distributed damage. Meanwhile, the strain rate field of the flawed sample has the strain localization around the flaw. In section 2 and section 3, the pre-existing flaw has made all curves of the statistics fluctuate with some regular patterns.
4. By taking full advantage of the volumetric strain rate field, the status of the damage evolution of the rock-like material sample can be analyzed and evaluated. The field contour maps of the MC strain rate can tell the damage position; the average deviation curve can judge the crack increment; the region average of the variation coefficient

can indicate the damage section. This strain rate field can provide assistance on the prediction of catastrophic failure of rock-like material.

Author Contributions: Conceptualization, J.J. and P.C.; methodology, J.J. and J.Z.; software, J.J., P.C. and C.M.; validation, J.L.; resources, J.J., P.C., J.Z. and X.B.; data curation, J.J., Y.W. and J.L.; writing original draft preparation, J.J.; funding acquisition, J.J., Y.W. and C.M.; supervision, X.B. All authors have read and agreed to the published version of the manuscript.

Funding: This research was financially supported by Fundamental Research Program of Shanxi Province (Program Nos. 202303021212078 and 202303021224001); the National Natural Science Foundation of China (NSFC) (No. 52178341); the Research Project of Science and Technology Department of Shanxi Province (No. 202203021222428).

Institutional Review Board Statement: Not applicable.

Informed Consent Statement: Not applicable.

Data Availability Statement: The data used to support the findings of this study are available from the corresponding author upon request.

Acknowledgments: We thank Central South University for their technical support.

Conflicts of Interest: Authors Jin Jin, Jun Zhang, Yanchao Wang and Chenxi Miao were employed by the company Shanxi Province Transportation Technology Research and Development Co., Ltd. The remaining authors declare that the research was conducted in the absence of any commercial or financial relationships that could be construed as a potential conflict of interest.

References

1. Cook, N.G.W. The failure of rock. *Int. J. Rock Mech. Min. Sci. Geomech. Abstr.* **1965**, *2*, 389–403. [CrossRef]
2. Ashby, M.F.; Sammis, C.G. The damage mechanics of brittle solids in compression. *Pure Appl. Geophys.* **1990**, *133*, 489–521. [CrossRef]
3. Martin, C.D.; Chandler, N.A. The progressive fracture of Lac du Bonnet granite. *Int. J. Rock Mech. Min. Sci. Geomech. Abstr.* **1994**, *31*, 643–659. [CrossRef]
4. Xu, X.; Ma, S.; Xia, M.; Ke, F.; Bai, Y. Damage evaluation and damage localization of rock. *Theor. Appl. Fract. Mech.* **2004**, *42*, 131–138. [CrossRef]
5. Liang, C.; Li, X.; Wang, S.; Li, S.; He, J.; Ma, C. Experimental investigations on rate-dependent stress-strain characteristics and energy mechanism of rock under uniaxal compression. *Yanshilixue Yu Gongcheng Xuebao/Chin. J. Rock Mech. Eng.* **2012**, *31*, 1830–1838.
6. Nikolic, M.; Ibrahimbegovic, A. Rock mechanics model capable of representing initial heterogeneities and full set of 3D failure mechanisms. *Comput. Methods Appl. Mech. Eng.* **2015**, *290*, 209–227. [CrossRef]
7. Schön, J.H. *Physical Properties of Rocks—Fundamentals and Principles of Petrophysics (first edition)*; Pergamon Press, Ltd.: Oxford, UK, 2004; pp. 3–4.
8. Soloviev, S.P.; Spivak, A.A. Electromagnetic signals generated by the electric polarization during the constrained deformation of rocks. *Izv. Phys. Solid Earth* **2009**, *45*, 347–355. [CrossRef]
9. Larionov, I.; Malkin, E.; Uvarov, V. Deformation-Electromagnetic Relations in Lithospheric Activity Manifestations. *E3s Web Conf.* **2018**, *62*, 03002. [CrossRef]
10. Song, X.; Li, X.; Li, Z.; Cheng, F.; Zhang, Z.; Niu, Y. Experimental research on the electromagnetic radiation (EMR) characteristics of cracked rock. *Environ. Sci. Pollut. Res.* **2017**, *25*, 6596–6608. [CrossRef]
11. Carlà, T.; Intrieri, E.; Di Traglia, F.; Nolesini, T.; Gigli, G.; Casagli, N. Guidelines on the use of inverse velocity method as a tool for setting alarm thresholds and forecasting landslides and structure collapses. *Landslides* **2016**, *14*, 517–534. [CrossRef]
12. Liu, S.; Xu, Z.; Wei, J.; Huang, J.; Wu, L. Experimental Study on Microwave Radiation From Deforming and Fracturing Rock Under Loading Outdoor. *IEEE Trans. Geosci. Remote Sens.* **2016**, *54*, 5578–5587. [CrossRef]
13. Shi, D.; Wang, J.; Xiong, L. Study on Noise Correction Algorithm of Infrared Emissivity of Rock under Uniaxial Compression. *Sustainability* **2022**, *14*, 12769. [CrossRef]
14. Huang, L.-Q.; Wu, X.; Li, X.-B.; Wang, S.-F. Influence of sensor array on MS/AE source location accuracy in rock mass. *Trans. Nonferrous Met. Soc. China* **2023**, *33*, 254–274. [CrossRef]

15. Bai, Y.; Xia, M.; Ke, F.; Li, H. Damage Field Equation and Criterion for Damage Localization. In *IUTAM Symposium on Rheology of Bodies with Defects*; Solid Mechanics and its Applications; Wang, R., Ed.; Springer: Dordrecht, The Netherlands, 1999; Volume 64, pp. 55–66. [CrossRef]
16. Hao, S.; Wang, H.; Xia, M.; Ke, F.; Bai, Y. Relationship between strain localization and catastrophic rupture. *Theor. Appl. Fract. Mech.* **2007**, *48*, 41–49. [CrossRef]
17. Cheng, J.; Yang, S.; Chen, K.; Ma, D.; Li, F.; Wang, L.M. Uniaxial experimental study of the acoustic emission and deformation behavior of composite rock based on 3D digital image correlation (DIC). *Acta Mech. Sin.* **2017**, *33*, 999–1021. [CrossRef]
18. Ma, S.; Zhou, H. Surface strain field evolution of rock specimen during failure process. *Chin. J. Rock Mech. Eng.* **2008**, *27*, 1667–1673.
19. Cheng, H.; Yang, X.; Zhang, C.; Ning, Z.; Li, Y. Characteristics of heterogeneous deformation and identification model of unstable damage state in pre-existing crack sandstone. *J. Min. Saf. Eng.* **2023**, *40*, 1290–1300. [CrossRef]
20. Song, H.; Zhang, H.; Fu, D.; Kang, Y.; Huang, G.; Qu, C.; Cai, Z. Experimental study on damage evolution of rock under uniform and concentrated loading conditions using digital image correlation. *Fatigue Fract. Eng. Mater. Struct.* **2013**, *36*, 760–768. [CrossRef]
21. Munoz, H.; Taheri, A.; Chanda, E.K. Pre-Peak and Post-Peak Rock Strain Characteristics During Uniaxial Compression by 3D Digital Image Correlation. *Rock Mech. Rock Eng.* **2016**, *49*, 2541–2554. [CrossRef]
22. Sun, Q.; Cai, C.; Zhang, S.; Tian, S.; Li, B.; Xia, Y.; Sun, Q. Study of localized deformation in geopolymer cemented coal gangue-fly ash backfill based on the digital speckle correlation method. *Constr. Build. Mater.* **2019**, *215*, 321–331. [CrossRef]
23. Salje, E.K.; Soto-Parra, D.E.; Planes, A.; Vives, E.; Reinecker, M.; Schranz, W. Failure mechanism in porous materials under compression: Crackling noise in mesoporous SiO_2. *Philos. Mag. Lett.* **2011**, *91*, 554–560. [CrossRef]
24. Nataf, G.F.; Castillo-Villa, P.O.; Baró, J.; Illa, X.; Vives, E.; Planes, A.; Salje, E.K. Avalanches in compressed porous SiO_2-based materials. *Phys. Review. E Stat. Nonlinear Soft Matter Phys.* **2014**, *902*, 022405. [CrossRef] [PubMed]
25. Zhao, Y.; Liu, H.; Xie, K.; Salje, E.K.; Jiang, X. Avalanches in Compressed Sandstone: Crackling Noise under Confinement. *Crystals* **2019**, *9*, 582. [CrossRef]
26. Ma, J.; Li, D.; Luo, P.; Zhang, C.; Gao, F. Dynamic Damage and Failure of Layered Sandstone with Pre-cracked Hole Under Combined Cyclic Impact and Static Loads. *Rock Mech. Rock Eng.* **2022**, *56*, 2271–2291. [CrossRef]
27. Gao, G.; Yao, W.; Xia, K.; Li, Z. Investigation of the rate dependence of fracture propagation in rocks using digital image correlation (DIC) method. *Eng. Fract. Mech.* **2015**, *138*, 146–155. [CrossRef]
28. Xianyu, W.; Lu, X.; Zhao, Q.; Han, G. Strength and deformation characteristics of rock with single pre-existing fissure under different loading conditions. *China Meas. Test* **2017**, *43*, 124–129.
29. Zhao, Y.; Gao, Y.; Wu, S.; Chen, L.; Zhang, C. Experimental and numerical study of failure characteristics of brittle rocks with single internal 3D open-type flaw. *Acta Geotech.* **2021**, *16*, 3087–3113. [CrossRef]
30. Ding, X.; Xiao, X.-C.; Lv, X.-F.; Wu, D.; Pan, Y. Analysis of Similarities and Differences between Acoustic Emission and Charge Signal Based on Fractal Characteristics of Coal Fracture. *Adv. Civ. Eng.* **2020**, *2020*, 8745039. [CrossRef]
31. Li, N.; Zhang, K.; Bao, R.; Fan, W. Quantitative description of fracture behavior and mechanical anisotropy of fractured rock mass: Sand 3D printing and DIC explorations. *Theor. Appl. Fract. Mech.* **2023**, *127*, 104065. [CrossRef]
32. Misra, S. Deformation localization at the tips of shear fractures: An analytical approach. *Tectonophysics* **2011**, *503*, 182–187. [CrossRef]
33. Zhou, X.; Zhang, Y.; Ha, Q.; Wang, J. Analysis of the localization of damage and the complete stress-strain relation for mesoscopic heterogeneous brittle rock subjected to compressive loads. *Appl. Math. Mech.* **2004**, *25*, 1039–1046. [CrossRef]
34. Wang, X.; Hou, W.; Dong, W. Assessment of the failure precursors of the strain fields for the coal specimens in the uniaxial compression based on the digital image correlation method. *J. Saf. Environ.* **2018**, *18*, 1237–1245. [CrossRef]
35. Zhang, K.; Li, N. Evolution law of strain field and precursor identification of flawed sandstone based on Digital Image Correlation method. *Hydrogeol. Eng. Geol.* **2021**, *48*, 150–156. [CrossRef]
36. Wang, P.; Liu, Q.; Zhang, Y.; Huang, Z. Identifying rock fracture precursor by multivariate analysis based on the digital image correlation technique. *Theor. Appl. Fract. Mech.* **2023**, *126*, 103987. [CrossRef]
37. Zhu, J.; Zhai, T.; Liao, Z.; Yang, S.; Liu, X.; Zhou, T. Low-Amplitude Wave Propagation and Attenuation Through Damaged Rock and a Classification Scheme for Rock Fracturing Degree. *Rock Mech. Rock Eng.* **2020**, *53*, 3983–4000. [CrossRef]
38. Zhang, L.; Yang, D.; Chen, Z.; Liu, A. Deformation and failure characteristics of sandstone under uniaxial compression using distributed fiber optic strain sensing. *J. Rock Mech. Geotech. Eng.* **2020**, *12*, 1046–1055. [CrossRef]
39. Hegger, S.; Vlachopoulos, N.; Poles, T.; Diederichs, M.S. Measuring the Full-Field Strain Response of Uniaxial Compression Test Specimens Using Distributed Fiber Optic Sensing. *Rock Mech. Rock Eng.* **2021**, *55*, 2615–2630. [CrossRef]

40. Jin, J.; Cao, P.; Chen, Y.; Pu, C.; Mao, D.; Fan, X. Influence of single flaw on the failure process and energy mechanics of rock-like material. *Comput. Geotech.* **2017**, *86*, 150–162. [CrossRef]
41. Sun, Q.; Liu, X.; Zhang, G.; Liu, C.; Jin, F. The mesoscopic structures of dense granular materials. *Adv. Mech.* **2017**, *47*, 263–308. [CrossRef]

applied sciences

MDPI

Article

Research on the Identification of Rock Mass Structural Planes and Extraction of Dominant Orientations Based on 3D Point Cloud

Jiarui Zhu [1], Yonghua Xia [1,2,*], Bin Wang [2,*], Ziliang Yang [3] and Kaihua Yang [3]

1 Faculty of Land and Resources Engineering, Kunming University of Science and Technology, Kunming 650093, China; 20222201124@stu.kust.edu.cn
2 City College, Kunming University of Science and Technology, Kunming 650051, China
3 Yunnan Geological Engineering the Second Investigation Institute, Kunming 650213, China; yzl284937289@outlook.com (Z.Y.); kaihuayang19@outlook.com (K.Y.)
* Correspondence: 20040063@kust.edu.cn (Y.X.); 20100058@kust.edu.cn (B.W.)

Abstract: The different spatial distribution forms of rock mass structural planes create weak zones in the rock mass, which is also a key factor in controlling rock mass stability. Accurately and efficiently identifying rock mass structural planes and obtaining their dominant orientations is critical for rock mass engineering design and construction. Traditional surveying methods for high and steep rock mass structural planes pose high safety risks, offer limited data, and make comprehensive statistical analysis difficult. This paper utilizes complex rock mass surface 3D point cloud data obtained through 3D laser scanning technology and uses the Hough space transform method to calculate the normal vectors of the 3D point cloud. Based on the difference in normal vectors and surface variation, region growing segmentation is applied to identify and extract rock mass structural planes. Additionally, the fast search and density peak clustering method (CFSFDP) is used for clustering analysis of the rock mass structural planes to obtain dominant orientations. This method was applied to a highway's high and steep rock slope, successfully identifying 281 structural planes and two sets of dominant structural planes. The orientation of the dominant structural planes identified through RocScience Dips 7.0 analysis showed a deviation of no more than ±3°, complying with engineering standards. The research results offer a feasible solution for the identification of high and steep rock mass structural planes and the extraction of the orientation of dominant structural planes.

Keywords: 3D point cloud; rock mass structural planes; dominant orientation; Hough space transform; region growing

1. Introduction

Rock mass structural planes refer to natural or artificial discontinuities in rock masses, including fractures, faults, joints, bedding, and shear zones. The structural planes developed within the rock mass are the weakest geological interfaces in terms of mechanical properties and are critical factors that determine the mechanical behavior and stability of the rock mass [1]. Accurate and efficient identification of rock mass structural planes and the extraction of dominant orientations have always been key concerns in the fields of geotechnical and geological engineering [2,3]. Traditional methods for surveying structural planes primarily rely on manual operations using simple tools such as measuring tapes and geological compasses for contact measurements. However, when dealing with high and steep rock masses or slopes, these methods are inefficient and carry significant safety risks [4]. Additionally, the data obtained from manual measurements are limited, and these discrete local data points are difficult to integrate for analysis, making comprehensive assessment of the entire rock mass or slope challenging.

With the development of non-contact measurement technologies, photogrammetry (UAV) and terrestrial laser scanning (TLS) have been widely applied in geological engineering. Photogrammetry can capture two-dimensional images of rock mass surfaces, and many researchers have used multi-view vision techniques to quickly reconstruct 3D models of the study area and obtain structural plane information from rock masses based on these 2D images [5–7]. However, photogrammetry is influenced by factors such as image resolution, camera parameters, 3D model construction algorithms, and external environmental conditions. As a result, the final 3D model often has low coordinate accuracy, low automation in structural plane identification, and significant errors in the extracted structural plane information. Terrestrial laser scanning technology overcomes the limitations of single-point measurements and can acquire high-precision, high-density 3D point cloud data that represent the structural features of the rock mass surface, providing fundamental data for structural plane identification and clustering analysis. As early as the late 20th century, Hoppe [8] began research on identifying specific planes within 3D point clouds by searching the neighborhood of each core point, applying the PCA method to fit local planes, and calculating the normal vectors of these planes to replace the core point's normal vector. This approach offered a new idea for accurately computing the normal vectors of point cloud data. S. Slob [9] constructed a triangulated mesh from point cloud data and calculated the orientation of each triangular facet, implementing a fuzzy K-means clustering algorithm to identify structural planes and calculate their spacing. Gigli G [10] set a coplanarity threshold and used a variable-sized search cube to select points that met the coplanarity threshold for plane fitting and developed the structural plane analysis software DiAna. Ge [11] processed point cloud data by meshing, set appropriate region-growing thresholds, and achieved intelligent identification of rock mass structural planes, summarizing the impact of threshold changes on identification results. Chen [12] applied an improved RANSAC algorithm and an enhanced Graham scan algorithm to identify rock mass structural planes and provide a refined description of structural plane boundaries. She developed the RDD (RANSAC discontinuity detection) program, which can extract orientation and size information of structural planes. Kong et al. [13] utilized the KNN algorithm to search for the optimal neighborhood, calculated the point cloud normal vectors based on coplanarity criteria and the least squares method, and then converted them into orientation information. Finally, the KDE and DBSCAN algorithms were combined to cluster the point cloud data, achieving the extraction of rock mass structural plane information. Gou [14] applied the Firefly Algorithm (FA) and Fuzzy C-means (FCM) algorithm for semi-automated clustering analysis of rock mass structural planes based on normal vectors. Gao [15] proposed a new method for identifying discontinuous sets of rock masses based on fast search and density peak identification and validated the accuracy of the method using regular geometric datasets. Cui, Chen, and Wang et al. [16–18] improved the traditional K-means algorithm using variable-length string genetic algorithms, maximum sample density, and the Artificial Fish Swarm Algorithm (AFSA), achieving clustering analysis of rock mass structural plane orientations. Kong et al. [19] used the fast search and density peak identification (CFSFDP) algorithm for clustering analysis of structural planes and used Fisher's K-index to filter out noise points, thus improving the accuracy of the clustering centers.

In summary, many scholars at home and abroad have long been dedicated to the automatic extraction of rock mass structural plane information from 3D point cloud data. Since 3D point cloud data is inherently discrete and lacks topological relationships, data mining methods are needed to extract characteristic values that represent rock mass structural planes, with the accurate calculation of normal vectors being particularly crucial [20–23]. Factors such as noise, point density, and the complexity of the rock mass surface can affect the accuracy of normal vector calculations. Therefore, this paper adopts the Hough space transform method based on statistical principles to accurately calculate the normal vectors of 3D point clouds and uses the differences in normal vectors between adjacent points for region-growing segmentation, achieving the automatic identification and information extraction of rock mass structural planes. The extraction of dominant orientations of structural planes plays an irreplaceable role in the stability analysis of rock masses. Dominant structural planes determine the sliding direction and failure modes of rock masses under different stress conditions. Accurate extraction of dominant orientations not only provides key input parameters for rock mass stability evaluation but also offers scientific guidance for support design and optimization of construction plans [24,25]. Existing clustering algorithms for rock mass structural planes are numerous but generally rely on trial and error and experience to determine the clustering centers and the number of clusters, making the results susceptible to subjective influences and limiting the amount of input data. In this paper, the sine squared value of the normal vectors is used as a similarity measure, and the fast search and density peak clustering method (CFSFDP) is applied for clustering analysis. This method automatically determines the clustering centers and numbers, avoiding subjective parameter settings, thereby improving the efficiency and accuracy of clustering and ensuring the precise extraction of the orientation of dominant structural planes [26–28].

2. Methods

Based on 3D point cloud data, this paper addresses the limitations in rock mass structural planes recognition and clustering methods. The Hough space transform method is employed to compute the normal vectors of the point cloud data. The differences in normal vectors are used as criteria for region growing segmentation, leading to the development of an effective and noise-resistant method for rock mass structural planes recognition and information extraction. Subsequently, the rock mass structural planes are used as clustering objects, and clustering analysis is performed using the Clustering by Fast Search and Find of Density Peaks (CFSFDP) method to extract the orientations of dominant structural planes. The specific workflow is shown in Figure 1.

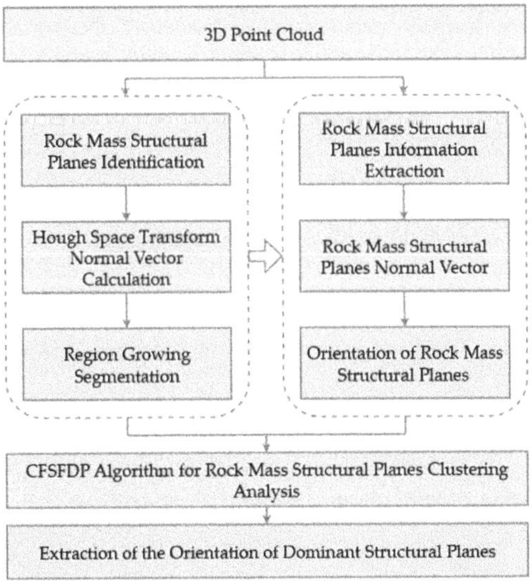

Figure 1. Flowchart of the Method for Identifying Rock Mass Structural Planes and Extracting Dominant Orientations.

2.1. *Identification and Information Extraction of Rock Mass Structural Planes*

2.1.1. Rock Mass Structural Planes Identification

The core idea of region growing is to use certain common features of point cloud data as the basis for judgment. By selecting arbitrary points and comparing them with neighboring points in one or multiple dimensions, points with similar features are segmented and merged. There are three key issues to consider during the region growing process [29].

1. Selection of Seed Points

Region growing is a seed-point-based point cloud segmentation algorithm. The initial seed points should be selected from areas with relatively high flatness to improve the computation speed. The surface variation SV_i of each point is calculated, where SV_i represents the surface roughness within the neighborhood. The points are sorted in ascending order based on their surface variation values, and the point with the smallest SV_i is chosen as the initial seed point.

2. Feature Judging Criteria

Rock mass structural planes are typically composed of relatively flat planes and sharp edges, with field investigations primarily using sharp edges as boundaries. In the identification of rock mass structural planes, the normal vector serves as a crucial feature for differentiating between various planes; points within the same rock mass structural plane should have similar or identical normal vector orientations. The difference in normal vectors between adjacent points is used as a criterion for region-growing segmentation of the point cloud data.

The Hough transform employs statistical principles to effectively enhance the noise resistance and computation of normal vectors on complex rock surfaces. The core of this method involves randomly selecting two points from the neighborhood and a seed point to form a plane, calculating the normal vector of that plane, and traversing all combinations to statistically identify the most frequently occurring normal vector as the final normal vector for the seed point. In the context of structural surfaces, any given point in space can be classified into three positional states: (1) away from the edge, (2) near the edge, and (3) on the edge. When the seed point is in states 1 or 2, two non-collinear points are randomly selected from the defined neighborhood K to form a plane with the seed point. The normal vectors calculated from all combinations will exhibit a "random" normal distribution, allowing for the determination of the most probable normal vector. When the point is in the third positional state (on the edge), due to the potential differences in point cloud density on either side of the edge, the method can still identify the most probable normal vector as the final normal vector for the seed point by appropriately setting the maximum number of plane combinations ($T_{\max}$) and confidence level. This approach effectively addresses the challenges posed by complex geometries and enhances the accuracy of rock structural surface extraction.

3. Stopping Conditions for Growth

When calculating the normal vector of the point cloud using the Hough space transform, if the K value is too large or the point cloud density within the neighborhood is too high, numerous planes will be generated during the combination process. Region growing stops when the number of planes exceeds the maximum plane count $T_{\max}$, or when the confidence level satisfies the requirement. If the number of clustered points in a structural plane is less than the minimum cluster point count $C_{\min}$, region growing will also stop.

The region growing process based on the normal vector differences using the Hough space transform is as follows:

(1) Set the neighborhood size K, calculate the surface variation SV_i of the point cloud, sort the point cloud data in ascending order by SV_i, and select the point with the smallest SV_i as the initial seed point.
(2) Randomly select two points from the neighborhood and the current seed point to form a plane T_i. Calculate the normal vector $\vec{N_i}$ of the plane T_i, then vote all normal vectors into the corresponding storage container. Region growing stops when the combination satisfies $T_{\max}$ and the confidence level is met. The mean of the normal vectors in the container with the highest vote count is selected as the final normal vector for the point.

(3) Set the normal vector angle threshold θ_{th}, calculate the angle $\Delta\theta_i$ between the normal vector of the neighboring point and the current seed point. If $\Delta\theta_i < \theta_{th}$, the neighboring point is added to the current region.
(4) Set the surface variation threshold SV_{th}, calculate the difference ΔSV_i between the surface variation of the neighboring point and the current seed point. If $\Delta SV_i < SV_{th}$, remove the current seed point and designate the neighboring point as the new seed point, continuing the growth process with the new seed point.
(5) Repeat steps (3) and (4). If the number of clustered points is less than C_{min}, the region growing process for that area is complete, and the region is added to the cluster array.
(6) Repeat step (5) until all points are traversed.

2.1.2. Extraction of Rock Mass Structural Planes Information

After region growing, the points within the same structural planes are grouped together to form a point set with different point sets assigned different colors to distinguish each structural plane. Based on the least squares principle, plane fitting is performed for each point set, and the normal vector of the plane is calculated. Using principles of analytical geometry, the normal vector is converted into the structural plane's orientation information, specifically dip direction and dip angle.

Let the normal vector of the structural planes be $\vec{N} = (a,b,c)^T$ with $c > 0$, and the formulas for calculating the dip direction α and dip angle β of the structural planes can be applied, as verified and used by Dong [30]. In the geographic coordinate system, assuming the positive direction of the Y-axis is north, the positive direction of the X-axis is east, and the positive direction of the Z-axis is upward, the formulas for calculating the dip direction α and dip angle β are as follows:

$$\begin{array}{c} \beta = \arccos(c) \\ if\ a \geq 0, b \geq 0, \alpha = \arcsin(a/\sin\beta) \\ if\ a < 0, b < 0, \alpha = 360^\circ - \arcsin(-a/\sin\beta) \\ if\ a \geq 0, b \geq 0, \alpha = 180^\circ - \arcsin(a/\sin\beta) \\ if\ a \geq 0, b > 0, \alpha = 180^\circ + \arcsin(-a/\sin\beta) \end{array} \tag{1}$$

2.1.3. Application Analysis of Rock Mass Structural Planes Identification and Information Extraction

4. Visualization Analysis of Normal Vectors

To further demonstrate the advantage of the Hough space transform in calculating the normal vectors of point clouds near the edges of structural planes, this paper uses both the Hough space transform and the PCA method to calculate the normal vectors of point cloud data for a cube. The normal vectors are then visualized using the PCL point cloud library. If the normal vector of each point is displayed, the result would be too cluttered to discern the distribution of the normal vectors at the edges of structural planes. Therefore, a normal vector is displayed every 10 points, and the length of each normal vector is set to 0.5. The visualization results are shown in Figure 2.

In the figure, the green parts represent the point cloud, and the white line segments represent the normal vectors. The direction of each white line indicates the normal direction at that point. Each face of the cube represents a structural plane, and each edge represents the boundary of a structural plane. As shown in Figure 2, in flat areas, the normal vectors calculated by both methods perform well. In Figure 2b, the normal vectors near the edges are divergent, and the change in the direction of the normal vectors is relatively smooth. However, in Figure 2d, at the same locations, the normal vectors are perpendicular to the planes on both sides of the edges, and the distribution shows a more distinct variation trend. The Hough space transform method significantly outperforms the PCA method in calculating normal vectors near the edges of structural planes. The main reason is that the PCA method calculates the normal vector of a core point using the spatial relationship of all points in its neighborhood. When the core point is near or on the edge of a structural plane, points from another structural plane can affect the accuracy of the core point's normal vector. In contrast, the Hough space transform method randomly selects two points in the neighborhood to form a plane with the core point and calculates its normal vector. The most voted normal vector is then selected as the core point's normal vector, effectively eliminating the influence of points from other structural planes on the calculation and providing excellent noise resistance.

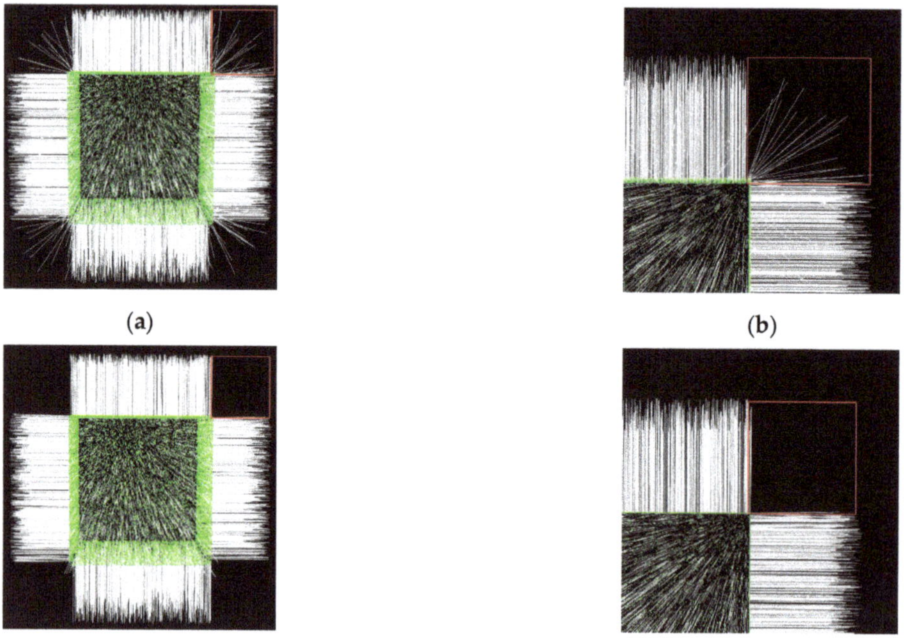

(a) (b) (c) (d)

Figure 2. Visualization of Normal Vectors on a Cube. (**a**,**b**) PCA method; (**c**,**d**) Hough space transform method.

5. Analysis of Region Growing Results

The normal vector difference-based region growing algorithm using the Hough space transform involves several key parameters: neighborhood size K, normal vector angle threshold θ_{th}, surface variation threshold SV_{th}, maximum number of planes $T_{\max}$, confidence level, and minimum number of cluster points $C_{\min}$. K represents the number of points in the neighborhood of point P_i; the smaller the K value, the stronger the seed point's characteristics, and the more fragmented the structural plane's segmentation becomes. The normal vector angle threshold θ_{th} and the surface variation threshold SV_{th} are limiting parameters. The $\theta_{th} \in [0, \pi]$ is set too high, points with small feature differences will be merged, reducing the number of identified structural planes. The $SV_{th} \in [0, 5]$ is used for seed point determination, and the smaller the value, the longer the algorithm takes, though it has minimal effect on the structural planes recognition result. The maximum number of planes $T_{\max}$ and the confidence level control the balance between computation time and accuracy. Generally, when $T_{\max} = 800$, the confidence level exceeds 0.95. $C_{\min}$ represents the minimum number of points within structural planes after region growing, and in practical applications, this value should be determined based on the point cloud density, typically ranging from 50 to 150.

After visualizing the normal vectors in the previous section, we further validate the reliability of this method using point cloud data generated for a regular hexagonal pyramid through the Open3D library. The hexagonal pyramid has a height of 1 m and a base edge length of 0.5 m, with an average point spacing of 4 mm. The number of points on each face is known, with a total of 146,913 points. Each face of the hexagonal pyramid represents a structural plane, and two opposite faces form a group, resulting in three groups of structural planes. First, we use both the PCA method and the Hough space transform to calculate the normal vectors of the hexagonal pyramid point cloud data. The same region growing parameters are set, and then region growing segmentation is performed on the point cloud data of the hexagonal pyramid. The results of the region growing process are shown in Figure 3.

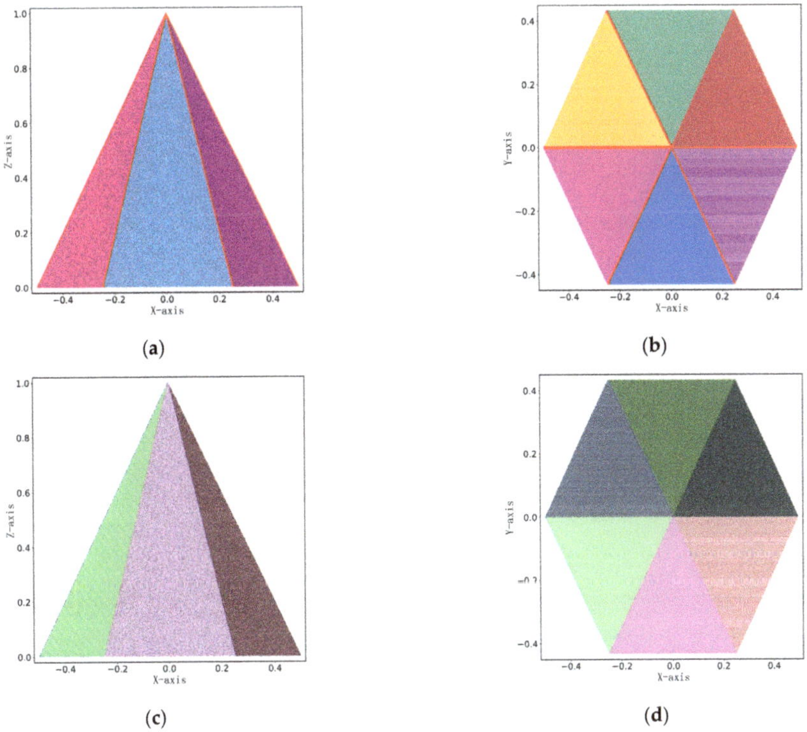

Figure 3. Region Growing Result Graph for Hexagonal Pyramid. (**a**) Front View using PCA; (**b**) Top View using PCA; (**c**) Front View using Hough space transform; (**d**) Top View using Hough space transform.

After region growing, the point cloud data within the same plane is assigned random colors and projected onto the XOZ and XOY planes to draw the front views (a), (c) and top views (b), (d) of the region growing results using the PCA method and Hough space transform. As shown in Figure 3, it is clear from the front and top views that each side of the regular hexagonal pyramid has been successfully identified and segmented. In Figure 3a,b, the point clouds along the edges and vertices of the hexagonal pyramid are red, indicating points that were not successfully identified. In contrast, in Figure 3c,d, only a small number of points in the same areas remain unrecognized. According to the study by Guo Ge [19], to quantify the effect of region growing, statistical analysis was performed on the number of identified and unidentified points within each plane. The recognition accuracy of region growing is measured by the identification rate ω, which is calculated using the following formula:

$$\omega = \frac{J_n^i}{J_n} \tag{2}$$

In the formula: J_n represents the total number of points within the structural plane, and J_n^i represents the number of successfully identified points within the structural plane.

Table 1 first lists the total number of points within the six structural planes of the hexagonal pyramid before region growing. Then, it shows the number of identified points within the six structural planes after region growing using the PCA method and the Hough space transform. The identification rate ω for both methods is calculated using Formula (2). The lowest identification rate using the PCA method was 94.45%, and the highest was 97.23%. In contrast, the lowest identification rate using the Hough space transform was 99.12%, and the highest was 99.65%. For each structural plane, the identification rate of the Hough space transform was higher than that of the PCA method. In terms of overall identification rate, the Hough space transform improved the identification rate by 3.9%.

Table 1. Region Growing Identified Point Count Table.

Structural Plane ID	Total Points (Count)	Identified Points (Count)		Unidentified Points (Count)		Identification Rate ω (%)	
		PCA	Hough	PCA	Hough	PCA	Hough
J1	28,559	27,218	28,459	1341	100	95.30	99.65
J2	28,479	27,142	28,356	1337	123	95.31	99.57
J3	25,292	23,889	25,070	1403	222	94.45	99.12
J4	21,995	21,095	21,853	900	142	95.91	99.35
J5	21,475	20,880	21,378	595	97	97.23	99.55
J6	21,113	20,099	20,935	1014	178	95.20	99.16
Total	146,913	140,323	146,051	6590	862	95.51	99.41

Based on the visual results of region growing and the identification rate statistics in Table 1, both methods achieved similar identification accuracy in the central regions of the structural planes. However, the Hough space transform was less affected by sharp spatial changes, resulting in higher identification accuracy at the edges of the structural planes.

2.2. Cluster Analysis of Rock Mass Structural Planes

Cluster analysis of rock mass structural planes involves grouping multiple structural planes to extract the orientation information of dominant structural planes. Referring to the Clustering by Fast Search and Find of Density Peaks (CFSFDP) method [31], the sine square value of the normal vectors of the structural planes is used as the metric for measuring the similarity between different structural planes. The decision value for each structural plane is calculated, and a decision graph is plotted to automatically identify the cluster centers and the number of clusters. The remaining non-central structural planes are assigned to clusters, and based on local density, the structural planes are divided into core and outlier planes. This method is suitable for large-scale cluster analysis of high and steep rock mass structural planes. The basic steps of the CFSFDP clustering algorithm are as follows:

1. Calculate the sine square value of the angle between the normal vectors of different structural planes, which serves as the clustering distance $d_{dist}(n_i, n_j)$;
2. Set the cutoff distance d_c and calculate the local density ρ_i, which can effectively identify outlier structural planes;
3. Arrange the local density ρ_i in descending order and calculate the control distance δ_i;
4. Plot the decision graph using the local density ρ_i as the horizontal axis and the control distance δ_i as the vertical axis and identify the cluster centers and the number of clusters;
5. Assign the remaining points (non-central points) to the nearest category with a higher density than the point, completing the assignment of all non-central points in one iteration;
6. Set the local density percentage, calculate the boundary density, and divide the structural planes into core and boundary points to identify outliers, thus improving clustering accuracy.

In Section 2.1.3, the region growing algorithm based on Hough space transform normal vector differences is used to identify the structural planes. By combining Formula (1), the dominant orientation information of the structural planes can be further extracted, and the CFSFDP clustering algorithm is applied to perform cluster analysis of the structural planes to obtain the orientation of dominant structural planes.

According to the spatial relationships of each face of the regular hexagonal pyramid, the six faces can be divided into three groups of structural planes with the same dip angle and dip directions differing by 180°. If conventional metrics such as Euclidean distance or Manhattan distance are used as the similarity measure between two structural planes, they would be classified into different categories. However, by using the sine square value of the normal vector angle as the clustering distance, this issue can be effectively resolved. The clustering distance calculation results for some structural planes are shown in Figure 4a. Additionally, as shown in Figure 4c,d, the number of clusters for the structural planes of the regular hexagonal pyramid is three, with the cluster centers of the three groups of structural planes represented by red, yellow, and blue. The extracted orientation information of the structural planes was also imported into RocScience Dips 7.0 to generate the polar density map, and the grouping results are shown in Figure 4b. The clustering results from both methods are consistent.

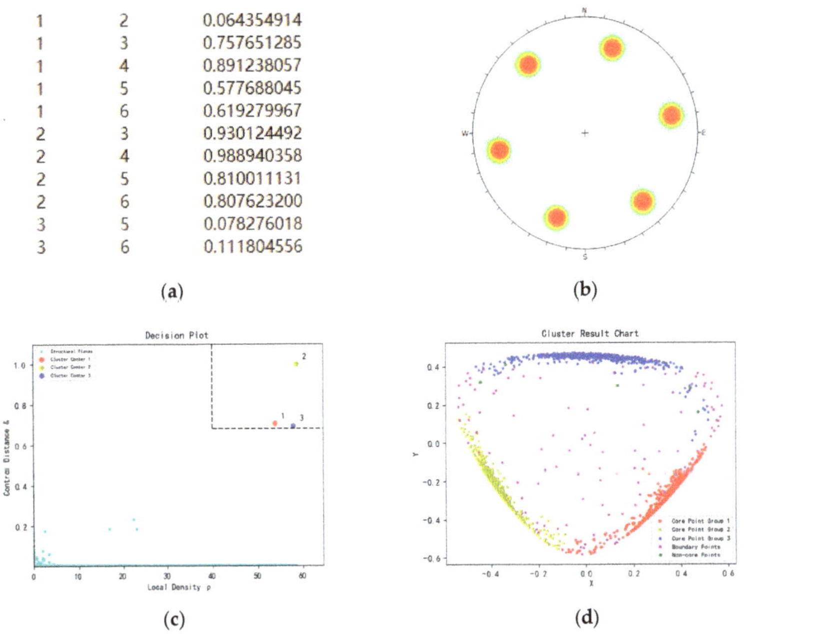

1	2	0.064354914
1	3	0.757651285
1	4	0.891238057
1	5	0.577688045
1	6	0.619279967
2	3	0.930124492
2	4	0.988940358
2	5	0.810011131
2	6	0.807623200
3	5	0.078276018
3	6	0.111804556

(a) (b) (c) (d)

Figure 4. Cluster Analysis Graph of the Regular Hexagonal Pyramid Structural Planes. (**a**) Clustering distance calculation result graph; (**b**) Polar density map; (**c**) Decision graph; (**d**) Clustering result graph.

3. Example Analyses

The data used in this paper was obtained from Slob's personal page on Research Gate, available at: https://www.researchgate.net/publication/289523409_raw_point_cloud_data_-ascii_x_y_z_intensity-_metadata, accessed on 27 May 2024 [32]. The data was collected along the slopes of highway TP 7403 km 06, near the village of Torroja, Spain. The study area consists entirely of Carboniferous rocks, primarily composed of shales and sandstones, which are commonly interbedded and resemble flysch deposits. These sediments have undergone low-grade metamorphism, resulting in the formation of slates, metasandstones, and greywackes, reflecting their complex geological history. The thickness of the Carboniferous sediments can reach up to 2000 m and experienced folding during the Hercynian orogeny, leading to the development of distinct layering and structural surfaces. The well-developed structural surfaces of the Carboniferous rocks make the exposures within these rocks highly suitable for demonstrating the methods proposed in this thesis.

The research data were obtained using the ILRIS_3D three-dimensional laser scanner from Optech, with an average point spacing of 6 mm. The data is stored in PCD format and includes three-dimensional coordinates (X, Y, Z) and RGB attribute information. Figure 5a shows a field image of the scanned area. This study only selected a portion of the point cloud data from the slope, indicated by the black rectangular box in Figure 5a, which is enlarged in Figure 5b. The overall dimensions of the selected research area are 3.8 m × 3 m. The point cloud data for the selected area underwent filtering to remove vegetation points or isolated points. The final number of points in the selected research area is 30,633, as illustrated in Figure 5.

(a) (b)

Figure 5. Field images of the study area. (**a**) Schematic diagram of the study area location; (**b**) Local magnification of the study area.

Figure 6b shows the identification results of rock mass structural planes after applying the Hough space transformation and normal vector difference region growing method. Different colored areas represent different rock mass structural planes, with a total of 281 planes identified. For the regular planar rock mass structural planes, the region growing identification performs well, with only a few points at the edges of the planes not successfully segmented. The black rectangular box in Figure 6 highlights the locations of non-planar rock mass structural planes or complex areas of rock surfaces, where the segmentation of rock mass structural planes is fragmented. This fragmentation is due to the inherent randomness when using only the normal vector as the criterion for region growing. After successfully identifying the rock mass structural e planes, the clustering distances (sine squared values), local density, and control distances between any two the rock mass structural planes are calculated, followed by clustering analysis using the CFSFDP clustering algorithm.

As shown in Figure 7a,b, the number of clustered structural planes in the study area is two, indicating the presence of two dominant structural plane groups. Additionally, the 281 identified structural planes are imported into RocScience Dips 7.0 software to generate poles and equal density plots. The number of dominant structural plane groups and their orientations are obtained, and a comparative analysis of the orientations derived from the two methods is conducted.

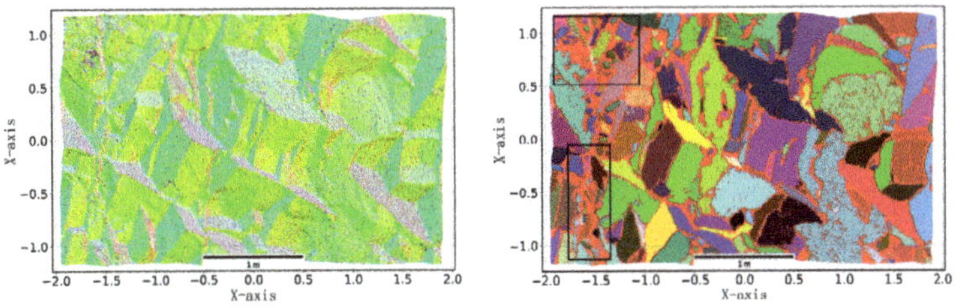

(**a**) (**b**)

Figure 6. Point Cloud Data of High and Steep Highway Slopes. (**a**) before region growing; (**b**) after region growing.

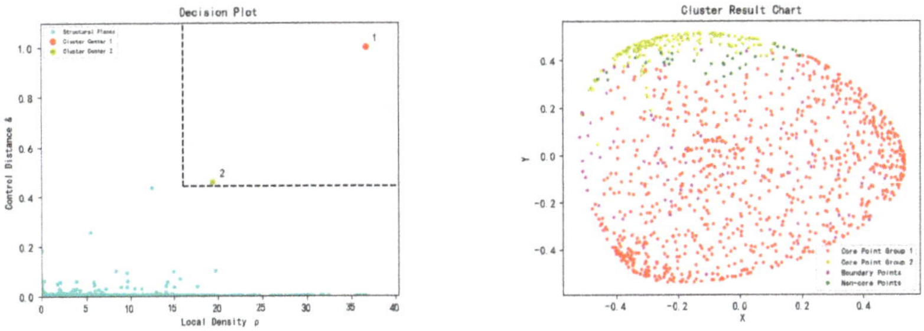

(**a**) (**b**)

Figure 7. Clustering Graph of Rock Mass Structural Planes on Highway Slope. (**a**) Decision Tree graph; (**b**) Clustering Result graph.

As shown in Figure 8, the number of dominant structural plane groups obtained using RocScience Dips 7.0 software is still two, consistent with the clustering analysis results of this study. The orientations of the dominant structural planes obtained by both methods are presented in Table 2. Compared to the results from the open-source geological analysis software RocScience Dips 7.0, the orientation errors for the J1 and J2 dominant structural plane groups are within 3°. This demonstrates that the method proposed in this study not only provides rapid and accurate identification of rock mass structural planes but also automatically determines the cluster centers and the number of clusters, thus obtaining the orientations of dominant structural planes. This method offers significant advantages over traditional geological exploration methods.

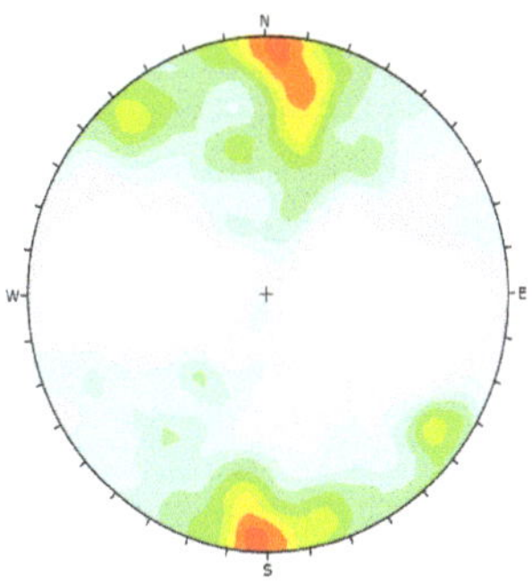

Figure 8. Pole Density Map of Rock Mass Structural Planes in Highway Slope.

Table 2. Statistical Table of the Orientations of Dominant Structural Planes in Highway Slope.

Method	J1	J2
	Dip Direction < Dip Angle	**Dip Direction < Dip Angle**
RocScience Dips 7.0	182.76° < 81.07°	282.42° < 33.13°
Proposed Method	183.08° < 82.32°	284.65° < 33.95°

4. Discussion

In the complex environments of various engineering projects, there are often many slopes of high and steep gradients formed either naturally or by human activity, making it impossible for geological survey personnel to conduct investigations using traditional contact measurement methods. This paper focuses on high and steep highway slopes, utilizing non-contact measurement methods to obtain 3D point cloud data from the study area. Based on this 3D point cloud data, we propose a method for rock mass structural plane identification using Hough space transformation and normal vector interpolation region growing, along with a CFSFDP clustering method for rock mass structural planes. This method is applied to a regular assembly model to verify its effectiveness. Below are three points of discussion regarding the research methods and results:

This paper employs the Hough space transform and vector interpolation region growing algorithm for rock mass structural plane identification and information extraction. The algorithm consists of two parts: Hough space transform for point cloud normal vector calculation and vector interpolation region growing. Both the Hough space transform and PCA methods calculate normal vectors based on neighborhood point relationships. PCA treats all points within a neighborhood as a whole, constructs a covariance matrix, performs eigenvalue decomposition, and finds the corresponding eigenvectors. The eigenvector with the largest variance represents the normal vector of the core point. When the core point is near or on the edge of a structural plane, points from non-core surfaces may be included in the covariance matrix, leading to inaccurate normal vector calculations. The core idea of the Hough space transform is to randomly select two points within the neighborhood to form a plane with the core point and calculate its normal vector. The normal vectors are stored in an accumulator container through a voting process. To ensure computational speed, a maximum number of planes is set; once this number is reached, it is assumed that sufficient confidence is obtained to ensure the accuracy of the point cloud normal vector. The normal vector with the highest number of votes is selected as the core point normal vector. This approach reduces the impact of points from non-core surfaces, making the Hough space transform method robust to noise and accurate in calculating the normal vectors at the edges of rock mass structural planes. In the cube point cloud normal vector visualization analysis experiment, the normal vectors calculated using the Hough space transform were perpendicular to the planes on both sides of the edges, showing a clear trend in normal vector changes. This demonstrates that the method has distinct advantages in calculating normal vectors for complex rock mass surface point clouds. Region growing, which only uses normal vector differences as the growth criterion, exhibits significant randomness. In practical applications, it is also evident that the identification of non-planar rock mass structural planes or complex areas of rock surfaces shows a decline in performance. To address this issue, future research could explore other descriptors that can represent the spatial characteristics of point clouds. Establishing a multi-dimensional feature fusion mechanism could replace the single-feature criterion with multi-dimensional feature values as criteria for region growing, thereby enhancing the effectiveness of rock mass structural planes identification.

The Fast Search and Find of Density Peaks (CFSFDP) algorithm is simple in steps, can automatically find clustering centers and determine the number of clusters, and does not rely on complex parameter tuning. Although the overall efficiency is high, calculating the clustering distance $d_{dist}(n_i, n_j)$ remains complex. Direct clustering of massive point cloud data requires calculating the clustering distance between each point and every other point, which involves a large amount of computation. This study first uses the region growing algorithm for rock mass structural plane identification and information extraction. After region growing, the point cloud data are segmented or merged into many blocks, each representing a rock mass structural plane. The number of structural planes is much smaller than the number of point clouds, making rock mass structural planes suitable for clustering and effectively addressing the problem of CFSFDP's high clustering distance complexity, which is only suitable for small datasets. In terms of computational resources, this study utilized a platform equipped with two Intel Xeon E5-2630 v4 processors, each having 10 cores and 20 threads, along with a system memory of 32 GB and a 1 TB mechanical hard drive. The total runtime for the computational process was approximately 15 min, demonstrating good efficiency. To effectively extend this method to broader research areas, we suggest that future work consider employing hierarchical processing and parallel computing strategies, which would further enhance the applicability and computational efficiency of the method in larger study regions.

In real scenarios, rock mass surfaces often have structural planes with the same dip angle but a 180° difference in dip direction. Using traditional Euclidean distance, Manhattan distance, Chebyshev distance, or spherical distance as clustering distances would misclassify such structural planes into different groups. By using the sine squared value of the angle between the normal vectors of different structural planes as the clustering distance $d_{dist}(n_i, n_j)$, this type of clustering error can be effectively avoided. Additionally, a cutoff distance d_c is set to effectively remove outlier structural planes, which helps handle noise points in the data and maintain the stability and accuracy of the clustering results.

5. Conclusions

Based on 3D point cloud data, this study utilizes Hough space transform to calculate point cloud normal vectors and employs vector differences as criteria for region growing, effectively achieving the identification and information extraction of rock mass structural planes. In the experimental validation using a hexagonal pyramid, six structural planes were successfully identified. The accuracy of the region growing was evaluated using visualization methods and recognition rates ω. The number of rock mass structural planes identified by this method matched the theoretical results, and the precision of the extracted structural plane orientations also met engineering specification requirements.

After completing the identification and information extraction of the mass structural planes, these surfaces were used as clustering objects, with the sine squared value of the normal vector differences serving as the clustering distance $d_{dist}(n_i, n_j)$. A cutoff distance d_c was set to filter out outlier surfaces, and the Fast Search and Find of Density Peaks (CFSFDP) clustering method was employed for clustering analysis of the identified surfaces. This approach not only addresses the issue of high complexity in calculating clustering distances with CFSFDP, which is suitable only for small datasets, but also improves clustering accuracy and noise resistance of the method. When applied to high and steep highway slope, the method successfully identified two dominant structural planes groups, with results closely matching those from RocScience Dips 7.0 clustering, and the orientation error of the dominant structural planes was within 3°. Overall, the method demonstrated strong practical value for large-scale geological surveys of high and steep rock masses. Future research will focus on further improving algorithm efficiency and handling larger datasets.

Author Contributions: Conceptualization, J.Z., Y.X., B.W., Z.Y. and K.Y.; methodology, J.Z., Y.X., B.W., Z.Y. and K.Y.; software, J.Z.; validation, J.Z., Y.X., B.W., Z.Y. and K.Y.; resources, Y.X., B.W., Z.Y. and K.Y.; data curation, J.Z., Z.Y. and K.Y.; writing—original draft preparation, J.Z.; funding acquisition, J.Z., Y.X., B.W., Z.Y. and K.Y. All authors have read and agreed to the published version of the manuscript.

Funding: This research was financially supported by the National Natural Science Foundation of China (No. 41861054).

Institutional Review Board Statement: Not applicable.

Informed Consent Statement: Not applicable.

Data Availability Statement: The data used to support the findings of this study are available from the first author upon request.

Acknowledgments: We thank the Second Survey Institute of Yunnan Geological Engineering for their technical support. We also thank the reviewers and editors of the Journal of Applied Sciences for their hard work.

Conflicts of Interest: Authors Ziliang Yang and Kaihua Yang were employed by the Yunnan Geological Engineering the Second Investigation Institute. The remaining authors declare that the research was conducted in the absence of any commercial or financial relationships that could be construed as a potential conflict of interest.

References

1. Zhu, H.H.; Pan, B.G.; Wu, W.; Yong, H.; Liu, J.; Li, H.M. Review on Collection and Extraction Methods of Rock Mass Discontinuity Information. *J. Basic Sci. Eng.* **2023**, *31*, 1339–1360.
2. Sun, G.Z. On the Theory of Structure-Controlled Rock Mass. *J. Eng. Geol.* **1993**, *1*, 14–18.
3. Ge, Y.F.; Tang, H.M.; Huang, L.; Wang, L.Q.; Sun, M.J.; Fan, Y.J. A New Representation Method for Three-Dimensional Joint Roughness Coefficient of Rock Mass Discontinuities. *Chin. J. Rock Mech. Eng.* **2012**, *31*, 2508–2517.
4. Wu, X.; Wang, F.Y.; Wang, M.C.; Zhang, X.Q.; Wang, Q.; Zhang, S. A New Method for Automatic Extraction and Analysis of Discontinuities Based on TIN on Rock Mass Surfaces. *Remote Sens.* **2021**, *13*, 2894. [CrossRef]
5. Wang, F.Y.; Chen, J.P.; Fu, X.H.; Shi, B.F. Study on Geometrical Information of Obtaining Rock Mass Discontinuities Based on VirtuoZo. *Chin. J. Rock Mech. Eng.* **2012**, *33*, 169–175.
6. Wang, S.H.; Mu, J.J.; Zhang, H.; Zhang, S.C. Spatial Modeling of Refined Structural Plane of Rock Mass and Block Stability Analysis. *J. Northeast. Univ. (Nat. Sci.)* **2019**, *40*, 1186–1189.
7. Li, S.C.; Liu, H.L.; Li, L.P.; Shi, S.S.; Zhang, Q.Q.; Sun, S.Q.; Hu, J. A Quantitative Method for Rock Structure at Working Faces of Tunnels Based on Digital Images and Its Application. *Chin. J. Rock Mech. Eng.* **2017**, *36*, 1–9.
8. Hoppe, H.; DeRose, T.; Duchamp, T.; McDonald, J.; Stuetzle, W. Surface Reconstruction from Unorganized Points. In Proceedings of the 19th Annual Conference on Computer Graphics and Interactive Techniques, New York, NY, USA, 1 July 1992; pp. 71–78.
9. Slob, S.; Van Knapen, B.; Hack, R.; Turner, K.; Kemeny, J. Method for Automated Discontinuity Analysis of Rock Slopes with Three-Dimensional Laser Scanning. *Transp. Res. Rec.* **2005**, *1913*, 187–194. [CrossRef]
10. Gigli, G.; Casagli, N. Semi-Automatic Extraction of Rock Mass Structural Data from High Resolution LIDAR Point Clouds. *Int. J. Rock Mech. Min. Sci.* **2011**, *48*, 187–198. [CrossRef]
11. Ge, Y.F. Intelligent Identification and Extraction of Geometric Properties of Rock Discontinuities Based on Terrestrial Laser Scanning. *J. Rock Mech. Eng.* **2017**, *36*, 3050.
12. Chen, N.; Cai, X.M.; Xia, J.W.; Zhang, S.; Jiang, Q.; Shi, C. Intelligent Interpretation of Rock Mass Discontinuity Based on Three—Dimensional Laser Point Cloud. *Earth Sci.* **2021**, *46*, 2351–2361.
13. Kong, X.L.; Xia, Y.H.; Wu, X.Q.; Wang, Z.H.; Yang, K.H.; Yan, M.; Li, C.; Tai, H.Y. Discontinuity Recognition and Information Extraction of High and Steep Cliff Rock Mass Based on Multi-Source Data Fusion. *Appl. Sci.* **2022**, *12*, 11258. [CrossRef]
14. Guo, J.T.; Liu, S.J.; Zhang, P.N.; Wu, L.X.; Zhou, W.H.; Yu, Y.A. Towards Semi-Automatic Rock Mass Discontinuity Orientation and Set Analysis from 3D Point Clouds. *Comput. Geosci.* **2017**, *103*, 164–172. [CrossRef]
15. Gao, F.; Chen, D.P.; Zhou, K.P.; Niu, W.J.; Liu, H.W. A Fast-Clustering Method for Identifying Rock Discontinuity Sets. *KSCE J. Civ. Eng.* **2019**, *23*, 556–566. [CrossRef]
16. Cui, X.J.; Yan, E.C.; Chen, W. Cluster analysis of discontinuity occurrence of rock mass based on improved genetic algorithm. *Rock Soil Mech.* **2019**, *40*, 374–380.
17. Chen, J.Q.; Zhu, H.H.; Li, X.J. Automatic Extraction of Discontinuity Orientation from Rock Mass Surface 3D Point Cloud. *Comput. Geosci.* **2016**, *95*, 18–31. [CrossRef]
18. Wang, S.H.; Ren, Y.P.; Chen, J.Z.; Zhang, Z.S. An Improved Fish Swarm Clustering Algorithm for Structural Grouping. *J. Northeast. Univ. (Nat. Sci.)* **2019**, *40*, 420.
19. Kong, D.H.; Wu, F.Q.; Saroglou, C. Automatic Identification and Characterization of Discontinuities in Rock Masses from 3D Point Clouds. *Eng. Geol.* **2020**, *265*, 105442. [CrossRef]
20. Daghigh, H.; Tannant, D.D.; Daghigh, V.; Lichti, D.D.; Lindenbergh, R. A Critical Review of Discontinuity Plane Extraction from 3D Point Cloud Data of Rock Mass Surfaces. *Comput. Geosci.* **2022**, *169*, 105241. [CrossRef]
21. Hu, L.; Xiao, J.; Wang, Y. Efficient and Automatic Plane Detection Approach for 3-D Rock Mass Point Clouds. *Multimed. Tools Appl.* **2020**, *79*, 839–864. [CrossRef]
22. Zhang, J.H. Intelligent Identification of Rock Mass Structural Plane and Stability Analysis of Rock Slope Block. *Sci. Rep.* **2022**, *12*, 16745. [CrossRef] [PubMed]
23. Zhao, X.D.; Liu, J.; Zhang, H.X.; You, W.; Liu, F.T. Rock mass Structural Plane Digital Recognition and Stope Stability Classification Based on the Photographic Surveying Method. *J. Min. Saf. Eng.* **2014**, *31*, 127.

24. Sheng, Y.; Song, T.J.; Chen, J.P.; Zhang, W. Clustering Analysis of Dominative Attitudes of Rock Mass Structural Plane Based on Firefly Algorithm. *J. Northeast. Univ. (Nat. Sci.)* **2015**, *36*, 284.
25. Feng, Y.; Ma, F.S.; Gong, C.C.; Guo, J.; Wang, S.F.; Liu, Z.C. Data Analysis Method for Optimized and Dominant Orientations of Joints in Rock Mass. *J. Eng. Geol.* **2011**, *19*, 887–892.
26. Ai, C.M.; Lu, Y.; Sun, P.P. Dynamic Cluster Grouping Analysis of Occurrence of Structural Plane in Natural Rock Mass. *J. Yangtze River Sci. Res. Inst.* **2023**, *40*, 106.
27. Song, J.L.; Huang, R.Q.; Pei, X.J. Particle swarm optimization algorithm based fuzzy c-means cluster analysis for discontinuities occurrence in rock mass. *J. Eng. Geol.* **2012**, *20*, 591–598.
28. Deng, J.H.; Chen, B.L.; Wu, X.M.; Xie, G.H. Self-Organized Cluster Analysis on Texture Plane Occurrence of Rock Mass. *J. Yangtze River Sci. Res. Inst.* **2011**, *28*, 50.
29. Guo, G. Study on Rock Mass Structural Plane Identification and Rock Mass Integrity Evaluation Based on 3D Laser Scanning Technology. Master's Thesis, Three Gorges University, Yichang, China, 2023.
30. Dong, X.J. The Three-Dimensional Laser Scanning Technique and Research on Its Engineering Application. Master's Thesis, Chengdu University of Technology, Chengdu, China, 2007.
31. Rodriguez, A.; Laio, A. Clustering by Fast Search and Find of Density Peaks. *Science* **2014**, *344*, 1492–1496. [CrossRef]
32. Slob, S. *Automated Rock Mass Characterization Using 3-D Terrestrial Laser Scanning*; ResearchGate: Berlin, Germany, 2010.

applied sciences

MDPI

Article

Shear Mechanical Behaviours and Size Effect of Band–Bedrock Interface: Discrete Element Method Simulation Insights

Hao Wang [1], Xueyan Guo [1], Xinrong Liu [1,2,3,*], Xiaohan Zhou [1,2,3] and Bin Xu [4]

1 School of Civil Engineering, Chongqing University, Chongqing 400045, China; cquwanghao@126.com (H.W.); 18569939178@163.com (X.G.); cqzhouxhan@126.com (X.Z.)
2 National Joint Engineering Research Center for Prevention and Control of Environmental Geological Hazards in the TGR Area, Chongqing University, Chongqing 400045, China
3 Key Laboratory of New Technology for Construction of Cities in Mountain Area of the Ministry of Education, Chongqing University, Chongqing 400045, China
4 School of River and Ocean Engineering, Chongqing Jiaotong University, Chongqing 400074, China; geotechnicale2016@163.com
* Correspondence: liuxrong@126.com

Abstract: The shear band is a prominent feature within the Banbiyan hazardous rock mass located in the Wushan section of the Three Gorges Reservoir area. This band constitutes a latent risk, as the potential for the rock mass to slide along the region threatens the safety of lives and property. Presently, the understanding of the shear mechanisms and the impact of shear band size on the band–bedrock interface is incomplete. In this study, based on band–bedrock shear laboratory tests, DEM simulation is used to investigate the shear-induced coalescence mechanism, stress evolution, and crack-type characteristics of the band–bedrock interface. In addition, the shear mechanical properties of samples considering specimen size, rock step height, and step width are further studied. The results show that the crack initiation and failure crack types observed in the first rock step are predominantly tensile. In contrast, the failure cracks in the remaining rock slabs and steps are primarily characterised by shear mode in addition to other mixed modes. The stress condition experienced by the first step is very near to the position of the applied point load, whereas the stress distribution across the remaining steps shows a more complex state of compressive–tensile stress. The relationship between shear parameters and sample size is best described by a negative exponential function. The representative elementary volume (REV) for shear parameters is suggested to be a sample with a geometric size of 350 mm. Notably, the peak shear strength and shear elastic modulus demonstrate a progressive increase with the rise in rock step height, with the amplifications reaching 91.37% and 115.83%, respectively. However, the residual strength exhibits an initial decline followed by a gradual ascent with increasing rock step height, with the amplitude of reduction and subsequent amplification being 23.73% and 116.94%, respectively. Additionally, a narrower rock step width is found to diminish the shear parameter values, which then tend to stabilise within a certain range as the step width increases.

Keywords: shear band–bedrock interface; shear mechanical behaviour; size effect; DEM simulation

1. Introduction

Tectonic discontinuity is the result of geotectonic movement and is widely distributed in rock mass, which leads to uncertainty regarding the stability of engineering geological bodies [1]. There are special engineering geological conditions in the Three Gorges Reservoir (TGR) area [2–5]. After many years of geological evolution, a large number of engineering rock masses with discontinuities have been formed. Located in the Wushan section of the TGR area, a special filling discontinuity exists in the Banbiyan dangerous rock mass (BDRM), as shown in Figure 1. The filling of this discontinuity is a shear band composed of rock blocks, which gradually formed in the process of rock mass unloading and structural movement. The failure mode of the BDRM is most likely to be of the slip type [6]. This kind of failure mainly depends on the material properties and structural form of the fillings. Once the BDRM becomes unstable, the safety of the Yangtze River waterway and surrounding people and property will be threatened. Therefore, further research on the shear mechanical behaviour and failure mechanism of the shear band has great significance for the stability control of the BDRM.

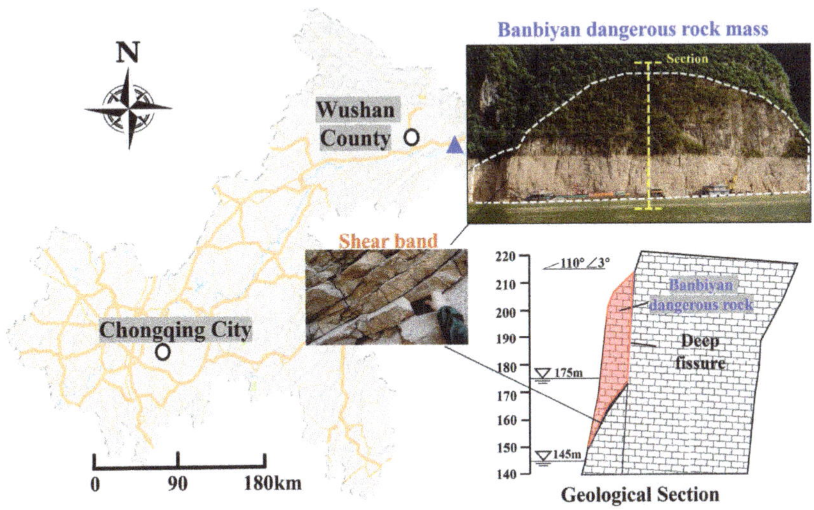

Figure 1. Geographical location and characteristics of shear band in Banbiyan dangerous rock mass.

As common special structures in natural rock mass, the shear mechanical properties of discontinuities have been widely researched. In the field of shear strength, the bilinear shear strength model was first proposed by Patton [7]. Subsequently, Barton et al. [8] developed the JRC-JCS model of discontinuity shear strength based on a large number of shear laboratory tests. Since then, numerous tests on the shear strength of discontinuities have been carried out, and a variety of peak strength models have been put forward, including the JRC-JMC model that considers the structural plane coincidence coefficient [9], the three-parameter hyperbolic nonlinear failure criterion [10], the strength model based on quantified surface descriptions [11], the peak strength criterion that considers morphology [12], and the brittle drop mode damage-based constitutive model [13]. The shape of a discontinuity is a considerable factor used to determine shear strength and failure, so research examining the shear behaviour of discontinuities with different morphological structural planes has received extensive attention. Liu et al. [14] studied the shear mechanical behaviour of rock joints under different morphologies, and the result showed that the peak shear stress approximately linearly increased and the peak shear displacement decreased with the increase in first-order asperities. According to an anisotropic plaster shear test, Liu et al. [15] found that with an increment of the inclination angle of the critical waviness, both the normal displacement and sheared-off asperity mass increased, following power law functions. Zhang et al. [16] conducted a sandstone discontinuity shear test under water deterioration and determined that the chemical action of water reduces the structural plane roughness and leads to a decrease in the shear strength. Le et al. [17] focused on the failure characteristics of rock-like samples with different geometric steps, and the influence of normal stress and step height on peak strength was identified. In addition to the above research, morphology characteristics [18–21] and lithological differences between upper and lower plates [22–24] were also considered during research involving discontinuity shear testing. From the aforementioned studies, it is evident that research on the shear mechanical behaviour of discontinuities has primarily focused on joints with natural roughness and rock bridges between joints, as these represent common models of discontinuities found in nature. However, special discontinuities such as the shear band in the Banbiyan dangerous rock mass, which are formed due to unique tectonic movements, are rarely encountered in either engineering or research. In particular, interfaces with rock steps have not been a focus of discontinuity research. As a result, their mechanical properties are often overlooked. In this context, there is a scarcity of in-depth studies on how the size of prominent steps affects the mechanical properties of such interfaces, especially regarding the size effect. Hence, it is clear that novel research on step-like band–bedrock interfaces needs to be conducted.

Understanding crack propagation during the shear process may help to identify the shear failure path in engineering rock mass. However, due to limitations in test conditions, it is difficult to clearly and accurately record some shear failure processes. In recent years, the development of numerical methods has solved this problem to some extent. The main advantage of the discrete element method (DEM) is to obtain information on the dynamic failure process and crack propagation of the specimen, and to monitor information that cannot be obtained from a variety of laboratory tests. Therefore, the DEM has been widely used in geotechnical engineering when numerical simulation is required. Bahaaddini et al. put forward the shear box genesis approach for generating the upper and lower plates in direct shear simulation [25], solving the problem of particle interlocking in the bond removal method. Cement–rock interface shear simulation tests were conducted by Yang et al. [26], and the evolution law of the sample force chain during the shear process was presented. Cao et al. [27] explored the shear behaviour of 3D nonpersistent jointed rock-like specimens using PFC software (Version 5.0) and obtained the structural plane stress variation law, which cannot be observed in experimental testing. The energy evolution process and the change in the crack number of samples with nonpersistent joints were clarified by Tai et al. [28] through simulation. Moreover, Chen et al. [29] summarised the crack types in rock containing nonpersistent joints under uniaxial compression by analysing the particle relative displacement. Song et al. [30] utilised the discrete element method to simulate the mechanical properties of sandstone under multi-level cyclic loading, revealing that models with higher heterogeneity in tensile strength and cohesion can better reproduce the stress–strain relationship patterns. In addition to reproducing the experimental results, numerical simulation is often used for extended learning due to its convenience. Observational studies have been carried out based on the DEM that examined the effect of size [31–34], random discrete fracture [35–37], fracture numbers, and geometry size [38,39] on the mechanical behaviour of rock or rock-like samples. It follows that the discrete element method may be helpful for the study of the micro-mechanical properties of shear bands in the BDRM. At the same time, the challenge of large workloads in size effect laboratory tests will be effectively solved through numerical simulation.

In summary, with the help of existing survey data and related references on the BDRM, a series of laboratory experiments and numerical simulations were carried out on the shear mechanical behaviour of the band–bedrock interface generalised from the BDRM. This paper is structured as follows. In Section 2, the geological conditions of the BDRM are briefly described. In Section 3, the establishment and schemes of the experiment and numerical simulation tests are introduced. Then, in Section 4, the shear coalescence mechanism, stress evolution, and crack-type characteristics of the band–bedrock interface are analysed. The effect of the sample and step geometry size on the shear mechanical properties of the band–bedrock interface is discussed in Section 5. Finally, the conclusions are presented in Section 6. It is hoped that this paper may provide some help in understanding the instability and failure mechanisms of the BDRM, contributing to the mitigation of geohazards in regions such as the Three Gorges Reservoir area.

2. Characteristics of Shear Bands

The Banbiyan dangerous rock mass, situated along the right bank of the Wushan section within the Three Gorges Reservoir area, is geographically positioned with approximate coordinates of latitude N29° to N32° and longitude E109° to E111°, as depicted in Figure 1. The principal lithology of this rock mass comprises medium- to thickly bedded marlstone, argillaceous limestone, and thin- to medium-bedded limestone, which are predominantly found in members 2 to 3 of the Lower Triassic Jialingjiang Formation. The trailing edge of the dangerous rock mass has been severed by profound fissures, effectively disconnecting it from the mother rock. Within the BDRM, two distinct shear bands of considerable width are present. Notably, Shear Band 1 is located above the waterline at 145 m and exhibits macroscopic fragmentation. Locally, there are scattered fragments with diameters ranging from approximately 8 cm to 45 cm. The lateral surface of Shear Band 1 is exposed within the hydro-fluctuation zone, making it particularly susceptible to erosion by river water. The interstitial material between these fragments is predominantly of low mud content and exhibits low cementation. The disorderly arrangement of the blocks leads to an interlocking effect. The presence of deep and large fissures significantly influences the stability of the BDRM, with the characteristics of the shear bands playing a pivotal role in determining the overall stability of the rock mass.

3. Laboratory and DEM Modelling Establishment

3.1. Sample Preparation

The fundamental materials for laboratory experiments were sourced from a location adjacent to the BDRM, ensuring that the stratigraphic lithology matched that of the BDRM, with the native material being limestone, as illustrated in Figure 2a. Through meticulous procedures of cutting, grinding, and other preparatory steps, the rock samples were fashioned into standard cylindrical specimens with precise geometric dimensions of 100 mm in height and 50 mm in diameter, as depicted in Figure 2b. These prepared samples were then subjected to a series of uniaxial compression tests to ascertain the fundamental mechanical properties of the rock, including, but not limited to, elastic modulus and compressive strength, as outlined in Figure 2c. The macroscopic mechanical parameters of the rock, as detailed in Table 1, served as a crucial foundation for all subsequent analyses and simulations.

(a) (b) (c)

Figure 2. Field sampling and parameter measurement: (**a**) Field sampling; (**b**) Sample processing; (**c**) Parametric measurement.

Table 1. Basic physical and mechanical parameters of limestone.

Limestone	Density (kg/m^3)	Elastic Modulus (GPa)	Uniaxial Compressive Strength (MPa)	Poisson's Ratio
Value	2580	13.99	131.74	0.24

Combining geological data with on-site survey results (Figure 3a), it can be observed that the shear band and the bedrock exhibit the characteristics of layered rock masses. The interface is stepped, and the stability is maintained by the interlocking of more regular rock slabs within the shear band and the recessed parts in the bedrock. Based on the characteristics of the layered rock mass, considering the size limitations of the servo direct shear testing machine, the precision of the experimental processing, and the proportion of the on-site block size within the shear band, the shear band was generalised into rock slabs of uniform thickness but varying heights, and the bedrock was a complete rock block with protruding steps. The step height was set at 10 mm, and the width at 22 mm. The rock slabs and the bedrock accurately formed a cube with dimensions of 200 mm. The generalised model and dimensions of the entire specimen are shown in Figure 3a. It is worth noting that regular stratified rock masses are generally isotropic in the horizontal plane; thus, the three-dimensional stratified rock mechanics problem can be treated as a planar problem perpendicular to the stratification. This is the reason why the interface roughness is considered a simple two-dimensional geometric problem. However, there are certain limitations to this kind of model generalisation, as the step size at the interface actually has some spatial variability, making it difficult to assess the mechanical properties of the interface. Despite this limitation, in this study, the mechanical shear behaviour of the interface with a stepped shape was the focus, and the non-uniform step size within the same specimen was not considered a harsh initial condition. This was essentially a preliminary exploratory study, hence the step size being unified. This model generalisation method can be applied to shear zones formed by the lateral unloading of layered slopes in steep gorge areas due to river incision or open-pit mining excavation. This is because the fracturing of layered rock masses will inevitably lead to the formation of interlayer fractures, resulting in this stepped interlocking structure.

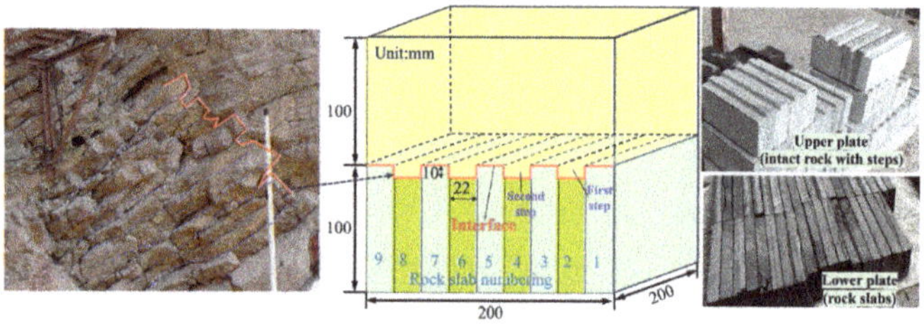

(a)

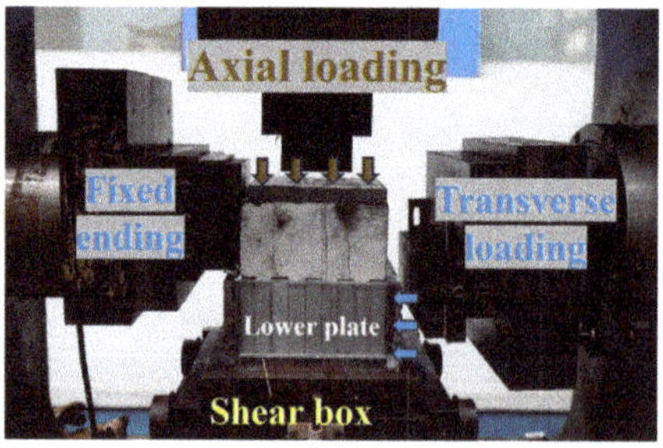

(b)

Figure 3. Experimental model and loading methods: (**a**) Experimental sample design; (**b**) Experimental loading methods.

3.2. Testing System and Procedure

The exposure of the shear band does not impose any additional stress constraints, thus allowing for a straightforward setting of boundary conditions in the shear tests. Consequently, the boundary condition for the shear tests was established as constant normal stress. The height of the dangerous rock mass situated above the shear band is approximately 110 m. Considering the height and density of the overburden rock, three distinct constant normal stress levels were set: 1 MPa, 2 MPa, and 4 MPa. The experimental procedures were conducted using a WDAJ-600 microcomputer-controlled electro-hydraulic servo rock shear rheological testing machine, as shown in Figure 3b. In these tests, the upper plate remains stationary, while the lower plate is put into the shear box. Normal stress is applied through a stress servo mechanism, and the shear load is incrementally applied to the lower plate in a leftward direction. The tangential loading is applied with a displacement-controlled loading speed of 0.5 mm/min.

3.3. DEM Modelling Establishment

PFC (Particle Flow Code) (Version 6.0 used in this study) is a DEM simulation analysis software program developed by Itasca, a company in the United States. In this software, the smallest units of the simulated object are circular particles (2D) or spheres (3D), and the particles (or spheres) are connected to each other through contacts [40]. Contacts can transmit forces and moments, and are updated and calculated at each timestep [40]. Therefore, PFC software can be used to simulate the complex mechanical behaviour of granular materials or blocks, including particle flow, breakage, deformation, and other large deformation issues [41–43]. However, due to the mechanical calculations and updates of tens of thousands of particles and contacts at each timestep, the computational efficiency and accuracy significantly increase with the increase in particle numbers. For complex models, modelling and computational time and accuracy have become its inevitable shortcomings. However, for rock or rock mass simulation tests, while ensuring the size effect, the number of particles can be controlled within an effective range. Therefore, PFC software has been proven to be an effective means of numerically simulating rock mechanics [15,22,27–29,41–43]. It can be seen from Figure 3a that the design of the on-site rock mass and indoor experiment has certain planar problem characteristics because the interface between the rock slab and the dangerous rock mass only has changes in two directions. In light of this, the three-dimensional characteristics of the interface roughness can be simplified into a planar problem. Considering the calculation speed and accuracy, the 2D version of the PFC software (PFC2D) was chosen for numerical simulations in this study. It is important to ensure that the dimensions of the simulated samples align precisely with those of the laboratory test specimens, as depicted in Figure 4a. The samples were established using a shear box genesis approach [25]. Particles were generated within the upper plate and the rock slabs of the lower plate, thereby maintaining the structural integrity and independence of each component. The particle radius in the simulation varies, with the maximum being 1.05 mm and the minimum 0.7 mm, resulting in a total of 13,812 particles that constitute the sample. As shown in Figure 4b, the simulation loading methodology involves the application of shear stress through the movement of the lower wall from left to right. The movement velocity of the walls is set to 0.05 m/s. Under this velocity, the calculation timestep is automatically set to 2.01×10^{-7} s/step. It needs 0.995×10^5 steps to move 1 mm. Referring to the related study results in reference [44], this loading velocity adheres to the quasi-static requirements necessary for the experimental testing.

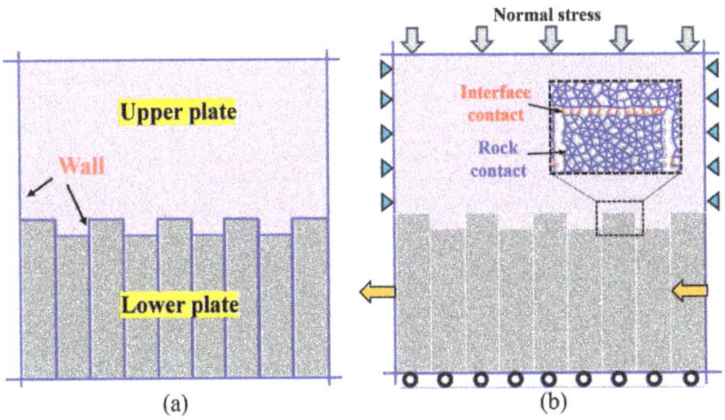

Figure 4. Numerical model establishment and loading methods: (**a**) Model establishment; (**b**) Loading methods.

3.4. Contact Model and Parameter Calibration

In PFC software, compared with the parallel-bond contact model, using a flat-joint model can address the issues of particle interlocking and the mismatched tensile-to-compressive strength ratio [45,46]. Hence, it is considered to be a constitutive model that can better simulate the failure and matching strength of rock samples and has been widely used in a number of studies [47–49]. In view of this, in the simulation test in this study, the flat-joint model was selected as the contact model for the rock and rock slab particles. The smooth-joint model is mainly used to simulate the friction action of rock discontinuities [25,50]. However, in addition to friction action, force transmission also occurs at the interface position during the loading process, and the use of the flat-joint model can give consideration to these two actions. Therefore, the flat-joint model was also chosen as the contact model for the interface.

The "trial and error" approach is the most commonly used method of parameter calibration [27,29,51,52]. With a given set of microscopic parameters, the simulated samples are subjected to loading conditions that mirror those of the laboratory tests. If the simulation results fall within an acceptable error range and show good correspondence with the experimental macroscopic parameters and failure modes, the microscopic parameters are deemed suitable for other simulation tests involving the same rock type. Consequently, in this study, the results of the uniaxial and laboratory shear tests on limestone were reproduced through numerical simulation. To maintain uniformity and accuracy in the simulation parameters, the uniaxial simulation sample was cut directly from the shear sample, as depicted in Figure 5a. Through a process of constant trial and error, results that closely correspond with the experimental data were obtained, as evidenced in Figure 5. Table 2 gives the microscopic parameters used in this study. A quantitative comparison of the mechanical characteristics between the simulation and experimental results is shown in Table 3. The findings further validate the accuracy of the parameter calibration by showing that the errors between the two are within a reasonable range.

Table 2. Microscopic parameters used in simulation.

Microscopic Parameters	Limestone	Interface
Particle minimum radius Rmin (mm)	0.70	/
Particle maximum radius Rmax (mm)	1.05	/
Deformability modulus E^* (GPa)	10.20	0.11
Normal to shear stiffness ratio k^*	2.00	0.50
Tensile strength fj_ten (MPa)	18.50	0.00
Bond strength fj_coh (MPa)	60.00	0.00
Friction angle fj_fa (°)	35.00	55.00
Friction coefficient fj_fric	0.45	0.80

Table 3. Parameter value comparison between experiment and simulation.

Mechanical Parameters	Experimental Result	Simulation Result	Error
Elastic Modulus (GPa)	13.99	13.64	2.50%
UCS (Mpa)	131.74	128.51	2.45%
Poisson's Ratio	0.24	0.25	4.17%

3.5. Experimental and Numerical Simulation Schemes

Besides the three fundamental laboratory shear tests, this study also explored how changes in the sample and step geometry size affect shear mechanical behaviour. Across all models, the microscopic parameters are held constant, following those listed in Table 2, to ensure the consistency and comparability of the results. The specific conditions for both the laboratory tests and numerical simulations are detailed in Table 4, along with their corresponding descriptions. The abbreviations "L" and "S" denote the basic laboratory and simulation tests, respectively. To probe the effects of varying sample sizes, step heights, and step widths, the simulation tests are categorised as "SS" for sample size, "SH" for step height, and "SW" for step width, with the subsequent numbers representing the specific altered values.

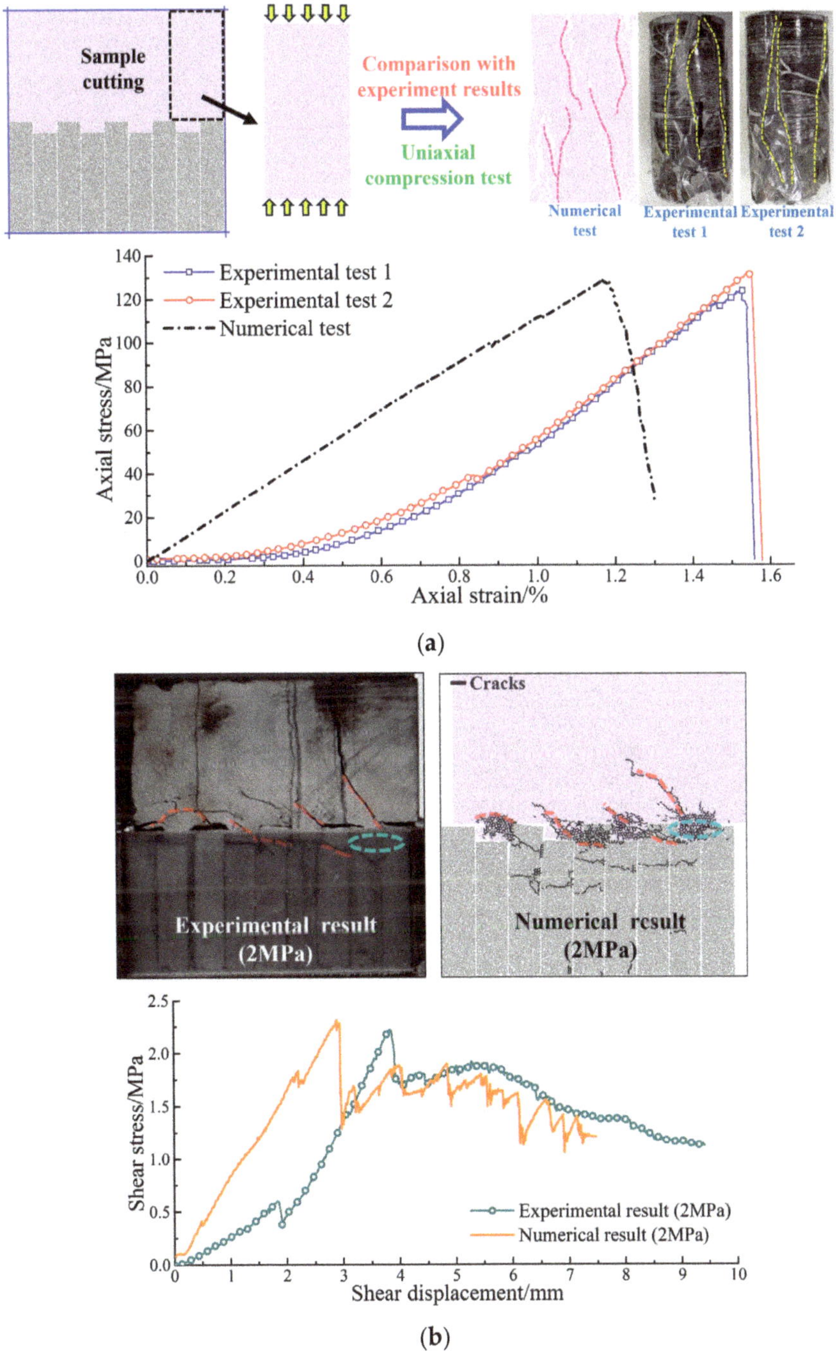

Figure 5. Parameter calibration results comparison: (**a**) Uniaxial compression test; (**b**) Shear test.

Table 4. Test scheme designs in this paper.

Scheme Name	Description	Scheme Name	Description
L-1	normal stress 1 MPa model size 200 mm step height 10 mm step width 22 mm	L-2	normal stress 2 MPa model size 200 mm step height 10 mm step width 22 mm
L-4	normal stress 4 MPa model size 200 mm step height 10 mm step width 22 mm	S-1	normal stress 1 MPa model size 200 mm step height 10 mm step width 22 mm
S-2	normal stress 2 MPa model size 200 mm step height 10 mm step width 22 mm	S-4	normal stress 4 MPa model size 200 mm step height 10 mm step width 22 mm
SS-100	model size 100 mm normal stress 1, 2, 4 MPa	SS-300	model size 300 mm normal stress 1, 2, 4 MPa
SS-350	model size 350 mm normal stress 1, 2, 4 MPa	SS-400	model size 400 mm normal stress 1, 2, 4 MPa
SH-3	Step height 3 mm normal stress 2 MPa	SH-5	Step height 5 mm normal stress 2 MPa
SH-7	Step height 7 mm normal stress 2 MPa	SH-13	Step height 13 mm normal stress 2 MPa
SH-15	Step height 15 mm normal stress 2 MPa	SW-16	Step width 16 mm normal stress 2 MPa
SW-18	Step width 18 mm normal stress 2 MPa	SW-20	Step width 20 mm normal stress 2 MPa
SW-24	Step width 24 mm normal stress 2 MPa	SW-26	Step width 26 mm normal stress 2 MPa
SW-28	Step width 28 mm normal stress 2 MPa		

4. Results Analysis

4.1. Failure Mode and Curve Characteristics

Figure 6a,b display the experimental and numerical shear stress–shear displacement and normal displacement–shear displacement curves for the "L" and "S" samples, respectively. The results from both the laboratory and simulation tests exhibit a strong correlation, indicating good agreement between the two sets of data. The shear stress curves reveal several distinct stages in the shear behaviour. The initial stage, marked by contact gap closure and micro-crack compaction, is unique to the experimental shear stress curves. The second stage corresponds to elastic behaviour, followed by the pre-peak elastic–plastic stage in the third stage. A noticeable stress drop is observed after reaching the peak (first peak) stress point, which is the fourth stage. In the post-peak stage, the shear stress initially increases slightly, and then gradually decreases, ultimately reaching a residual value. Notably, for samples "L-4" and "S-4", the post-peak shear stress value surpasses the first peak value, indicating high residual strength. This phenomenon can be explained by the heightened frictional force caused by the increased normal stress.

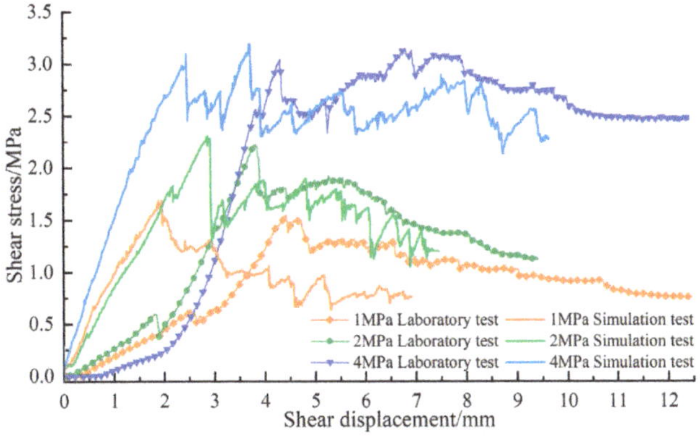

(a)

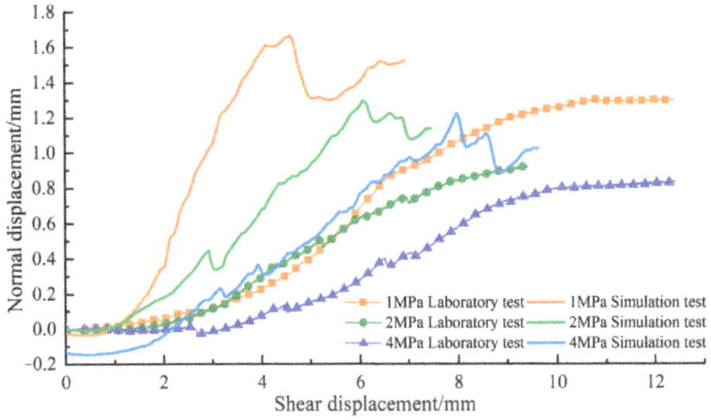

(b)

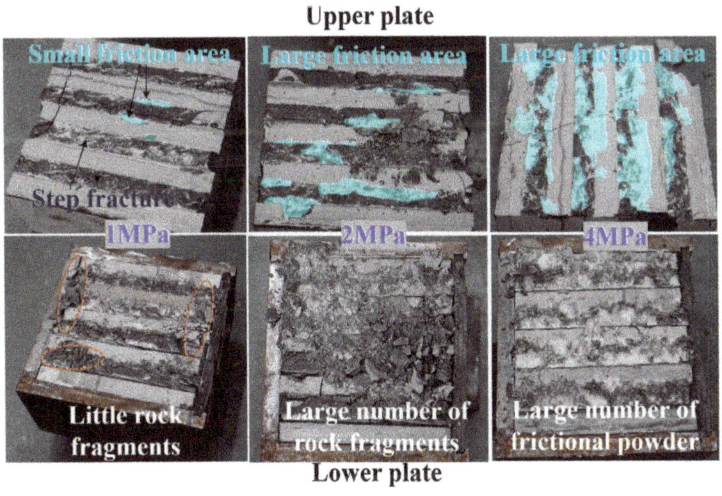

(c)

Figure 6. *Cont.*

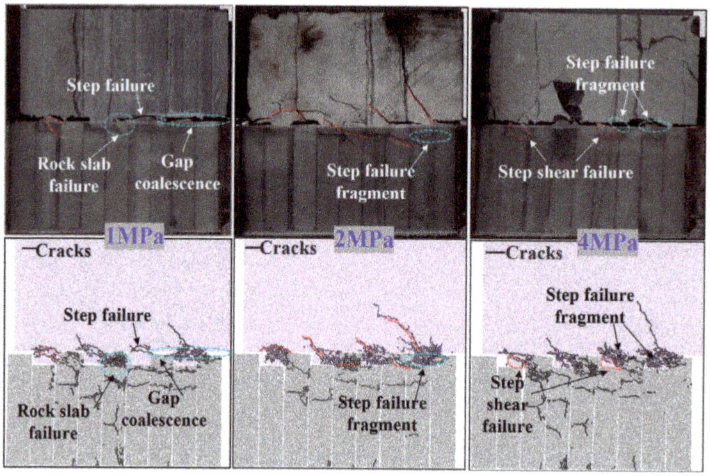

(d)

Figure 6. Curve characteristics and failure modes of sample "L" and "S": (**a**) Shear stress–shear displacement curves; (**b**) Normal displacement–shear displacement curves; (**c**) Interface failure; (**d**) Failure mode comparison.

The normal displacement–shear displacement curves highlight the samples' dilatation effects. The experimental curves typically exhibit no growth during the compaction stage, stable growth during the failure stage, and gradual growth as they approach the residual value. Due to limitations in particle radius and the accuracy of stress servo, the residual part of the simulation curves exhibits some fluctuation, with the normal displacement value in the residual region being marginally larger than that observed in the experimental results. Nevertheless, the overall trend of the simulation curves closely mirrors that of the experimental curves.

Figure 6c,d illustrate the failure modes observed in the tests. As the normal stress increases, the interface strength increasingly relies on both the strength of the steps and the frictional action at the interface, with the contribution of friction to the shear strength being progressively enhanced. Under low normal stress, specimen failure is predominantly characterised by step fracture, with a limited frictional area and minimal rock fragments at the interface. With intermediate normal stress, the failure mode transitions to step fragmentation accompanied by a significant accumulation of rock fragments. At high normal stress, the steps break, the rock fragments are ground into powder, and a large frictional area is formed at the interface.

4.2. Failure and Crack Evolution Process Analysis

Using sample "S-2" as a representative case, Figure 7 illustrates the progression of failure and crack evolution. At a shear displacement of 0.48 mm, cracks initiate from the root of the first rock step to the upper right place (point A). As the displacement continues to increase, cracks spread over a small area, with a few emerging in the lower rock slab and the first rock step (point B). Upon reaching the peak stress point, the first step experiences failure, marked by a sudden increase in the number of cracks and an instantaneous drop in stress. With the broken step no longer able to provide effective shear resistance, the second step immediately assumes the primary load-bearing role at point C, leading to the initiation of cracks within it. From point C to D, during the post-peak stage, transversely propagating cracks begin to appear within the rock slabs. Another stress drop is observed at point E, signalling the failure of the second rock step and the coalescence of numerous cracks. Subsequently, the shear stress starts to decline gradually. At point F, cracks in the two steps on the right side propagate transversely. Finally, at the point of ultimate failure (point H), both the upper rock step and the lower rock slab near the interface are severely damaged, with cracks coalescing horizontally across the interface.

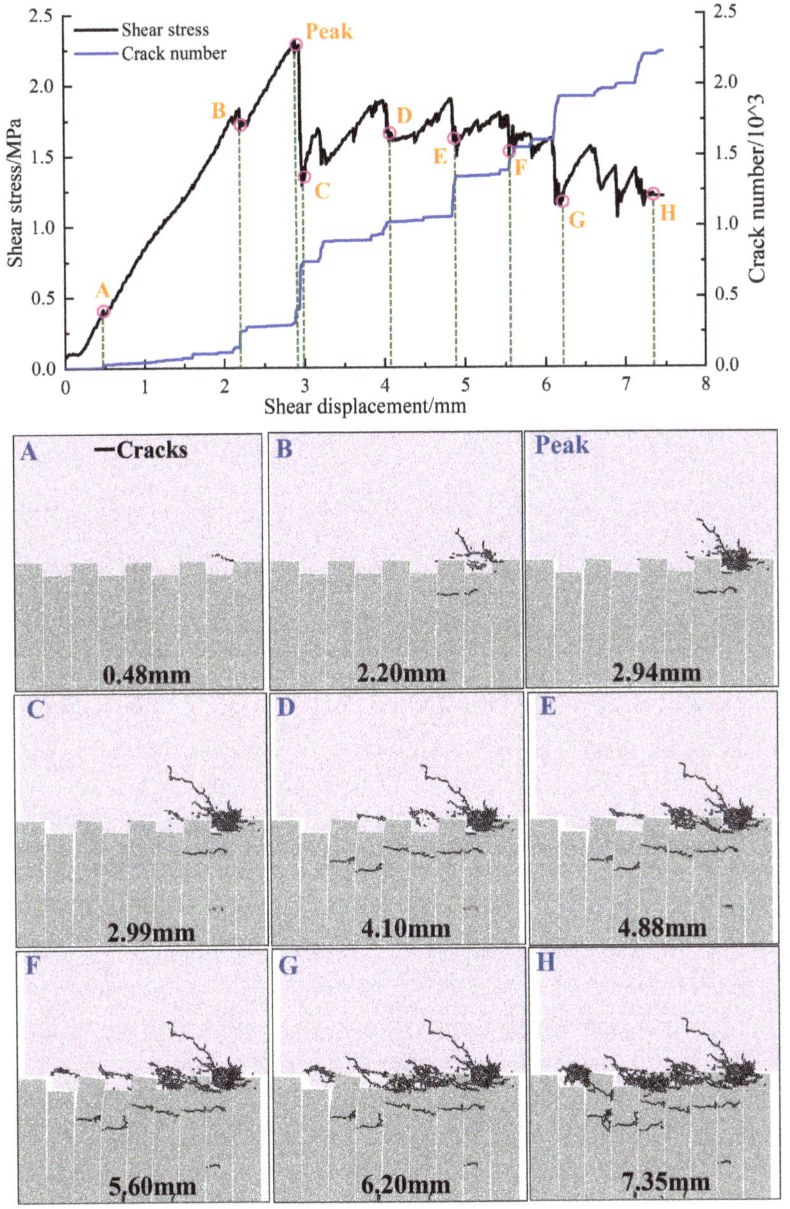

Figure 7. Failure process and crack evolution of sample under 2 MPa normal stress (Points A to H are different state points in the whole failure process).

A comparative analysis of the simulation results for samples "S-1", "S-2", and "S-4" (with the results for "S-1" and "S-4" not presented due to length constraints) reveals variations in the degree of damage and the timing of failure events among the three samples. Despite these differences, the overall failure process remains consistent across the samples, characterised by progressive failure from right to left.

4.3. Displacement Field and Crack Type Analysis

4.3.1. Description of Crack Types

In the discrete element method, the excessive relative displacement between particles can lead to broken bonds, as reflected in the contacts within the PFC software. This process is instrumental in simulating the formation of cracks within the material. The classification of crack types is typically based on the relative normal and tangential displacements of the particles [29,53–56], and they are broadly categorised into three main types: tension, shear, and mixed failure. The types of cracks are described as follows and plotted in Figure 8:

(1) Direct tensile (DT). This type of crack forms when the displacement directions of two particles are opposite and parallel to the line connecting their centres.
(2) Relative tensile (RT). The displacement directions of the two particles are aligned and parallel to the line connecting their centres; however, the magnitudes of the displacements differ.
(3) Direct shear (DS). The directions of the two particles' displacement are opposite and perpendicular to the line connecting their centres.
(4) Relative shear (RS). The directions of the two particles' displacement are uniform and perpendicular to the line connecting their centres, but with differing displacement magnitudes.
(5) Mixed failure (MF). This mode of failure is characterised by the simultaneous presence of relative displacements in both the normal and tangential directions for the two particles.

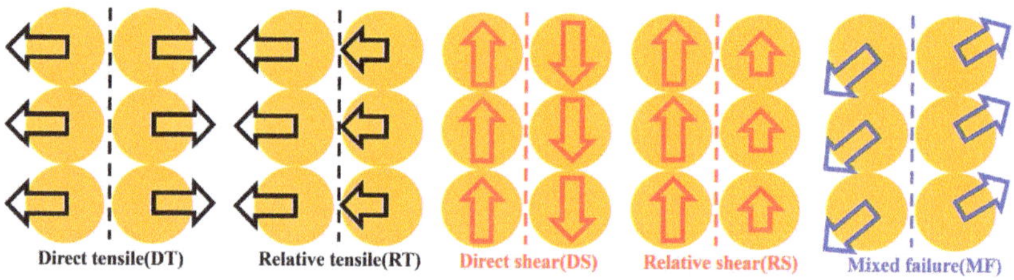

Figure 8. Schematics of different crack types (modified from reference [56]).

4.3.2. Displacement Field Evolution and Crack Types

Figure 9 presents a displacement field cloud chart of sample "S-2" throughout its failure process, where the orientation of the displacement vectors is indicated by arrows, and the presence of cracks is denoted by short black lines. To facilitate a more nuanced understanding of the crack types, several magnified local views are included.

At point A, the root of the first rock step exhibits RT cracks, indicating that the initial failure is primarily influenced by tensile stress. Upon reaching the peak stress point, both DT and RT cracks are observed within the first step, while the cracks in the rock slab are predominantly DT in nature. This further confirms that the failure of the first rock step is largely driven by tensile stress. At point D, the crack propagation patterns in the second and third rock steps begin to deviate from those observed in the first rock step. Notably, the cracks initiating at the left root are of the mixed failure (MF) type, and those propagating to the right are of the relative tensile (RT) type. During this stage, shear stress starts to exert a more pronounced effect, particularly evident as direct shear (DS) cracks at the bottom of the second step. By point F, RS and MF cracks are observed in rock slab 3, signifying a transition to shearing as the dominant mode of failure. Similarly, the failure of the second rock step is primarily attributed to shear stress, with the cracks therein being predominantly of the RS and MF types. Upon the ultimate failure of the sample, the fourth rock step is characterised by the predominance of RS cracks, which are largely influenced by shear stress. Additionally, the third rock step and rock slab 5 exhibit the presence of both RS and mixed failure (MF) cracks. The interplay of these crack types within these regions indicates a complex interaction of tensile and shear stresses leading to the comprehensive failure of the rock mass.

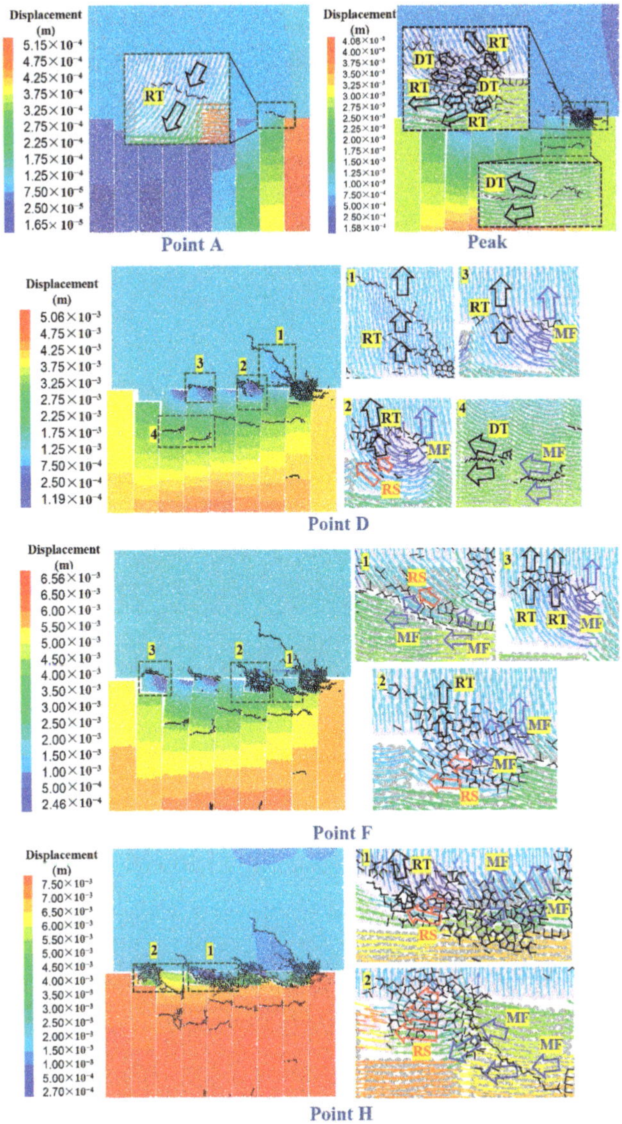

Figure 9. Displacement field and crack type of sample "S-2" during failure process.

In summary, the initiation and failure of the first rock step are predominantly influenced by tensile stress, resulting in tensile-type cracks. For the subsequent rock steps, failure occurs after the peak stress point. The lateral and vertical movement of the particles gradually increases, leading to the progressive enhancement of the shear stress influence. As the deformation continues to accumulate, the sample transitions to a state where shear stress becomes the predominant factor in the failure mechanism.

4.4. Stress Evolution and Failure Mechanism

The visualisation of the stress field is an invaluable technique for observing the failure process and analysing crack behaviour [57]. Principal stress axes serve as an effective tool for depicting the state of stress and the orientation of the principal stress directions, which can greatly aid in the analysis of failure mechanisms [58,59]. To capture the stress field within the sample, a total of 1727 measurement circles were arranged, as depicted in Figure 10. These measurement circles were designed with an overlap to ensure the acquisition of more detailed and comprehensive stress data. The horizontal, vertical, and shear stress values monitored by these measurement circles were extracted and utilised to calculate the maximum and minimum principal stresses. Furthermore, the "Tensor" tool within the PFC software was employed to plot the principal stress axes, which were drawn by importing the stress values. As presented in Figure 11, the long line at the stress cross in the figure signifies the direction of the maximum principal stress, while the short line indicates the direction of the minimum principal stress. Compressive stress is shown as positive and tensile stress as negative in the colour representation of the stress values. For an in-depth examination, five different states of sample "S-2" were chosen in order to obtain better knowledge of the failure process.

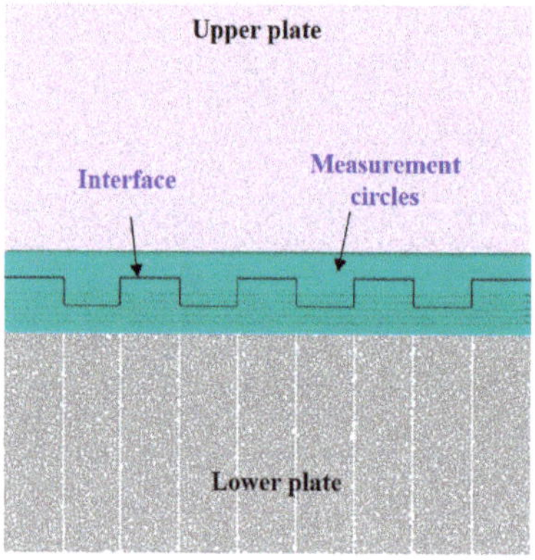

Figure 10. Arrangement of measurement circles.

Point B (Figure 11a): The maximum value of the maximum principal stress is observed at the left root of the first step. In this step, the minimum principal stress is nearly zero, and the direction of the maximum principal stress deflects counterclockwise by approximately 60 degrees (noting that the maximum principal stress direction is vertical during the servo stage). The first step is in a point-load stress state, which could lead to splitting failure. Because of its biaxial compression, the rock slab is the primary vertical load-bearing element.

Point C (Figure 11b): Following the attainment of bearing capacity limitations, splitting cracks appear in the first step (refer to Figure 7). The propagation path of these cracks aligns with the direction of the maximum principal stress identified at Point B. A pronounced concentration of the maximum principal stress is evident on both sides of the first step root and the left root of the second step. The horizontal direction in rock slab 2, the leftward deflection direction in rock slab 3, and the counterclockwise deflection in the second rock step make up a whole maximum primary stress transmission path.

Point D (Figure 11c): The direction of the maximum principal stress in the first rock step gradually deflects towards the vertical direction. This indicates that the first rock step has lost its bearing capacity, which is also evident from the significant stress concentration at the left root of the second rock step. The two sides of rock slab 2 continue to bear the majority of the minimal principal stress. At the interface of rock slab 3 and the second rock step, there is slight tensile stress.

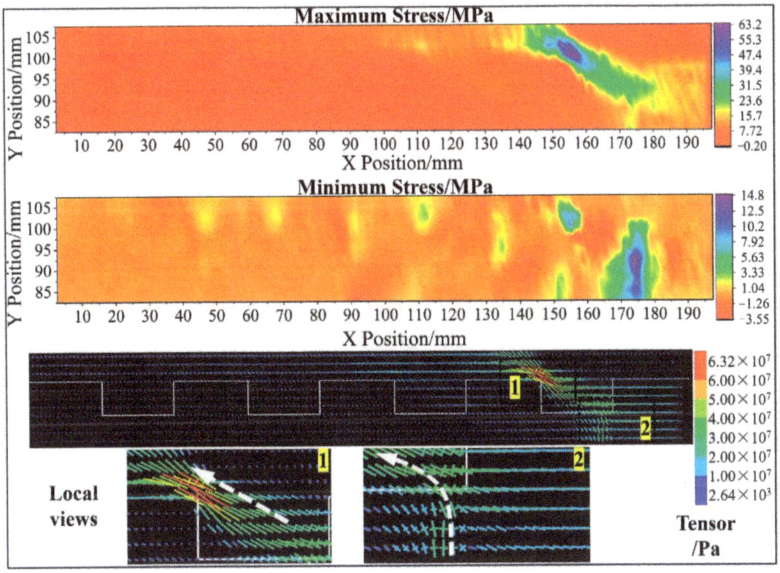

(a)

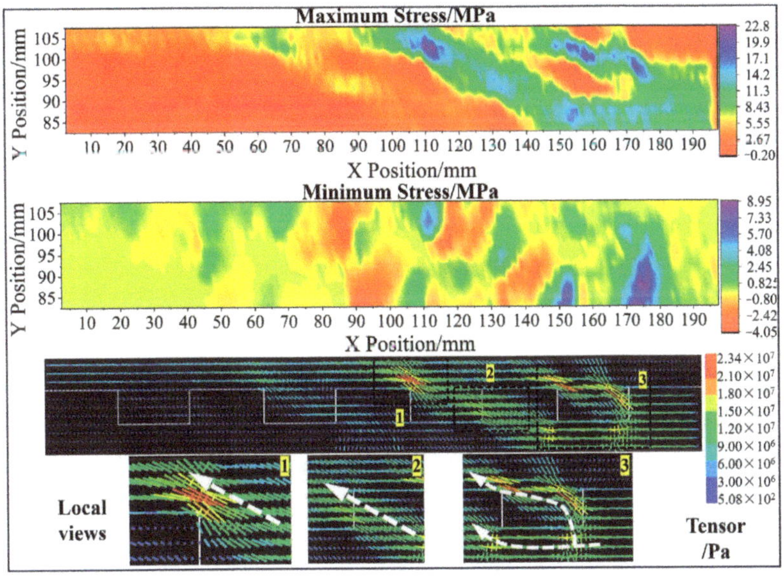

(b)

Figure 11. *Cont.*

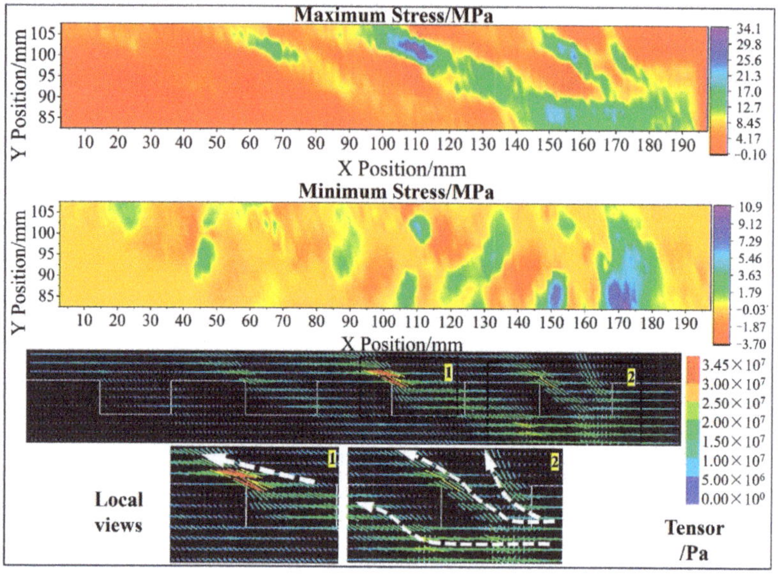

(c)

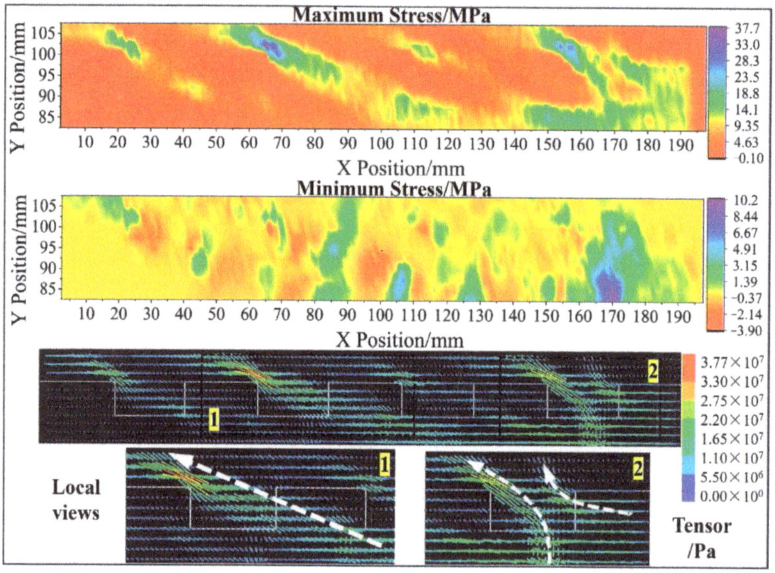

(d)

Figure 11. *Cont.*

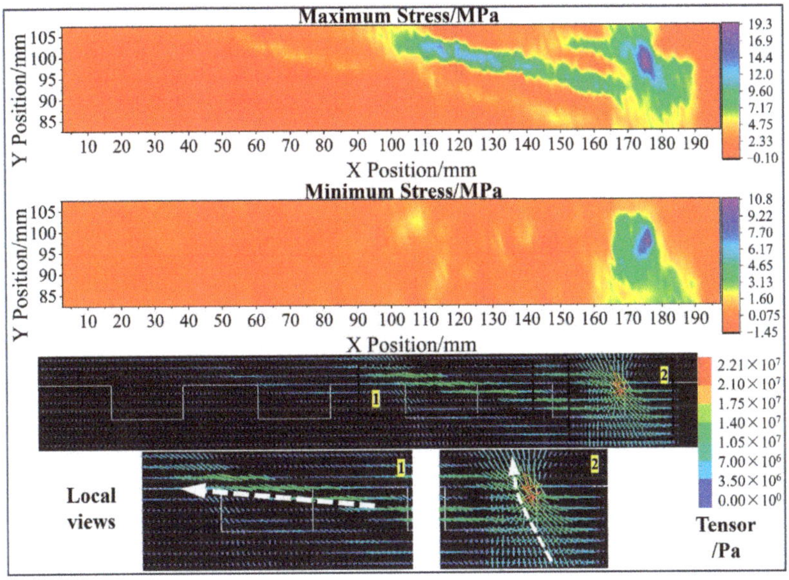

(e)

Figure 11. Maximum and minimum stress evolution of sample "S-2" during failure process: (**a**) Point B; (**b**) Point C; (**c**) Point D; (**d**) Point F; (**e**) Point H.

Point F (Figure 11d): As the shear displacement increases, the particles in the first rock step are compressed. The stress concentration in the first rock step reemerges due to particle re-contact. The left root of the third rock step becomes the primary location of the stress concentration. The fracturing of the lower rock slabs results in the disrupted continuity of the stress distribution.

Point H (Figure 11e): After all rock steps have failed, the maximum principal stress in the first rock step deflects to a near-vertical direction, making it the main location to bear normal stress at the residual stage. The direction of the maximum principal stress in the second rock step and rock slab 3 is near horizontal, with the minimum principal stress being almost zero, indicating that this area is the origin of residual shear stress.

Based on the analysis of crack development and stress evolution throughout the failure process of the sample, the following failure mechanism can be summarised. The failure of the first rock step occurs before the peak, and the crack initiation is mainly driven by the tensile stress. A point-load stress state is experienced by the initial rock step, which frequently leads to splitting failure. The interlocking characteristics of the rock steps allow the initially fractured step to continue bearing a portion of the shear force. The stress distribution among the remaining rock steps is then changed by this initial collapse, putting them in a complicated compressive–tensile stress condition. The rock steps' mode of failure shifts to a combination of shear and tensile processes in the post-peak stage as displacement grows and the impact of shear friction becomes more noticeable.

5. Discussion

5.1. *Effect of Model Size on Shear Characteristics*

Figure 12 shows the failure modes and the associated shear stress–shear displacement curves for the sample group "SS". With the increase in the sample size, the post-peak plasticity becomes more obvious in the curves. In particular, strain softening behaviour is apparent in the curves when the sample size surpasses 200 mm, and it is difficult for the stress values to rebound after the peak stress point (as demonstrated in the 200 mm sample). This is primarily attributed to the diminishing interlocking of particles as the sample size expands. Larger sample sizes lead to both an increase in the quantity of cracks generated during the shear process and an improvement in the degree of crack propagation. Under a normal stress of 1 MPa, large-sized samples exhibit a failure pattern similar to that of smaller-sized samples, with shear dilatancy being a common phenomenon. However, under higher normal stresses of 2 MPa and 4 MPa, the occurrence of shear dilatancy is less frequent in the failure of large-sized samples. Rather, there is more damage near the interface, especially to the steps and rock slabs.

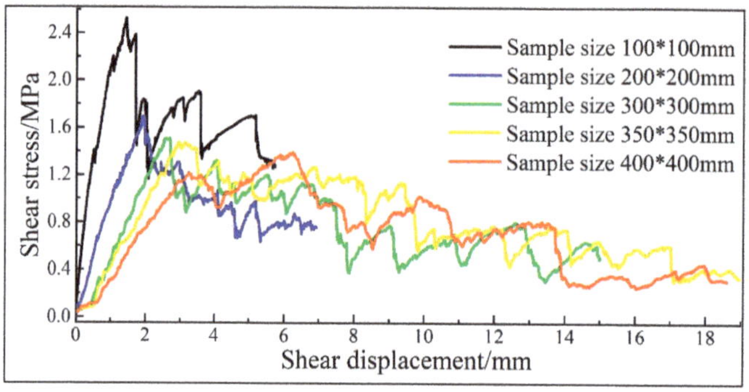

(**a**)

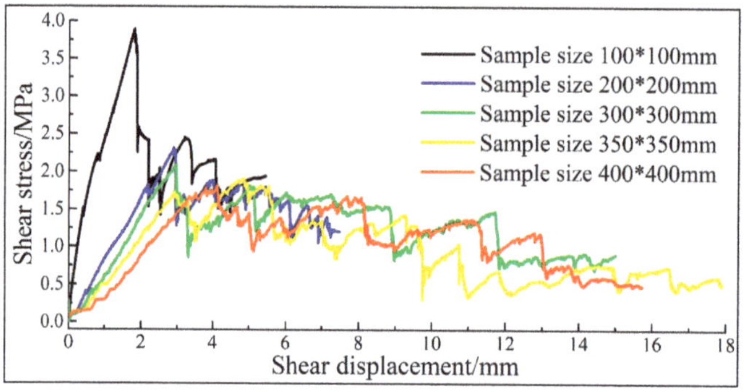

(**b**)

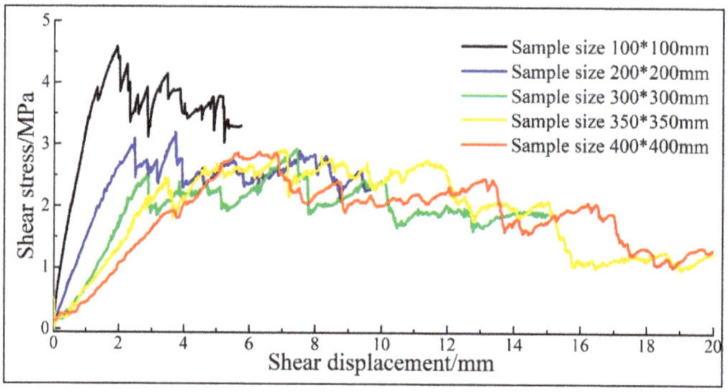

(**c**)

Figure 12. *Cont.*

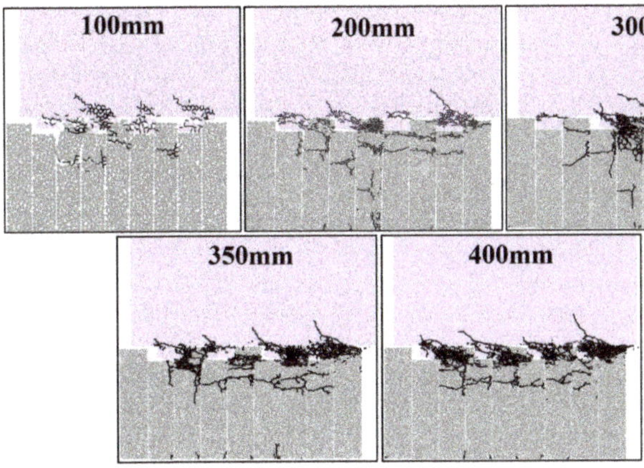

(d)

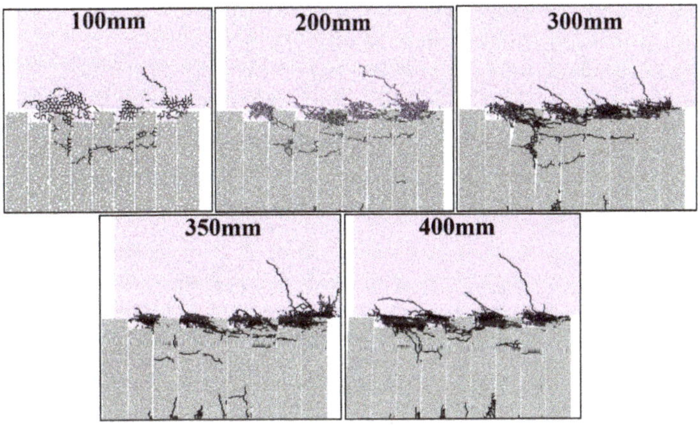

(e)

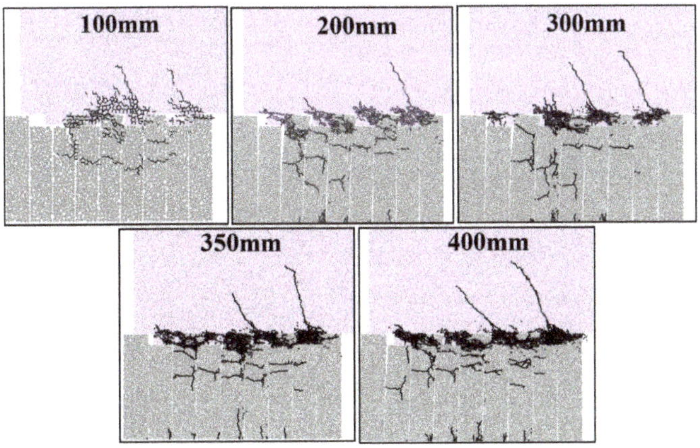

(f)

Figure 12. Size effect on stress–displacement curves and failure modes: (**a**) Curve 1 MPa; (**b**) Curve 2 MPa; (**c**) Curve 4 MPa; (**d**) Failure mode 1 MPa; (**e**) Failure mode 2 MPa; (**f**) Failure mode 4 MPa.

The shear mechanical parameters of samples varying in size under different stress conditions were calculated, with the results presented in Figure 13. Residual strength is a value that falls within a specified interval because, during the residual stage, the curves show oscillations within a set range. To determine this range value, the residual strength is calculated as the average of the maximum and minimum values within the fluctuation range, with the standard deviation used to construct the error bars for graphical representation. As can be seen in Figure 13, a distinct size effect is observable in the variation of the three shear parameters (peak shear strength, shear elastic modulus, and residual strength) as the sample size increases. When the model size expands from 100 mm to 200 mm, there is a significant decrease in the values of all three shear parameters. However, as the sample size increases from 200 mm to 400 mm, the rate of decrease for these shear parameters begins to slow down, indicating a less pronounced attenuation. This change in the shear parameters can be characterised by a negative exponential function. To describe the variation law of the shear mechanical parameters with sample size more clearly, fitting functions were employed, which are listed in Equations (1)–(3).

$$1\ \text{MPa}\begin{cases} \tau_p = 4.128exp(-d/76.639) + 1.405 \\ E_{shear} = 0.449exp(-d/74.657) + 0.198 \\ \tau_r = 2.114exp(-d/175.356) + 0.126 \end{cases} \quad (1)$$

$$2\ \text{MPa}\begin{cases} \tau_p = 8.331exp(-d/70.807) + 1.857 \\ E_{shear} = 1.829exp(-d/34.974) + 0.220 \\ \tau_r = 3.197exp(-d/273.597) - 0.286 \end{cases} \quad (2)$$

$$4\ \text{MPa}\begin{cases} \tau_p = 9.213exp(-d/59.403) + 2.874 \\ E_{shear} = 0.389exp(-d/729.927) \\ \tau_r = 6.377exp(-d/544.967) - 1.981 \end{cases} \quad (3)$$

Here, τ_p, E_{shear}, τ_r, and d are the shear peak strength (MPa), shear elastic modulus (GPa), residual strength (MPa), and sample size (mm), respectively.

As the sample size exceeds 350 mm, the variation in the shear parameter values becomes minor with further increases in sample size. Consequently, for the band–bedrock interface under investigation in this study, under the given stress conditions and step height-to-width ratio, the representative elementary volume (REV) for the shear parameters is determined as a sample with a geometry size of 350 mm.

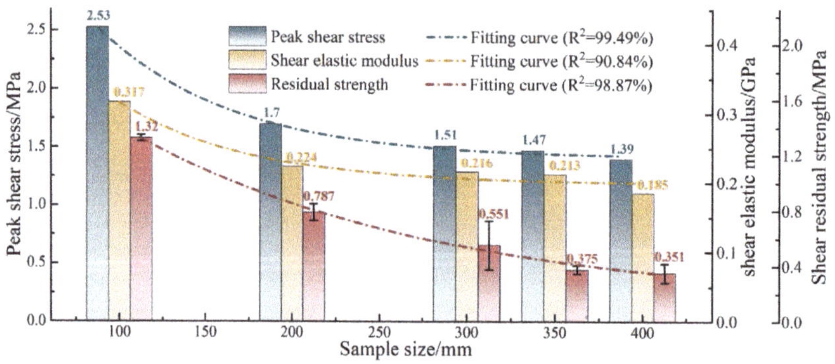

(a)

Figure 13. *Cont.*

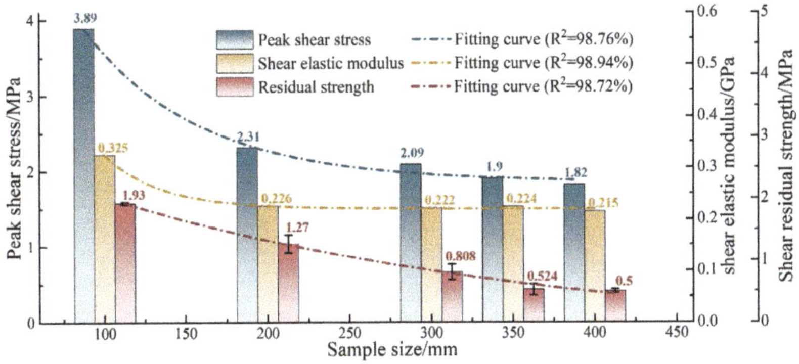

(**b**)

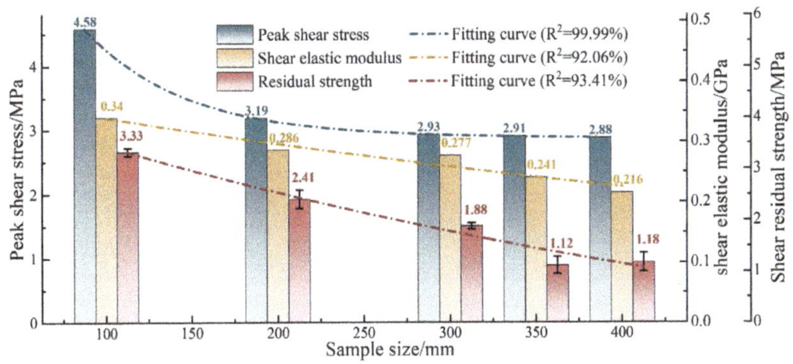

(**c**)

Figure 13. Size effect on shear mechanical parameters: (**a**) 1 MPa; (**b**) 2 MPa; (**c**) 4 MPa.

5.2. Effect of Rock Step Height on Shear Characteristics

Figure 14 illustrates the effect of rock step height on the shear characteristics of samples with a size of 200 mm, summarising three typical shear stress curves (Figure 14a) and failure modes (Figure 14b). When the rock step height is small, the stress curve transitions from strain hardening to strain softening following the elastic stage. The characteristics of the curves and failure modes for moderate rock step height were discussed in Section 4 and are not reiterated here. With large rock step height, a plastic stage emerges prior to the peak, and the strain softening stage occurs immediately after the peak.

During the shear process of the samples with a small rock step height, the degree of stress concentration is high due to the limited contact area. After failure, the reduction in the contact area exacerbates the stress concentration, which is the primary cause of strain hardening. Once the steps have broken to a certain extent, shear resistance is primarily provided by particle friction, leading to shear contraction within the sample and resulting in strain softening. Conversely, when the rock step height is large, the increased contact area mitigates the degree of stress concentration, making the shear behaviour of the specimen more prone to plastic deformation. Shear dilation is less noticeable near the beginning of the steps, where failure usually starts. It is evident that the rock step height regulates the failure mode and the characteristics of the stress curve by influencing the level of stress concentration and the degree of shear dilation.

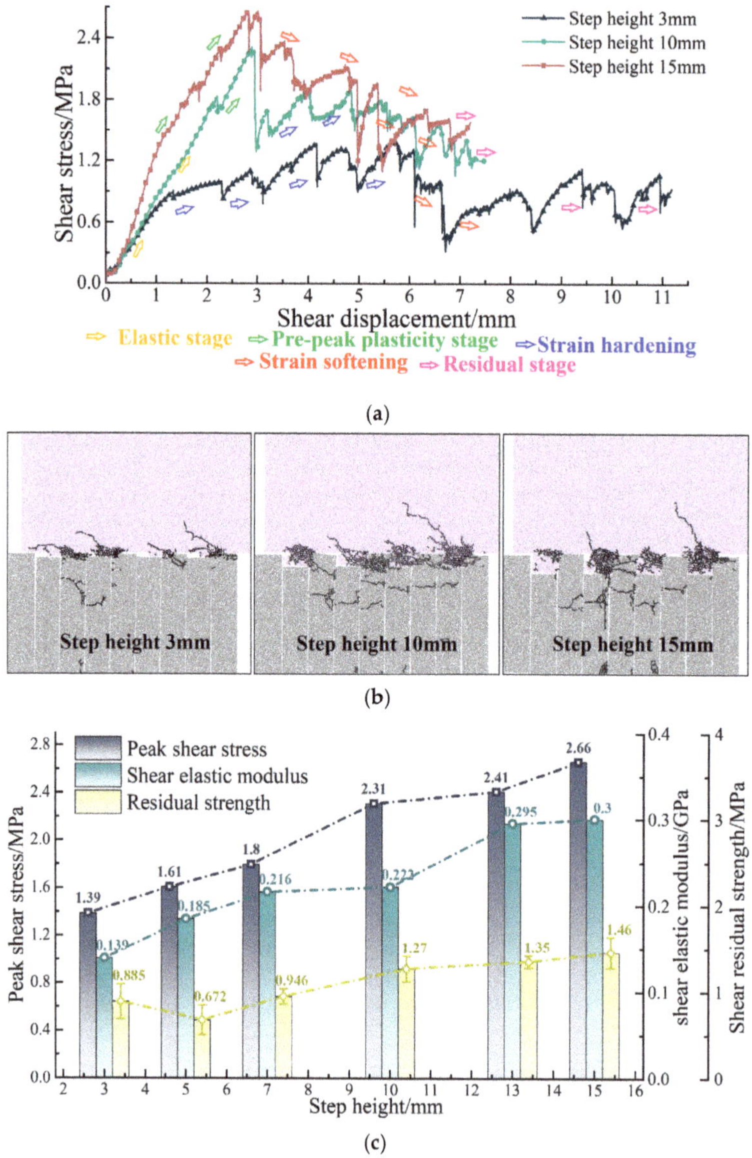

Figure 14. Effect of step height on sample shear characteristics with geometry size 200 mm: (**a**) Three typical shear–displacement curves; (**b**) Three typical failure modes; (**c**) Shear parameters.

The variation in shear mechanical parameters influenced by rock step height is depicted in Figure 14c. Both the peak shear strength and shear elastic modulus incrementally rise with increasing rock step height. At a step height of 3 mm, the values of these parameters are 1.39 MPa and 0.139 GPa, respectively. At a rock step height of 15 mm, they escalate to 2.66 MPa and 0.30 GPa, with amplifications of 91.37% and 115.83%, respectively. The residual strength initially declines from 0.885 MPa (at a rock step height of 3 mm) to 0.672 MPa (at a rock step height of 5 mm) and then gradually ascends to 1.46 MPa (at a step height of 15 mm), with amplitude reductions and amplifications of 23.73% and 116.94%, respectively.

5.3. Effect of Step Width on Shear Characteristics

The influence of rock step width on the failure mode and curve type of the samples is pronounced, as depicted in Figure 15a,b. With a narrow rock step width (16 mm), the rock slab exhibits lower bending and shear stiffness, leading to the formation of a multitude of fractures within the slab during the shearing process. Upon the failure of the rock slabs, they tend to incline along the fracture lines, resulting in point contacts between the fractured blocks and the step blocks, which helps to sustain shear resistance. This mechanism is reflected in the shear stress curve, which exhibits an approximately flat segment. As the stress concentration at the point contacts gradually increases, damage to the steps becomes evident, causing a drop in the stress curve before it enters the residual stage. Conversely, as the step width increases, stress concentration is less likely due to the reduced inclination of the rock slab. This results in a more uniform distribution of stress, and the interface damage tends to progress in a relatively smooth manner.

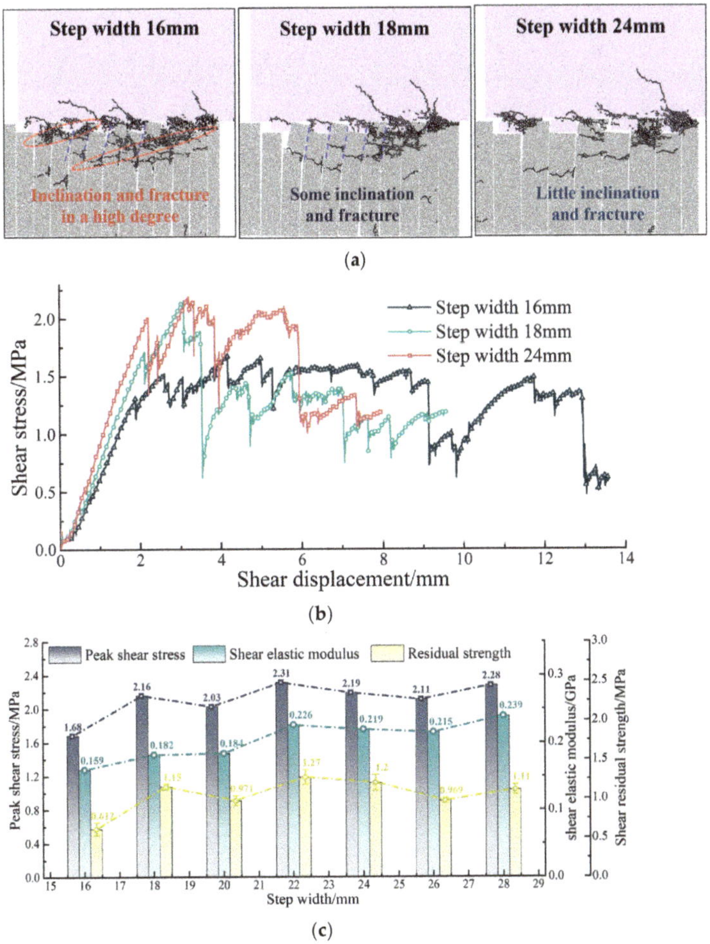

Figure 15. Effect of step width on sample shear characteristics with geometry size 200 mm: (**a**) Three typical failure modes; (**b**) Shear–displacement curves; (**c**) Shear parameters.

Figure 15c shows the effect of rock step width on the shear mechanical parameters. An analysis of the figure reveals that the minimum values for peak shear stress, shear elastic modulus, and residual strength, which are observed in the 16 mm sample, are 1.68 MPa, 0.159 GPa, and 0.612 MPa, respectively. When the layer thickness ranges from 18 to 28 mm, these parameters exhibit irregular fluctuations, with ranges of 2.03–2.31 MPa, 0.182–0.239 GPa, and 0.969–1.27 MPa, respectively. It can be deduced that decreasing the step width will result in a drop in the shear parameter values, which then stabilise within a given range when the step width increases.

6. Conclusions

In this study, numerical simulation was used to examine the shear coalescence mechanism, stress evolution, and crack type characteristics of the band–bedrock interface. Moreover, the effects of the sample and step geometry size on the shear mechanical properties of the band–bedrock interface were identified. The specific conclusions are as follows:

(1) The initiation and failure cracks in the first rock step are predominantly tensile. As the induced shear displacement increases, the influence of shear stress becomes more pronounced, leading to shear and mixed types of failure cracks in the subsequent rock slabs and steps.

(2) The failure of the first rock step occurs before the peak, primarily due to tensile stress, resulting in a point-load stress state that induces splitting failure. The remaining steps transition into a complex compressive–tensile stress state, with their failure occurring progressively.

(3) With the increase in the sample size, the post-peak plasticity of the stress curves with strain softening becomes the predominant characteristic when the sample size exceeds 200 mm. The variation in the shear parameters with sample size can be described by a negative exponential function. Once the sample size surpasses 350 mm, the shear parameter values approach consistency. The representative elementary volume of the shear parameters can be considered as a sample with a geometry size of 350 mm.

(4) The rock step height significantly influences the failure mode and stress curve characteristics by affecting the stress concentration and shear dilation. The peak shear strength and shear elastic modulus increase with rock step height, with respective amplifications of 91.37% and 115.83%. The residual strength initially decreases and then gradually increases with step height, exhibiting amplitude reductions and amplifications of 23.73% and 116.94%.

(5) A smaller step width results in a higher number of fractures during shearing, maintaining shear resistance through point contacts between rock slabs and step blocks. As the rock step width increases, the stress concentration is mitigated due to the reduced inclination of the broken rock slab, leading to the stabilisation of the shear parameter values.

Author Contributions: Conceptualization, H.W., X.G., X.L. and X.Z.; methodology, H.W., X.G. and X.Z.; software, H.W. and X.G.; validation, X.L., X.Z. and B.X.; formal analysis, H.W. and X.G.; investigation, H.W., X.G. and B.X.; resources, X.L. and B.X.; data curation, X.L., X.Z. and B.X.; writing—original draft preparation, H.W. and X.G.; writing—review and editing, X.G., X.L., X.Z. and B.X.; visualization, X.G. and X.Z.; supervision, X.L. and B.X.; project administration, X.L. and B.X.; funding acquisition, X.L. and B.X. All authors have read and agreed to the published version of the manuscript.

Funding: This research was funded by the Chongqing Talents and Out-standing Scientists Project (cstc2024ycjh-bgzxm0032), the National Natural Science Foundation of China (41972266), the China Postdoctoral Science Foundation (2023M730432) and the Special Funding for Chongqing Postdoctoral Research Project (2022CQBSHTB1010). And The APC was funded by Liu, X. and Xu, B.

Data Availability Statement: The original contributions presented in the study are included in the article, further inquiries can be directed to the corresponding author.

Acknowledgments: The authors are very grateful for the editor and the anonymous reviewers' valuable comments. The authors are also very grateful for the Suliman's work on language proofreading which improves readability of this paper.

Conflicts of Interest: The authors declare that they have no conflicts of interest which may affect paper publication.

References

1. Shang, J.; West, L.; Hencher, S.; Zhao, Z. Geological discontinuity persistence: Implications and quantification. *Eng. Geol.* **2018**, *241*, 41–54. [CrossRef]
2. Chen, C.; Yang, H.; Song, K.; Liang, D.; Zhang, Y.; Ni, J. Dissolution feature differences of carbonate rock within hydro-fluctuation belt located in the Three Gorges Reservoir Area. *Eng. Geol.* **2023**, *327*, 107362. [CrossRef]
3. Tang, H.; Wasowski, J.; Juang, C. Geohazards in the three Gorges Reservoir Area, China—Lessons learned from decades of research. *Eng. Geol.* **2019**, *261*, 105267. [CrossRef]
4. Yin, Y.; Huang, B.; Wang, W.; Wei, Y.; Ma, X.; Ma, F.; Zhao, C. Reservoir-induced landslides and risk control in Three Gorges Project on Yangtze River, China. *J. Rock Mech. Geotech. Eng.* **2016**, *8*, 557–595. [CrossRef]
5. Liu, X.; Wang, Y.; Xu, B.; Zhou, X.; Guo, X.; Miao, L. Dynamic damage evolution of bank slopes with serrated structural planes considering the deteriorated rock mass and frequent reservoir induced earthquakes. *Int. J. Min. Sci. Technol.* **2023**, *33*, 1131–1145. [CrossRef]
6. Yin, Y.; Wang, L.; Huang, B.; Zhang, Z.; Dai, Z. Evolution analysis of the Banbiyan dangerous rock mass in the three gorges reservoir area. *Georisk-Assess. Manag. Risk Eng. Syst. Geohazards* **2023**, *17*, 376–386. [CrossRef]
7. Patton, F. Multiple modes of shear failure in rock. In *Proceeding of the Congress of 1st International Society of Rock Mechanics, Lisbon, Portugal, 25 September–1 October 1966*; International Society for Rock Mechanics and Rock Engineering: Lisbon, Portugal, 1966; pp. 509–513.
8. Barton, N.; Choubey, V. The shear strength of rock joints in theory and practice. *Rock Mech. Rock Eng.* **1977**, *10*, 1–54. [CrossRef]
9. Zhao, J. A new JRC-JMC shear strength criterion for rock joint. *Chin. J. Rock Mech. Eng.* **1998**, *17*, 349–357.
10. Maksimović, M. The shear strength components of rough rock joint. *Int. J. Rock Mech. Min. Sci. Geomech. Abstr.* **1996**, *33*, 769–783. [CrossRef]
11. Xia, C.; Tang, Z.; Xiao, W.; Song, Y. New peak shear strength criterion of rock joints based on quantified surface description. *Rock Mech. Rock Eng.* **2014**, *47*, 387–400. [CrossRef]
12. Yang, J.; Rong, G.; Hou, D.; Peng, J.; Zhou, C. Experimental study on peak shear strength criterion for rock joints. *Rock Mech. Rock Eng.* **2016**, *49*, 821–835. [CrossRef]
13. Liu, X.; Li, P.; Guo, X.; Luo, X.; Zhou, X.; Miao, L.; Zhou, F.; Wang, H. A statistical damage-based constitutive model for shearing of rock joints in brittle drop mode. *Int. J. Min. Sci. Technol.* **2024**. [CrossRef]
14. Liu, X.; Xu, B.; Huang, J.; Lin, G.; Zhou, X.; Wang, J.; Xiong, F. Macro-meso shear mechanical behaviors of coalescent rock joints with different morphologies. *Chin. J. Geotech. Eng.* **2021**, *43*, 406–415.
15. Liu, R.; Lou, S.; Li, X.; Han, G.; Jiang, Y. Anisotropic surface roughness and shear behaviors of rough-walled plaster joints under constant normal load and constant normal stiffness conditions. *J. Rock Mech. Geotech. Eng.* **2022**, *12*, 338–352. [CrossRef]
16. Zhang, G.; Zhang, J.; Zou, B.; Zou, J.; Xiong, F.; Lin, Z.; Zhang, Z. An Empirical Study on Water-Induced Shear Weakening Behaviors of Rough-Walled Sandstone Joints. *Rock Mech. Rock Eng.* **2023**, *56*, 3475–3497. [CrossRef]
17. Le, H.; Wei, J.; Sun, S.; Zhang, C. Cracking behaviors and strength of rock-like samples with steps of different geometries under shear stress. *J. Mt. Sci.* **2021**, *18*, 1352–1370. [CrossRef]
18. Zhang, S.; Wang, G.; Jiang, Y.; Wang, C.; Xu, F. Shear behaviours and roughness degeneration based on a quantified rock joint surface description. *Int. J. Min. Sci. Technol.* **2023**, *33*, 1301–1316. [CrossRef]
19. Jiao, F.; Xu, J.; Peng, S.; Hei, M.; Zhang, X.; Cheng, L. Experimental study on shear mechanical properties and damage evolution of artificial structural plane under constant normal stiffness. *J. China Coal Soc.* **2023**, *48*, 4065–4077.
20. Bao, H.; Hu, Q.; Lan, H.; Yan, C.; Sun, W.; Liu, S.; Ma, K.; Wang, J. Spatio-temporal evolution law of anisotropic shear damage on rock mass joint surface affected by joint morphology. *Eng. Fail. Anal.* **2024**, *165*, 108820. [CrossRef]
21. Zhang, Q.; Luo, Z.; Shen, D.; Shi, Z.; Pan, Q. Shear stress relaxation behavior of rock joints considering 3D morphological characteristics. *Bull. Eng. Geol. Environ.* **2023**, *84*, 429. [CrossRef]
22. Song, L.; Jiang, Q.; Li, Y.; Yang, C.; Liu, C.; Yang, B.; Li, S. Improved JRC-JCS shear strength formula for soft-hard natural joints. *Rock Soil Mech.* **2017**, *38*, 2789–2798.
23. Ghazvinian, A.; Taghichian, A.; Hashemi, M.; Mar'ashi, S. The shear behaviour of bedding planes of weakness between two different rock types with high strength difference. *Rock Mech. Rock Eng.* **2010**, *43*, 69–87. [CrossRef]
24. Hu, Z.; Wang, Q.; Ma, Y.; Lv, H.; Liu, W.; Yan, R.; Wang, K.; Shao, T.; Sun, Y. Study on shear failure characteristics of fiber-reinforced shotcrete-granite interface based on surface scanning. *Case Stud. Constr. Mater.* **2024**, *21*, e03486. [CrossRef]
25. Bahaaddini, M.; Sharrock, G.; Hebblewhite, B. Numerical direct shear tests to model the shear behaviour of rock joints. *Comput. Geotech.* **2013**, *51*, 101–115. [CrossRef]

26. Yang, K.; Hu, Q.; Zhao, H.; Zhao, M.; Zhou, S. Numerical study on the shear behavior of concrete-rock joints with similar triangular asperities. *Comput. Geotech.* **2023**, *159*, 105468. [CrossRef]
27. Cao, R.; Yao, R.; Lin, H.; Lin, Q.; Meng, Q.; Li, T. Shear behaviour of 3D nonpersistent jointed rock-like specimens: Experiment and numerical simulation. *Comput. Geotech.* **2022**, *148*, 104858. [CrossRef]
28. Tai, D.; Qi, S.; Zheng, B.; Wang, C.; Guo, S.; Luo, G. Shear mechanical properties and energy evolution of rock-like samples containing multiple combinations of non-persistent joints. *J. Rock Mech. Geotech. Eng.* **2023**, *15*, 1651–1670. [CrossRef]
29. Chen, M.; Yang, S.; Ranjith, P.; Zhang, Y. Cracking behavior of rock containing non-persistent joints with various joints inclinations. *Theor. Appl. Fract. Mech.* **2020**, *109*, 102701. [CrossRef]
30. Song, Z.; Yang, Z.; Zhang, M.; Wang, F.; Herbst, M.; Konietzky, H. Effect of heterogeneity on mechanical and micro-seismic behaviors of sandstone subjected to multi-level cyclic loading: A discrete element method investigation. *J. Rock Mech. Geotech. Eng.* **2023**, *15*, 2556–2581. [CrossRef]
31. Shao, X.; Chi, S.; Tao, Y.; Zhou, X. DEM simulation of the size effect on the wetting deformation of rockfill materials based on single-particle crushing tests. *Comput. Geotech.* **2020**, *123*, 103429. [CrossRef]
32. Bahaaddini, M.; Hagan, P.; Mitra, R.; Hebblewhite, B. Scale effect on the shear behaviour of rock joints based on a numerical study. *Eng. Geol.* **2014**, *181*, 212–223. [CrossRef]
33. Lin, H.; Kang, W.; Oh, J.; Canbulat, I.; Hebblewhite, B. Numerical simulation on borehole breakout and borehole size effect using discrete element method. *Int. J. Min. Sci. Technol.* **2020**, *30*, 623–633. [CrossRef]
34. Sun, Y.; Kwok, C.; Duan, K. Size effects on crystalline rock masses: Insights from grain-based DEM modeling. *Comput. Geotech.* **2024**, *171*, 106376. [CrossRef]
35. Ye, Y.; Li, C.; Zeng, Y.; Tang, H. 3D DEM simulations of the variability of rock mechanical behaviour based on random rock microcracks. *Int. J. Rock Mech. Min. Sci.* **2023**, *167*, 105395. [CrossRef]
36. Wei, Y.; Feng, Y.; Tian, Z.; Yang, T.; Li, X.; Dai, Z.; Deng, J. Borehole stability in naturally fractured rocks with drilling mud intrusion and associated fracture strength weakening: A coupled DFN-DEM approach. *J. Rock Mech. Geotech. Eng.* **2024**, *16*, 1565–1581. [CrossRef]
37. Jiang, S.; Chen, J.; Wang, Z.; Zheng, J.; Huang, J.; Lv, Q. Three-Dimensional Discrete Element Analysis of Jointed Rock Slope Stability Based on the Universal Elliptical Disc Model. *Rock Mech. Rock Eng.* **2024**, *57*, 505–525. [CrossRef]
38. Vaziri, M.; Tavakoli, H.; Bahaaddini, M. Statistical analysis on the mechanical behaviour of non-persistent jointed rock masses using combined DEM and DFN. *Bull. Eng. Geol. Environ.* **2022**, *81*, 177. [CrossRef]
39. Zhang, Y.; Jiang, Y.; Asahina, D.; Wang, C. Experimental and numerical investigation on shear failure behavior of rock-like samples containing multiple non-persistent joints. *Rock Mech. Rock Eng.* **2020**, *53*, 4717–4744. [CrossRef]
40. ICG (Itasca Consulting Group). *PFC-Particle Flow Code Documentation, Version 7.0*; Itasca Consulting Group Inc.: Minneapolis, MN, USA, 2019.
41. Lei, X.; Gong, W.; Li, B.; Zhao, C.; Tang, H.; Wang, L. Probabilistic Analysis of Earthquake-induced Failure and Runout Behaviors of Rock Slopes with Discrete Fracture Network. *Comput. Geotech.* **2023**, *159*, 105487.
42. Ni, X.; Ma, J.; Zhang, F. Mechanism of the variation in axial strain of sand subjected to undrained cyclic triaxial loading explained by DEM with non-spherical particles. *Comput. Geotech.* **2024**, *165*, 105846. [CrossRef]
43. Rafiee, R.; Ataei, M.; KhalooKakaie, R.; Jalali, S.; Sereshki, F.; Noroozi, M. Numerical modeling of influence parameters in cavabililty of rock mass in block caving mines. *Int. J. Rock Mech. Min. Sci.* **2018**, *105*, 22–27. [CrossRef]
44. Zhang, X.; Wong, L. Loading rate effects on cracking behavior of flaw-contained specimens under uniaxial compression. *Int. J. Fract.* **2013**, *180*, 93–110. [CrossRef]
45. Xu, X.; Wu, S.; Gao, Y.; Xu, M. Effects of Micro-structure and Micro-parameters on Brazilian Tensile Strength Using Flat-Joint Model. *Rock Mech. Rock Eng.* **2016**, *49*, 3575–3595. [CrossRef]
46. Wu, S.; Xu, X. A Study of Three intrinsic problems of the classic discrete element method using flat-joint model. *Rock Mech. Rock Eng.* **2016**, *49*, 1813–1830. [CrossRef]
47. Yang, Y.; Qiao, W. Numerical investigation of the shear behavior of granite materials containing discontinuous joints by utilizing the flat-joint model. *Comput. Geotech.* **2018**, *104*, 69–80. [CrossRef]
48. Chen, Y.; Wong, L. A study on mechanical properties and fracturing behavior of Carrara marble with the flat-jointed model. *Int. J. Numer. Anal. Methods Geomech.* **2020**, *44*, 803–822. [CrossRef]
49. Cao, R.; Yao, R.; Dai, H.; Qiu, X.; Lin, H.; Li, K. Fracture behaviour of transversely isotropic rocks under pure mode III fracture: Experiment and numerical simulation. *Theor. Appl. Fract. Mech.* **2024**, *129*, 104208. [CrossRef]
50. Xie, S.; Lin, H.; Duan, H.; Liu, H.; Liu, B. Numerical study on cracking behavior and fracture failure mechanism of fractured rocks under shear loading. *Comput. Part. Mech.* **2024**, *11*, 903–920. [CrossRef]
51. Cao, R.; Cao, P.; Lin, H.; Ma, G.; Chen, Y. Failure characteristics of intermittent fissures under a compressive-shear test: Experimental and numerical analyses. *Theor. Appl. Fract. Mech.* **2018**, *96*, 740–757. [CrossRef]
52. Lin, Q.; Cao, P.; Meng, J.; Cao, R.; Zhao, Z. Strength and failure characteristics of jointed rock mass with double circular holes under uniaxial compression: Insights from discrete element method modelling. *Theor. Appl. Fract. Mech.* **2020**, *109*, 102692. [CrossRef]
53. Zhao, Y.; Li, B.; Wu, S.; Chen, L. Failure mechanism of brittle rock with 3D parallel preset flaws based on the particle displacement trend method. *Theor. Appl. Fract. Mech.* **2022**, *117*, 103193. [CrossRef]

54. Zhang, X.; Wong, L. Displacement field analysis for cracking processes in bonded-particle mode. *Bull. Eng. Geol. Environ.* **2014**, *71*, 13–21. [CrossRef]
55. Sharafisafa, M.; Aliabadian, Z.; Tahmasebinia, F.; Shen, L. Experimental and numerical investigations on crack development in 3D printed rock-like specimens with pre-existing flaws. *Eng. Fract. Mech.* **2021**, *241*, 107396.
56. Luo, X.; Cao, P.; Lin, Q.; Li, S. Mechanical behaviour of fracture-filled rock-like specimens under compression-shear loads: An experimental and numerical study. *Theor. Appl. Fract. Mech.* **2021**, *113*, 102935. [CrossRef]
57. Cai, W.; Li, Y.; Gao, K.; Wang, K. Crack propagation mechanism in rock-like specimens containing intermittent flaws under shear loading. *Theor. Appl. Fract. Mech.* **2022**, *117*, 103187. [CrossRef]
58. Zhao, Y.; Chen, C.; Qi, Q.; Wu, S. Rotation and deflection of 3D principal stress axes induced by a prefabricated single flaw in sandstone: A numerical investigation based on DEM. *Theor. Appl. Fract. Mech.* **2022**, *120*, 103430. [CrossRef]
59. Chen, C.; Zhao, Y.; Wu, S. Influences of true 3D twin flaws on principal stress status and crack occurrence of brittle rock-like samples under uniaxial compression: Insights from experimental and DEM studies. *Theor. Appl. Fract. Mech.* **2023**, *127*, 104057. [CrossRef]

applied sciences

MDPI

Article

Characterization of Seismic Dynamic Response of Uranium Tailings Dams Based on Discrete Element Method

Ming Lan [1,2], Hongyu Huang [2] and Yan He [2,*]

1 Deep Mining Laboratory of Shandong Gold Group Co., Ltd., Yantai 261400, China; lm1988@usc.edu.cn
2 School of Resource Environment and Safety Engineering, University of South China, Hengyang 421000, China; huanghongyu@stu.usc.edu.cn
* Correspondence: yanhe@usc.edu.cn

Abstract: Tailings dams play a critical role in ensuring the safety of mining operations. However, earthquakes can cause breaches in these dams, resulting in significant casualties and property damage. This study investigates the dynamic response characteristics of uranium tailings dams subjected to seismic loading, employing the discrete element method. It specifically analyzes how seismic wave amplitude, frequency, and the friction angle of tailings sand affect the dams' dynamic response. The results reveal that the peak ground acceleration ratio (PGAR) exhibits an increasing–decreasing–increasing pattern with elevation. When the friction angle of the tailings sand exceeds 35°, the overall stability of the dam improves. Conversely, a friction angle below 25° significantly increases the risk of dam failure. Additionally, the dam shows a reduced dynamic response to seismic waves with frequencies exceeding 15 Hz. At lower frequencies, deformation and damage are primarily concentrated on the slope face, while at higher frequencies, damage is predominantly located at the bottom of the model. These findings provide a theoretical foundation and reference for the safe operation of tailings dams, highlighting their practical significance.

Keywords: discrete element method; uranium tailings pond; dynamic characteristics

1. Introduction

Tailings dam facilities are used to store tailings and other industrial waste discharged from a mine after ore separation [1,2]. Uranium tailings ponds are an important component of uranium mine safety production due to their large capacity, high potential energy, and long storage time [3]. Currently, approximately 1 billion cubic meters of uranium tailings are produced worldwide [4,5]. A breach in a uranium tailings impoundment could result in severe environmental pollution and significant harm to downstream residents. Seismic activity is the primary cause of tailings pond failures [6,7]. Earthquakes can cause the liquefaction of tailings sands, weakening the strength of tailings materials and leading to static deformation of the dam. This can result in residual deformation in the dam structure, causing damage and reducing its overall stability [8,9]. Therefore, studying the dynamic response characteristics under seismic loading is of great practical significance.

Zhou, Jin, and Li [10–12] conducted shaking table tests to analyze the acceleration, pore pressure, and displacement of the dam body under seismic waves of varying peak ground accelerations. Their findings revealed that the acceleration peaks of the dam body exhibited distinct spatial effects. Moreover, they observed that the PGA amplification coefficients measured on the slope surface consistently surpassed those in the slope. Notably, they discovered that the dynamic amplification effect on the high side slopes correlated significantly with the slope's intrinsic frequency. However, laboratory experiments are typically on a small scale, and although shake table tests have their advantages, they also have clear limitations.

Model material parameters play a crucial role in numerical simulations, significantly impacting the authenticity, accuracy, and reliability of results [13,14]. Traditional trial-and-error methods for calibrating rock mechanical parameters are both complex and time-consuming. In contrast, using finite element analysis and machine learning offers an efficient way to accurately predict these parameters [15]. The Weibull distribution can describe the heterogeneity of rock hydraulics, providing a method for parameter distribution in heterogeneous rocks [16].

The dynamic response of the dam under earthquake was simulated using finite element software, and the seismic dynamic response of the dam was analyzed in terms of acceleration, stress, displacement, liquefied area, and stability [17,18]. The vertical dynamic response of a slope is less influenced by elevation, while the horizontal dynamic response has an obvious elevation amplification effect and topographic effect. The ground vibration amplification effect is related to the mechanical strength of the structural surface, as well as the waveform and spectral characteristics of the seismic waves [19]. Jin et al. [20] conducted model testing and numerical calculations using the finite element method to observe the gradual increase in the horizontal and vertical displacements of the monitoring points on the dam body with an increase in input peak acceleration. However, the acceleration amplification coefficient showed a decreasing trend. The collapse of the tailings dam did not exhibit clear evidence of a sliding surface as the input peak acceleration increased. Instead, the collapse mode presented as uniform sliding across the entire structure. Using the OpenSees software v.3.5.0, Vijayasri, T. [21] conducted an analysis on five distinct seismic motions. The study compared different boundary conditions in terms of displacement flow vectors, pore pressures, and stress–strain curves during the shaking period. The findings revealed that the viscous boundary conditions provided a more accurate representation of the actual field conditions in comparison to the fixed boundary conditions. Ferdosi Behnam et al. [22] conducted simulations and analyses to examine the effects of different waste rock reinforcement methods on the seismic performance of tailings dams. The study focused on the impact of seismic wave frequencies and found that low-frequency waves have a greater influence on tailings dams compared to high-frequency waves. The application of non-uniform seismic loads leads to a 10-30% decrease in the dynamic response of concrete panel rockfill dams compared to that of uniform loads. Furthermore, the edges of the concrete panels experience higher dynamic tensile stresses [23,24]. Prodromos N. Psarropoulos and Yiannis Tsompanakis employed an equivalent viscoelasticity model to simulate the dynamic response of different types of tailings dams to seismic forces. They considered the nonlinear properties of the foundation soil and the deposited tailings in order to identify the potential instability modes and seismic failure characteristics of these dams [25]. Most current studies on the dynamic characteristics of tailings dams rely on numerical simulation techniques like the limit equilibrium method and finite element method, which typically provide satisfactory results. Among the various types of numerical simulation methods, the discrete unit method can make up for some of the shortcomings of the finite element method when handling uranium tailings sand due to its discrete nature [26].

This paper examines the dynamic response characteristics of a uranium tailings dam under seismic load using the discrete element method. This study investigates the acceleration, displacement, and velocity change patterns of the uranium tailings dam. Furthermore, it quantitatively analyzes the factors influencing the dynamic response characteristics of the uranium tailings dam, such as the friction angle of the tailings sand and seismic wave amplitude. By controlling the variables, the study calculates the impact of these factors on the dynamic response of the tailings dam. The research findings can be utilized in the development of tailings ponds, providing a theoretical basis and reference for the safe operation of dams. This is of significant importance in safeguarding the safety of downstream residents and protecting the ecological environment.

2. Overview of Uranium Tailings Dams

A typical basin landform in southern China is formed by nine water-retaining dams and surrounding hills. This study focuses on a 300-meter-long section of one such uranium tailings dam. The initial dam of a tailings impoundment is a crushed, homogeneous earth dam with a height of 20 m, a downstream slope ratio of about 1:2.8, and a stacked dam with a slope ratio of about 1:3.4. The inner side of the top of the initial dam and the dam of the stacked dam platform are on bedrock. Table 1 provides an overview of the primary geotechnical parameters for this tailings impoundment.

Table 1. Geotechnical physical and mechanical parameters of tailings dams.

Materials	Density /($kg{\cdot}m^{-3}$)	Porosity	Internal Friction Angle/(°)	Cohesion /(kPa)	Young's Modulus /(GPa)	Poisson's Ratio
Uranium tailings sand	1850	0.5	30	6	0.36	0.3
Saturated uranium tailings sand	2050	0.5	28	8	0.36	0.27
Tail powder soil	2050	0.39	18	10	0.61	0.28
Initial dam	1860	0.39	19	61	0.28	0.32
Base rock	2040	0.43	20	60	1.05	0.28

3. Modeling of Dam Dynamics

Considering the geological complexity of the tailings dam, the dam body was simplified according to relevant data and field investigation, and the 3D dynamic analysis model shown in Figure 1 was established. In order to accurately represent the propagation of seismic waves, the size must be smaller than 1/8~1/10 of the minimum wavelength of the input wave, and the size of the model unit was optimized without affecting the calculation accuracy, and a total of 42,574 blocks were determined after trial calculations. Monitoring points are set up in the model to record the change in the velocity, displacement and the acceleration of key points under seismic load.

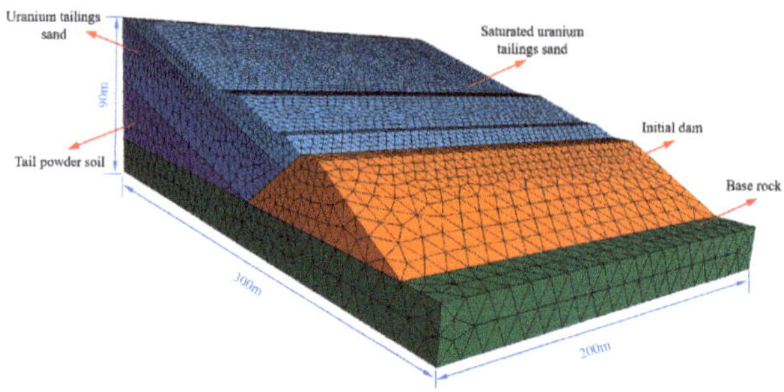

Figure 1. Three-dimensional uranium tailings dam model.

3.1. Loading Seismic Loads

There are two main types of seismic waves: artificial seismic waves and real seismic waves. In this paper, the simulation input seismic wave comes from the PEER ground motion database; the sampling interval is 0.005 s, its peak acceleration is 0.20 g, and the duration of the earthquake is 21.25 s. The time course curves of the acceleration, velocity, and displacement of the seismic wave are shown in Figures 2 and 3.

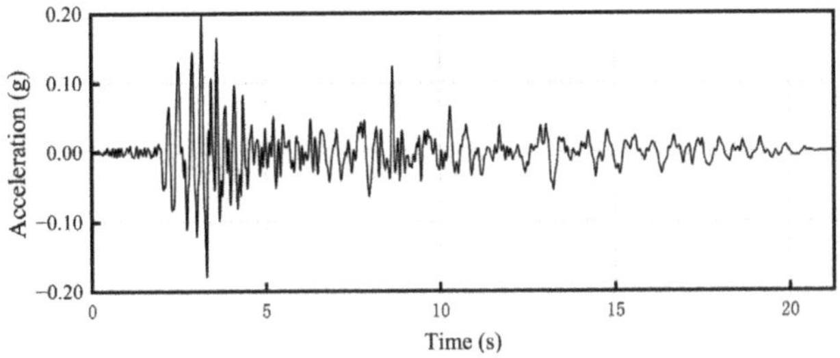

Figure 2. Accelerogram of input seismic wave in x-direction.

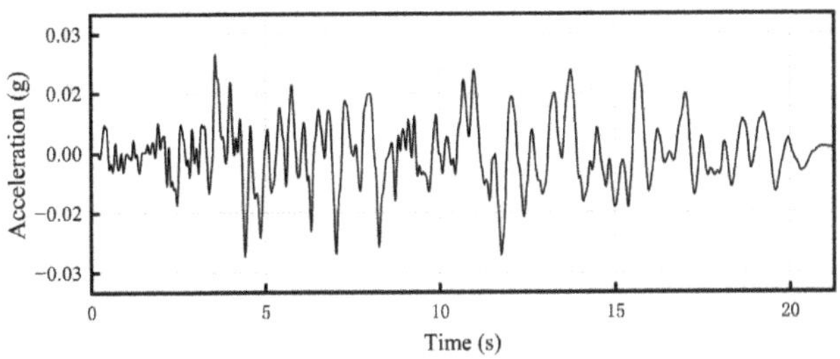

Figure 3. Accelerogram of input seismic wave in z-direction.

3.2. *Setting of Boundary Conditions and Damping*

In the static analysis, horizontal displacement constraints are applied to the model, while three-way displacement constraints are set on the bottom surface. The slope surface is left unconstrained, serving as a free surface. When conducting a dynamic analysis, the use of fixed or elastic boundaries can cause outward waves to reflect back into the model, resulting in distorted dynamic calculations. To address this issue, boundary conditions in 3DEC are primarily set as free-field boundaries and viscous boundaries [27,28]. The viscous boundary is positioned at the bottom of the model, surrounded by a free-field boundary (refer to Figure 4). Additionally, the seismic wave must be converted into a stress–time scale and applied to the bottom of the model using the transformation formula [16]:

$$\sigma_n = -2\rho c_p v_n \quad (1)$$

$$\tau_n = -2\rho c_s v_s \quad (2)$$

where: σ_n, σ_n are the normal and tangential stresses applied to the viscous boundary, respectively; ρ is the density; c_p, c_s are the longitudinal and transverse wave speeds, respectively; v_n, v_s are the normal and tangential velocity components on the model boundary.

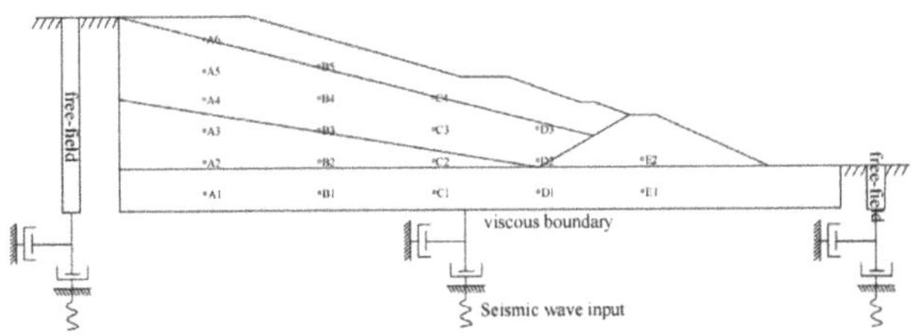

Figure 4. Uranium tailings dam profile and distribution of monitoring points.

In the dynamic computational analysis of 3DEC, three common forms of damping are Rayleigh damping, local damping, and Maxwell damping. However, local damping is not suitable for simulating dynamic calculations under seismic loading as it cannot effectively attenuate the high-frequency components of complex waveforms. Both Rayleigh damping and Maxwell damping effectively prevent self-oscillation in systems and manage complex waveforms. Maxwell damping schemes with three components that produce relatively constant damping over the frequency range are relevant to most seismic deformation analyses. The numerical computation time required for Maxwell damping is much less than that required for Rayleigh damping [29,30].

4. Analysis of Numerical Simulation Results

The computational analysis of discrete elements involves two steps. First, computation is performed under self-weight conditions to establish the initial stress field required for dynamic analysis (Figure 5). Then, the seismic wave is applied after resetting the initial velocity and displacement to conduct the dynamic computational analysis. Following the cessation of seismic vibrations, tailings dams may continue to deform. To obtain more reliable results, calculations should persist until the model reaches a state of rest after the seismic loading is completed. The permanent displacement map of the dam body in the x-direction (Figure 6) indicates that, under seismic loading, horizontal displacement is primarily concentrated at the slope face. This displacement is distributed in a stratified manner along a specific curved surface, with a maximum value of 0.26 meters. An analysis of Figure 7 reveals that the maximum permanent displacement in the z-direction is concentrated at the foot of the slope and the bottom of the dam body; however, the displacement at the bottom of the dam body is in the opposite direction to that at the foot of the slope. The cloud diagram of the maximum permanent displacement (Figure 8) illustrates a maximum value of 0.29 m. This instability is predominantly concentrated at the slope face and the foot of the slope, similar to the displacement observed in the x-direction. This suggests that horizontal seismic waves have a more significant impact on the stability of the dam body. Consequently, the instability of the dam body arises from the potential curved sliding surface along the slope, influenced by the superposition of seismic waves in both the horizontal and vertical directions.

Figure 5. Cloud view of the displacement of the dam body under gravity.

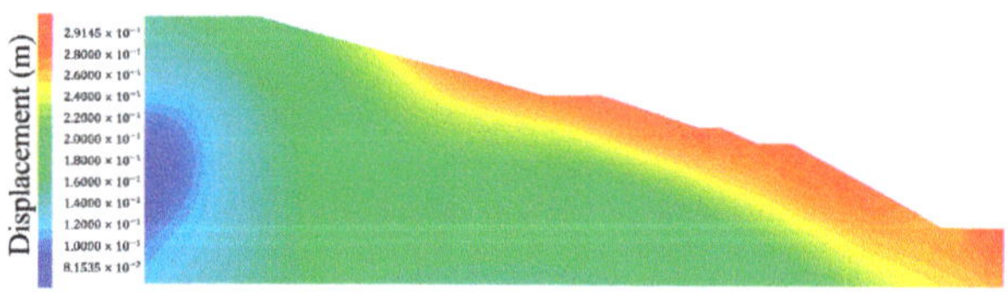

Figure 6. Cloud map of maximum permanent displacement in x-direction.

Figure 7. Cloud map of maximum permanent displacement in z-direction.

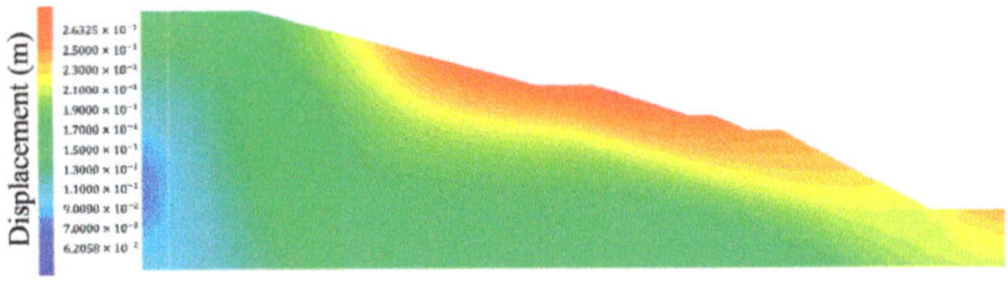

Figure 8. Maximum permanent displacement cloud map.

The time range curves of monitoring point A1 reveal a high degree of agreement between the slope velocity time range curves and the input displacement time range of the horizontal seismic wave (refer to Figure 9). This indicates that the boundary conditions of the dynamic model and the seismic loading are correct. The numerical calculations also recorded the peak values of the kinematic parameters and their amplification coefficients for each monitoring point simultaneously (Table 2).

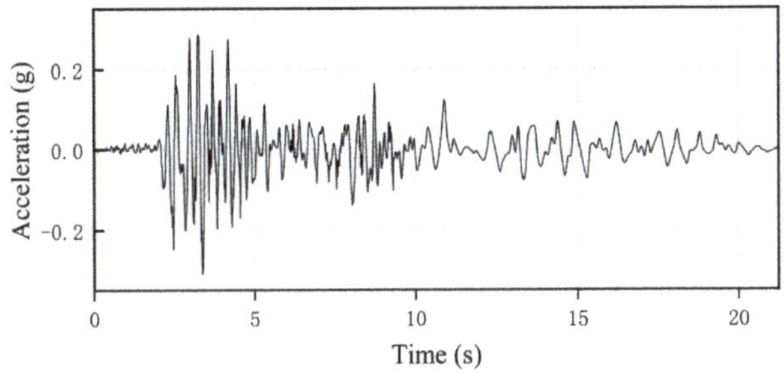

Figure 9. Time course curves of acceleration at monitoring point A1.

Table 2. Peak values and amplification factors of motion parameters at each monitoring point.

Monitoring Point	Elevation /m	Peak X-Velocity /($m \cdot s^{-1}$)	Horizontal Velocity Amplification Factor	Peak X-Displacement/ (cm)	X-Displacement Amplification Factor
A1	10	0.143	0.99	10.6	0.91
A2	25	0.156	1.08	11.3	0.97
A3	40	0.167	1.16	11.9	1.02
A4	55	0.188	1.31	12.3	1.05
A5	70	0.169	1.17	13.3	1.14
A6	85	0.154	1.07	15.2	1.30
B1	10	0.158	1.11	15.0	1.28
B2	25	0.148	1.03	16.4	1.40
B3	40	0.170	1.18	17.6	1.50
B4	55	0.177	1.23	18.9	1.62
B5	70	0.178	1.24	20.8	1.78
C1	10	0.147	1.02	15.4	1.32
C2	25	0.131	0.91	17.1	1.46
C3	40	0.145	1.01	19.2	1.64
C4	55	0.159	1.10	22.6	1.93
D1	10	0.130	0.90	15.8	1.35
D2	25	0.126	0.88	17.6	1.50
D3	40	0.170	1.18	21.2	1.81
E1	10	0.131	0.91	17.5	1.50
E2	25	0.146	1.01	20.7	1.77

5. Analysis of Response-Influencing Factors

5.1. Effect of Seismic Wave Amplitude on Dam Stability

The kinematic parameters of monitoring point C4 were observed at various amplitudes. Figure 10 illustrates that the displacement value at this monitoring point increases with amplitude, demonstrating a nearly linear relationship. Figure 11 reveals that the peak ground acceleration ratio (PGAR) does not increase linearly with elevation; instead, it exhibits a pattern of increasing, followed by decreasing, and then increasing again. This nonlinear behavior is attributed to the accumulation of shear strain in the tailings sand under seismic action, which reduces the stiffness and increases damping, thereby enhancing the filtering effects. Additionally, it was observed that the PGAR decreases as the seismic wave amplitude increases. This is because the filtering effect of the tailings sand becomes more pronounced at higher seismic wave amplitudes, leading to a reduction in the PGAR.

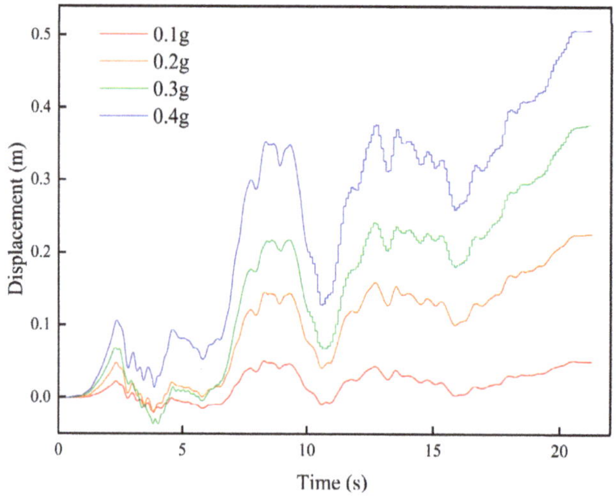

Figure 10. Time course of C4 displacement for different seismic wave amplitudes.

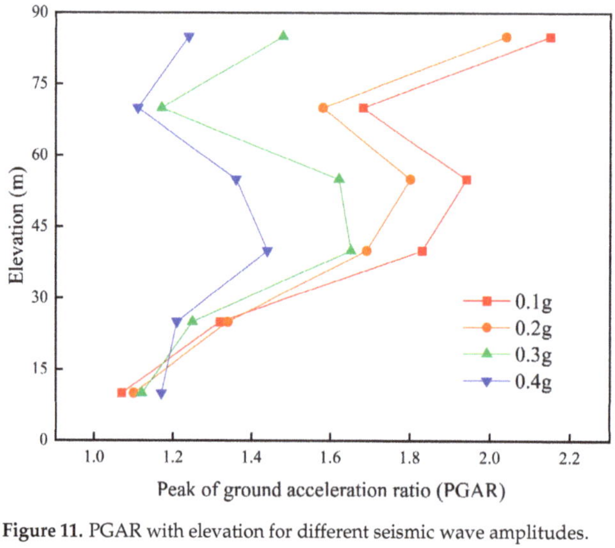

Figure 11. PGAR with elevation for different seismic wave amplitudes.

5.2. Effect of Tailings Sand Friction Angle on Dam Stability

During the calibration of the simulation parameters, it was discovered that the equivalent internal friction angle significantly impacts the simulation results [31,32]. Figure 12 shows the time course of the horizontal displacement at monitoring point C4 on the slope under different friction angles, and it can be seen that the permanent displacement value decreases significantly with the increasing friction angle, and the extreme value of displacement also decreases significantly, which indicates that its stability increases with the increase in the friction angle of the tailings sands. At friction angles of 35° and 40°, the differences in the horizontal displacement curves at the same point are relatively minor, and there is minimal accumulation of horizontal displacement. This observation suggests that the stability of the uranium tailings impoundment improves when the friction angle of the tailings exceeds 35°. The horizontal displacement curves at monitoring point C4 exhibit minimal variation for friction angles of 35° and 40°. Upon the application of a seismic load, the permanent horizontal displacement is markedly smaller compared to that observed at a friction angle of 30°, with the displacement values decreasing, which indicates that the tailings dam remains stable. This finding suggests that when the friction angle exceeds 35°, the stability of the tailings dam is effectively maintained. Furthermore, increasing the friction angle beyond 35° has a diminishing impact on the dam stability. However, it is important to note that during earthquakes, heavy rainfall tends to reduce the friction angle of the tailings sand. At a friction angle of 20°, both the peak and permanent displacement values increase significantly. Following the application of a seismic load, the horizontal displacement at the monitoring point reaches 0.47 m and continues to rise, signaling instability in the tailings dam.

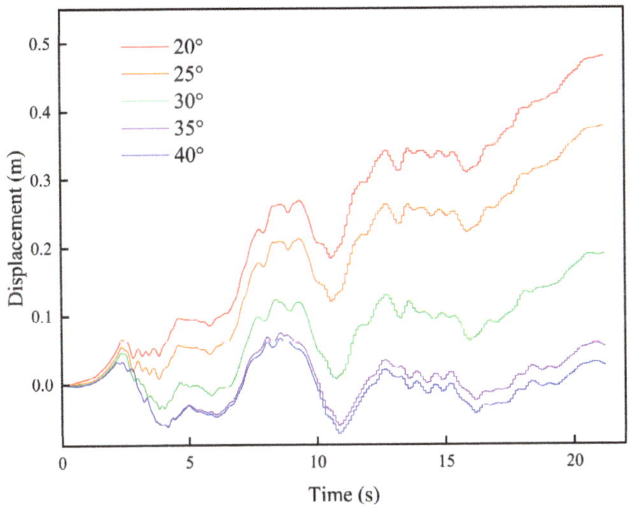

Figure 12. X-displacement time course of monitoring point C4 under different friction angles.

Figure 13 shows the change rule of the acceleration response along the vertical direction under different friction angles of the model: with the increase in the tailings sand friction angle, the acceleration amplification coefficient continues to increase. As the friction angle of the tailings increases, the acceleration amplification factor also rises, thereby enhancing the dynamic response of the tailings dam. The friction angle of the material is a crucial factor affecting the acceleration amplification factor, which tends to increase proportionally with the friction angle. This suggests that, within a certain range, increasing the friction angle of uranium tailings improves the overall stability of the dam.

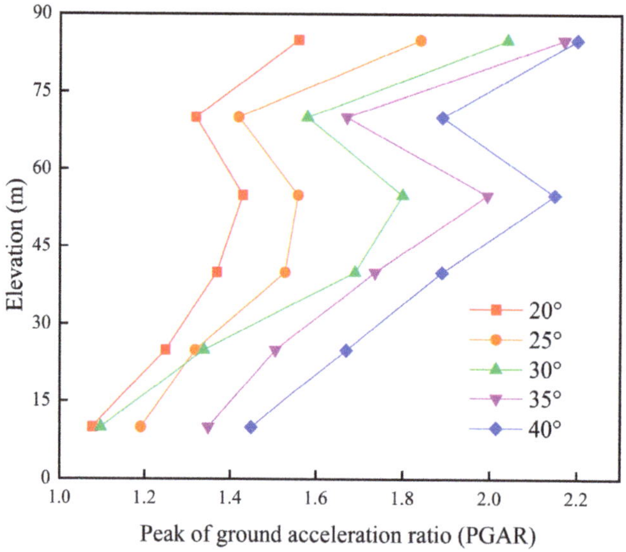

Figure 13. PGAR at different friction angles.

5.3. Effect of Seismic Loading Frequency on Dam Stability

The frequency of natural seismic waves typically ranges from 5 to 15 Hz. This study analyzed the dynamic response characteristics of the dam model using seismic wave frequencies of 5, 10, 15, 20, and 25 Hz. The acceleration response increases with elevation, exhibiting a significant amplification effect that peaks at the top of the slope. As the frequency of the input seismic wave increases, the amplification effect of acceleration with elevation tends to decrease (refer to Figure 14). At an input frequency of 5 Hz, the peak ground acceleration ratio (PGAR) amplification coefficient is at its highest as the dominant frequency is lowest at this point, resulting in a pronounced amplification effect on the slope body. Conversely, higher input frequencies contain more high-frequency energy, which enhances the filtering effect of the geotechnical body, thereby reducing the PGAR amplification coefficient response. Figure 15 illustrates the time course curves of displacement in the x-direction at monitoring point C4 for different frequencies. The maximum displacement value decreases as the frequency increases due to the reduced acceleration response of the higher-frequency seismic waves. When the seismic load frequency exceeds 15 Hz, the displacement value of the model due to the seismic load remains nearly unchanged, indicating that high-frequency waves have a lesser impact on the dam body.

The dynamic response of the model is characterized by its macroscopic deformation. According to the theory of elastic wave scattering, seismic stress waves split the wave field upon encountering a heterogeneous interface to maintain equilibrium. This phenomenon is primarily manifested through reflections from the free surface and both reflection and refraction at the slope discontinuity interface. Consequently, the seismic wave field near the interface becomes a complex combination of various wave types. As illustrated in Figure 16, the maximum permanent displacement in the horizontal direction of the tailings pond decreases with the increasing frequency of the input seismic wave. When the frequency exceeds 15 Hz, the displacement amplification ceases, likely due to the tailings' filtering effect and the dam's inherent inhibition.

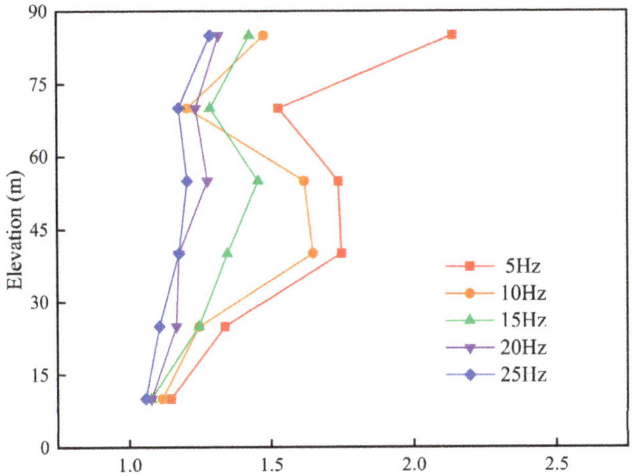

Figure 14. PGAR of C4 at various frequencies.

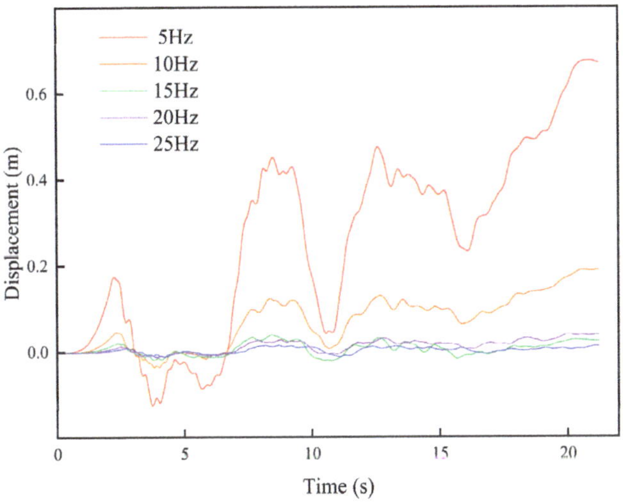

Figure 15. X-displacement time course of monitoring point C4 under different frequencies.

Figures 17–20 illustrate the maximum displacement cloud of the dam body at various frequencies. At lower frequencies, the maximum displacement of the dam body model is concentrated at the foot of the slope, with the maximum permanent displacement values distributed stratified along a specific radial pattern. As the frequency increases, the maximum displacement of the dam body model gradually shifts towards the top of the slope. Notably, at 25 Hz, the distribution of the maximum permanent displacement differs significantly from that observed at other frequencies. Figure 21 depicts the deformation damage at frequencies of 5 Hz and 25 Hz, respectively. At lower frequencies, the deformation damage to the dam model is concentrated in the shallow surface area and at the foot of the slope, resulting in localized cracking. In contrast, at 25 Hz, deformation damage primarily occurs at the bottom of the dam body, characterized by the predominance of lateral crack development, while the slope surface exhibits minimal damage.

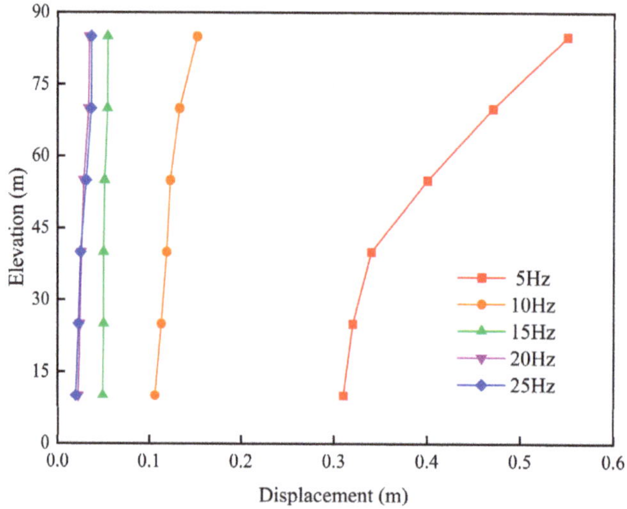

Figure 16. Change rule of maximum permanent displacement in x-direction with elevation at different frequencies.

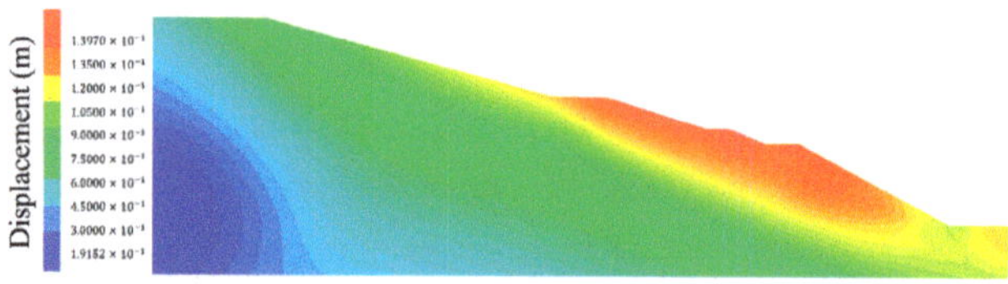

Figure 17. The maximum permanent displacement at a frequency of 5 Hz.

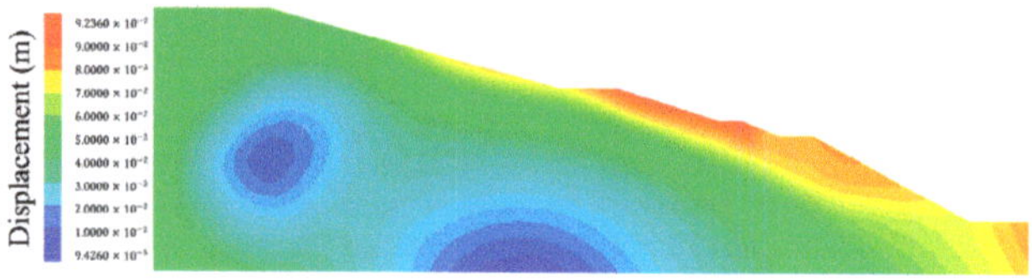

Figure 18. The maximum permanent displacement at a frequency of 15 Hz.

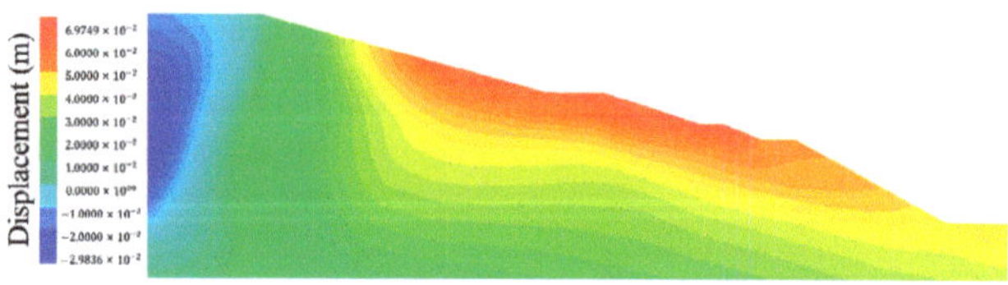

Figure 19. The maximum permanent displacement at a frequency of 20 Hz.

Figure 20. The maximum permanent displacement at a frequency of 25 Hz.

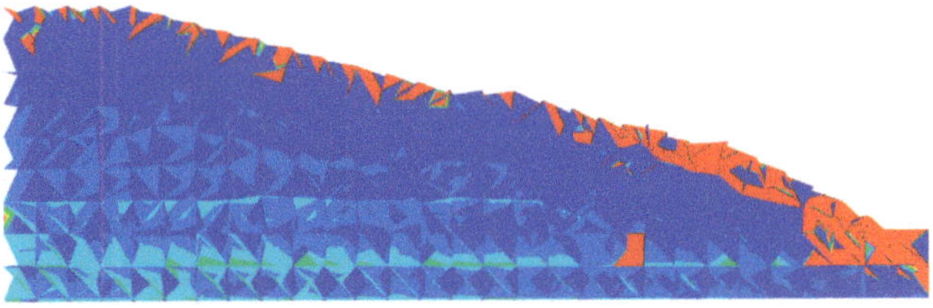

(**a**) 5 Hz.

(**b**) 25 Hz.

Figure 21. Fracture development at frequencies of 5 Hz and 25 Hz.

6. Conclusions

This paper investigates the dynamic response characteristics of a uranium tailings pond under seismic loads using the discrete element simulation software 3DEC v.5.2. It examines how the tailings sand friction angle, seismic wave amplitude, and frequency affect these characteristics. The main conclusions are as follows:

(1) The maximum permanent displacement was observed at the slope face and distributed in layers along a certain radian. The peak ground acceleration ratio (PGAR) does not increase linearly with altitude; rather, it follows an increase–decrease–increase pattern.

(2) As the amplitude of the seismic wave increases, slope displacement and velocity show an upward trend, while the PGAR exhibits a distinct decreasing trend.

(3) The internal friction angle of the dam body increases as the sand particles become coarser. This results in the stronger dynamic stability of the dam body. The displacement trend remains similar to that of the original distribution, but there is a significant reduction in the extreme value and permanent displacement. When the friction angle exceeds 35°, the overall stability of the dam is relatively good, and further increases in the friction angle have minimal impact on enhancing the stability of the tailings dam. Conversely, when the friction angle falls below 25°, the risk of dam failure increases significantly.

(4) Tailings sand serves as an effective filter for high-frequency seismic waves, significantly diminishing the dam's dynamic response to waves with frequencies exceeding 15 Hz. In contrast, low-frequency seismic waves are the predominant cause of damage to tailings ponds. When the frequency is low, the deformation damage to the model slope is concentrated on the slope face; conversely, when the frequency is high, the deformation damage is predominantly located at the bottom of the model.

Author Contributions: M.L.: Writing—original draft, Funding acquisition, Project administration, Supervision. H.H.: Data curation, Software, Validation. Y.H.: Writing—review & editing, Funding acquisition, Methodology, Resources. All authors have read and agreed to the published version of the manuscript.

Funding: This research was funded by National Natural Science Foundation of China (Grant Nos. 12005099 and 12205144), and Natural Science Foundation of Hunan Province (Grant Nos. 2024JJ6372 and 2023JJ40546).

Institutional Review Board Statement: Not applicable.

Informed Consent Statement: Not applicable.

Data Availability Statement: The datasets used and/or analyzed during the current study are available from the corresponding author on reasonable request.

Acknowledgments: The valuable support and assistance provided by Yang Liu during the course of this research are gratefully acknowledged.

Conflicts of Interest: Author Ming Lan was employed by the company of Dep Mining Laboratory of Shandong Gold Group Co. Ltd. The remaining authors declare that the research was conducted in the absence of any commercial or financial relationships that could be construed as a potential conflict of interest.

References

1. Rana, N.M.; Ghahramani, N.; Evans, S.G.; McDougall, S.; Small, A.; Take, W.A. Catastrophic mass flows resulting from tailings impoundment failures. *Eng. Geol.* **2021**, *292*, 106262. [CrossRef]
2. Agurto-Detzel, H.; Bianchi, M.; Assumpção, M.; Schimmel, M.; Collaço, B.; Ciardelli, C.; Barbosa, J.R.; Calhau, J. The tailings dam failure of 5 November 2015 in SE Brazil and its preceding seismic sequence. *Geophys. Res. Lett.* **2016**, *43*, 4929–4936. [CrossRef]
3. Dai, X.; Wu, X.; Hong, Y.; Xie, J.; Lin, D.; Chen, Y.; Liu, Y. Safety and stability evaluation of the uranium tailings impoundment dam: Based on the improved AHP-cloud model. *J. Radiat. Res. Appl. Sci.* **2022**, *15*, 21–31. [CrossRef]
4. Owen, J.R.; Kemp, D.; Lèbre, É.; Svobodova, K.; Pérez Murillo, G. Catastrophic tailings dam failures and disaster risk disclosure. *Int. J. Disaster Risk Reduct.* **2020**, *42*. [CrossRef]
5. Jiang, F.; Wu, H.; Liu, Y.; Chen, G.; Guo, J.; Wang, Z. Comprehensive evaluation system for stability of multiple dams in a uranium tailings reservoir: Based on the TOPSIS model and bow tie model. *R. Soc. Open Sci.* **2020**, *7*, 191566. [CrossRef]
6. Islam, K.; Murakami, S. Global-scale impact analysis of mine tailings dam failures: 1915–2020. *Glob. Environ. Change* **2021**, *70*. [CrossRef]
7. Lamour, R.J.A.; Wasserman, M.A.V.; Rochedo, E.R.R.; Lugon Junior, J. Assessment of the environmental radiological impact in the emergency exposure phase through the simulation of a uranium mining tailings dam breach. *Braz. J. Radiat. Sci.* **2023**, *11*, 1–20. [CrossRef]
8. Villavicencio, G.; Espinace, R.; Palma, J.; Fourie, A.; Valenzuela, P. Failures of sand tailings dams in a highly seismic country. *Can. Geotech. J.* **2013**, *51*, 449–464. [CrossRef]
9. Lyu, Z.; Chai, J.; Xu, Z.; Qin, Y.; Cao, J. A Comprehensive Review on Reasons for Tailings Dam Failures Based on Case History. *Adv. Civ. Eng.* **2019**, *2019*, 4159306. [CrossRef]
10. Jin, J.; Ding, Q.; Cui, H.; Zhang, P.; Xiao, X.; Lv, X. Dynamic response characteristics of a tailing dam determined by shaking-table tests. *Arab. J. Geosci.* **2020**, *13*, 897. [CrossRef]
11. Zhou, Z.; Ren, C.; Xu, G.; Zhan, H.; Liu, T. Dynamic Failure Mode and Dynamic Response of High Slope Using Shaking Table Test. *Shock. Vibration.* **2019**, *2019*, 4802740. [CrossRef]
12. Li, Q.; Ma, G.; Li, P.; Su, Z. Dynamic characteristics of tailings dam with geotextile tubes under seismic load. *Rev. Adv. Mater. Sci.* **2021**, *60*, 599–614. [CrossRef]
13. Lin, Q.; Zhang, S.; Liu, H.; Shao, Z. Water saturation effects on the fracturing mechanism of sandstone excavating by TBM disc cutters. *Arch. Civ. Mech. Eng.* **2024**, *24*, 154. [CrossRef]
14. Cao, R.-h.; Yao, R.; Meng, J.; Qibin, L.; Lin, H.; Su, L. Failure mechanism of non-persistent jointed rock-like specimens under uniaxial loading: Laboratory testing. *Int. J. Rock Mech. Min. Sci.* **2020**, *132*, 104341. [CrossRef]
15. Zhu, D.; Yu, B.; Wang, D.; Zhang, Y. Fusion of finite element and machine learning methods to predict rock shear strength parameters. *J. Geophys. Eng.* **2024**, *21*, 1183–1193. [CrossRef]
16. Itasca Consulting Group. *3DEC User Manual*; Itasca Consulting Group: Minneapolis, MN, USA, 2019. Available online: https://docs.itascacg.com/3dec700/3dec/docproject/source/3dechome.html (accessed on 7 August 2024).
17. Ren, Z.; Wang, K.; Zhang, Q.S.; Xu, Z.M.; Tang, Z.G.; Chen, J.P.; Yang, J.Q.; Xu, Z.H. Earthquake dynamic response behavior of Xiangchong valley type tailings impoundment in Yunnan, China. *J. Mt. Sci.* **2018**, *15*, 82–99. [CrossRef]
18. Chundi, S.; Baolin, X.; Wei, W. Analysis on seismic dynamic response and liquefaction area of tailings dam. *Int. J. Comput. Appl. Technol.* **2018**, *57*, 183–191. [CrossRef]
19. Xu, M.; Yu, X.; Pan, Y.; Liu, X.; Zhao, Y.; Hu, J. Analysis of the seismic dynamic response and failure mode of the Layue landslide. *Landslides* **2023**, *20*, 1135–1148. [CrossRef]
20. Jin, J.; Song, C.; Liang, B.; Chen, Y.; Su, M. Dynamic characteristics of tailings reservoir under seismic load. *Environ. Earth Sci.* **2018**, *77*, 654. [CrossRef]
21. Vijayasri, T. Dynamic characterization of tailing dam using fully coupled dynamic analysis with different boundary conditions—A case study. *Earthq. Eng. Eng. Vib.* **2023**, *22*, 995–1013. [CrossRef]
22. Ferdosi, B.; James, M.; Aubertin, M. Investigation of the Effect of Waste Rock Inclusions Configuration on the Seismic Performance of a Tailings Impoundment. *Geotech. Geol. Eng.* **2015**, *33*, 1519–1537. [CrossRef]

23. Xu, B.; Zhou, Y.; Zhou, C.G.; Kong, X.J.; Zou, D.G. Dynamic responses of concrete-faced rockfill dam due to different seismic motion input methods. *Int. J. Distrib. Sens. Netw.* **2018**, *14*, 1550147718804687. [CrossRef]
24. Yao, Y.; Wang, R.; Liu, T.Y.; Zhang, J.M. Seismic response of high concrete face rockfill dams subjected to non-uniform input motion. *Acta Geotech.* **2019**, *14*, 83–100. [CrossRef]
25. Psarropoulos, P.N.; Tsompanakis, Y. Stability of tailings dams under static and seismic loading. *Can. Geotech. J.* **2008**, *45*, 663–675. [CrossRef]
26. Bićanić, N. Discrete Element Methods. *Encycl. Comput. Mech.* **2004**.
27. Wu, J.; Wang, Y.; Dong, S.; Chen, Y.; Wang, L. Genetic mechanism and failure process of the Mogangling seismic landslide. *J. Geol. Soc. India* **2013**, *82*, 277–282. [CrossRef]
28. Wang, Y.; Wu, L.Z.; Gu, J. Process analysis of the Moxi earthquake-induced Lantianwan landslide in the Dadu River, China. *Bull. Eng. Geol. Environ.* **2019**, *78*, 4731–4742. [CrossRef]
29. Lemos, J.V.; Dawson, E.M.; Cheng, Z. Application of Maxwell damping in the dynamic analysis of masonry structures with discrete elements. *Int. J. Mason. Res. Innov.* **2022**, *7*, 663–686. [CrossRef]
30. Maxwell Damping: An Alternative to Rayleigh Damping. In *Geo-Extreme 2021*; American Society of Civil Engineers: Reston, VA, USA, 2021; pp. 34–45. [CrossRef]
31. Liu, B.; He, K.; Han, M.; Hu, X.; Wu, T.; Wu, M.; Ma, G. Dynamic process simulation of the Xiaogangjian rockslide occurred in shattered mountain based on 3DEC and DFN. *Comput. Geotech.* **2021**, *134*, 104122. [CrossRef]
32. Fan, X.; Xu, Q.; Scaringi, G.; Dai, L.; Li, W.; Dong, X.; Zhu, X.; Pei, X.; Dai, K.; Havenith, H.-B. Failure mechanism and kinematics of the deadly June 24th 2017 Xinmo landslide, Maoxian, Sichuan, China. *Landslides* **2017**, *14*, 2129–2146. [CrossRef]

MDPI AG
Grosspeteranlage 5
4052 Basel
Switzerland
Tel.: +41 61 683 77 34

Applied Sciences Editorial Office
E-mail: applsci@mdpi.com
www.mdpi.com/journal/applsci

www.ingramcontent.com/pod-product-compliance
Lightning Source LLC
Chambersburg PA
CBHW061438210926
51850CB00008B/1549
* 9 7 8 3 7 2 5 8 7 4 2 2 4 *